Mathematics and War

Edited by
Bernhelm Booß-Bavnbek and
Jens Høyrup

Springer Basel AG

Editors:

Bernhelm Booß-Bavnbek
Roskilde University
Department of Studies in Mathematics and Physics and their Functions in Education, Research and Applications - IMUFA
P.O. Box 260
4000 Roskilde
DENMARK

Email: booss@mmf.ruc.dk

Jens Høyrup
Roskilde University
Department of Psychology and of Philosophy and Science Studies

P.O. Box 260
4000 Roskilde
DENMARK

Email: jensh@ruc.dk

A CIP catalogue record for this book is available from the Library of Congress, Washington D.C., USA

Deutsche Bibliothek Cataloging-in-Publication Data
Die Deutsche Bibliothek lists this publication in the Deutsche Nationalbibliografie; detailed bibliographic data is available in the Internet at <http://dnb.ddb.de>.

ISBN 978-3-7643-1634-1 ISBN 978-3-0348-8093-0 (eBook)
DOI 10.1007/978-3-0348-8093-0

Ursprünglich erschienen bei Birkhäuser Verlag, Basel, Switzerland 2003

Printed on acid-free paper produced from chlorine-free pulp
Cover design: Micha Lotrovsky, Therwil, Schweiz

ISBN 978-3-7643-1634-1

9 8 7 6 5 4 3 2 1 www.birkhauser-science.com

Preface

In this moment of writing, the whole world is waiting for an announced war; at the moment of reading, there is scarce hope the war will not have taken place or be still in course. This is the *raison d'être* of the present volume.

"Mathematics and war" may seem a peculiar topic when war seems so much more important than mathematics. Importance, however, is concrete, and one of the facets which brings forth and shapes contemporary warfare is its profound mathematization. Therefore, a discussion of war which omits mathematics is bound to miss many points. Reversely, a discussion of mathematics which neglects its actual involvement in war and warfare may also miss quite a few essential points.

Mathematization, however, is also a global and rather abstract description, which on its part must be understood concretely. The papers collected in the volume therefore take up a number of different but mutually related issues, some of them historical, some of them philosophical, some of them concerned with the functioning of weapons and with the actual planning and conduct of war, some of them with the mental impact of the mathematization of fighting on the general population as well as on the fighting crews. Some of the contributions are written from a more or less outspokenly pacifist perspective, others are not. Some of the historically oriented papers deal with personal history, others with military history or the social and institutional history of mathematics; but the history focuses strongly on the last sixty or seventy years.

The reason for this is, on one hand, that our primary motivation has been ethical and aimed at understanding the actual state of the world – the main purpose of the book is not the gaining of pure theoretical insight but to gather valid knowledge that may be used in the situation where the human race is actually navigating; on the other, that the interaction between mathematics and warfare changed dramatically with World War II – indeed that the reasons that make understanding of this interaction practically urgent only came into being with World War II.

The general issues which are touched on are not covered broadly. Even this has its reasons. For one thing, "mathematics and war" is a relatively new interest, at least as a confined research field; broad generalizations must therefore wait until detailed studies are at hand (in the introduction, the editors permit themselves not to follow this wise rule). Moreover, each detailed study fills out a number of pages, and the volume can hold only a limited number of pages. For the purpose of understanding the world and reflecting upon what can be done *now*, the studies contained in the volume and the multitude of perspectives they represent should be useful as points of orientation in a domain that is in need of further development. For this purpose we see it as no disadvantage that many of the studies combine the presentation of factual information and reflection with moral or political stances. At least in their combination, they should approach Weberian *Wertfreiheit*.

Evidently, the editors have thought about which further perspectives could contribute to the global picture. One important theme, though touched at indirectly in several contributions, deserves in-depth investigation on its own: the relation between the public sphere and its institutions (including the media), general mental attitudes about supposedly clinically-rational technology, and the genuine or pretended mathematical rationality of modern warfare.

The volume grew out of an "International Meeting on Mathematics and War", Karlskrona, Sweden, 29–31 August 2002. We use the opportunity to thank Maurice and Charlyne de Gosson and the Blekinge Institute of Technology and its Mathematics Department for organizing splendidly this conference, which was also supported by Stig Andur Pedersen of The Danish Network for History and Philosophy of Mathematics (MATHNET) and Reiner Braun of The International Network of Engineers and Scientists for Global Responsibility (INES).

We also thank Birkhäuser Verlag for having the courage to risk its money on the book.

January 31, 2003

B. Booß-Bavnbek
J. Høyrup
Roskilde University

Postscript
During the war which *did* take place between the moment of writing the above and the moment of sending the proof-read pages to print, the very President of the American Mathematical Society took notice, not so much of the military role of mathematics in general as of mathematics as a key component in the preservation of US military superiority, implicitly promising the loyal assistance of his members (and asking for adequate payment): "[...] for a military commander to have secure communications in the field depends on fundamental advances in number theory that just a few years ago were touted as the sort of pure science that would never be applied. Future progress in this seemingly abstract area, by us or by hostile forces, could threaten the security of all these communications and would require fundamental new science to repair the damage." (David Eisenbud, April 9, 2003, to a US Congress subcommittee overseeing the funding of the US National Science Foundation; source: http://www.ams.org/notices/200306/inside.pdf).

Table of Contents

III Ethical Issues

IV Enlightenment Perspectives

Introduction

BERNHELM BOOSS-BAVNBEK and JENS HØYRUP*

Physicists, chemists, and biologists have a tradition of discussing meta-aspects of their subject – among which are the military use and misuse of the knowledge they produce. Concerns of the latter kind are rare among mathematicians.

No rule without exceptions. During the Vietnam war, a number of appeals were circulated among US mathematicians (with reverberations in particular in France and Japan and at the International Congress of Mathematicians in Moscow in 1966 and Nice in 1970) not to engage in war-related work. One such appeal was published in the *Notices of the American Mathematical Society* in January 1968. Alexandre Grothendieck's resigning from mathematics fell in the context of this debate. [Godement 1978], no longer debate but politico-economical analysis, was written from a mathematician's perspective even though it did not deal with mathematical research in particular. [Gross 1978], also written by a mathematician, was shorter but concentrated on mathematics.

In the new context of the euro-missile controversy of the early 1980s, military research came into the focus of debate at universities of Western Germany. [Booß & Høyrup 1984] was an offspring of this new discussion concentrating on mathematics; the broad discussion is reflected in [Tschirner & Göbel (eds) 1990]. The "Forum on Military Funding of Mathematics" published in the *Mathematical Intelligencer* 1987 no. 4 reflects problems arising for the US mathematical community from the "Strategic Defense Initiative" in the same phase.[1] Some more publications followed, mainly with historical emphasis.

As warfare is now again becoming an all-too-obvious aspect of our world and a no less obvious part of "Western" policies, time seemed ripe for taking up the issue anew. Just after the Kosovo War, *Zentralblatt für Didaktik der Mathematik* dedicated an issue to it (vol. 98 no. 3, June 1998); August 29–31, 2002, 42 mathematicians, historians of mathematics, military historians and analysts, and philosophers gathered in the historic Swedish military port of Karlskrona, to discuss a wide variety of questions, each representing its own perspective on or facet of the global topic "mathematics and war". Tentatively, the themes can be grouped as follows:

- To what extent has the military played an active part throughout history, and in particular since World War II, in shaping modern mathematics and the careers of mathematicians?
- Are mathematical thinking, mathematical methods, and mathematically supported technology[2] about to change the character and performance of modern warfare, and if so, how does this influence the public and the military?

* Both Roskilde University, 4000 Roskilde, Denmark. Email: booss@ruc.dk, jensh@ruc.dk

[1] Cf. also [Davis 1989].

- What were, in times of war, the ethical choices of outstanding individuals like the physicist Niels Bohr and the mathematician Alan Turing? To what extent can general ethical discussions provide guidance for working mathematicians?
- What was the role of mathematical thinking in shaping the modern international law of war and peace? Can mathematical arguments support actual conflict solution?

The present volume does not constitute the official proceedings of this meeting, but most contributions were presented there. This introductory essay represents the editors' personal attempt to synthesize the outcome of an inherently (and intentionally) very disparate exchange of information and views. Admittedly, the synthesis also draws on the material presented in [Booß & Høyrup 1984] and on the perspectives developed in that booklet. In order to balance this personal bias we refer in running footnotes to the contributions to the volume. Such references do not necessarily imply that the editors agree with all points of view expressed in the contribution in question, nor that the editors postulate the author's agreement with their interpretation of the subject-matter. In general, no other participants are of course responsible in any way for what we say in this introduction or for what we conclude from what they said; by participating in the endeavour, all of us – authors as well as editors – have agreed implicitly that we find civilized disagreement enlightening.

Perspectives from Mathematics

All mathematicians know the tales, reliable or not, about Archimedes and his defence of Syracuse. They may also have heard about early modern ballistics and fortification mathematics and the importance of trigonometry for navigation. All these cases of mathematics being entangled in conquest, warfare or preparation of war have one thing in common: that which was combined with technical and military knack was almost exclusively *already existing* mathematics. In this respect such examples do not differ from the use of simple accounting mathematics in logistics – which after all is likely to have been much more important from the military point of view. Mathematics served as a toolbox, and military officers may have been the largest group that received some general mathematical training; but the involvement of mathematics as a global endeavour with the military institution was not very intimate, and specifically military applications had no independent role as a shaping force for mathematics. Tartaglia's composition of straight lines and circles in ballistics was clearly inspired from gunnery and the war against the Turks. When Galileo introduced the parabolic law this origin was already left behind, and the theory was linked instead to the philosophical discussion of local motion – and was largely irrelevant for the firing of guns because of the influence of air resistance, as pointed out explicitly by Galileo.

[2] This "broad concept" of mathematics is the one that serves in the rest of this introductory essay; it also embraces computers and computer science.

Figure 1. "Noli turbare circulos meos": When Archimedes's city was conquered in spite of his astounding mathematical engineering, he pretended that he had only made pure mathematics – thus according to an anecdote that has remained popular since Roman Antiquity; the mosaic is an eighteenth-century forgery. [Courtesy: P. C. Bol, Städelsches Kunstinstitut, Frankfurt/M.]

Even to this rule ("the involvement of mathematics as a global endeavour ...") there is an exception. That part of the Sumero-Babylonian legacy which is most spoken of in general histories of mathematics – namely the invention and implementation of the place value system – may be a child of war. In c. 2074 BCE, king Shulgi organized a military reform in the Sumerian Empire, and the next year an administrative reform (seemingly introduced under the pretext of a state of emergency but soon made permanent) enroled the larger part of the working population in quasi-servile labour crews and made overseer scribes accountable for the performance of their crews, calculated in abstract units worth $^1/_{60}$ of a working day (12 minutes) and according to fixed norms. In the ensuing bookkeeping, all work and output therefore had to be calculated precisely and converted into these abstract units, which asked for multiplications and divisions *en masse*. Therefore, a place value system with base 60 was introduced for intermediate calculations.[3] Its functioning presupposed the use of tables of multiplication, reciprocals and technical constants and the training for their use in school; the implementation of a system whose basic idea had been "in the air" for some centuries therefore asked for decisions made at the level of the state and implemented with great force. Then as in many later situations, only war provided the opportunity for such social willpower.

Apart from that the conclusion stands that "the involvement of mathematics as a global endeavour with the military institution was not very intimate, and specifically military applications had no independent role as a shaping force for mathematics" until a century ago. Since around 1500 CE, as already mentioned, the employment of fortification mathematicians and the teaching of naval and artillery officers certainly played a social role for mathematics by providing job opportunities and a market for mathematics text books (copiously decorated with military symbols).

This we may regard as the past. The contemporary situation – the one that is our real interest – can be said to start around the First World War, and to reach full development during the Second World war.[4]

During World War I, two important new military technologies depended on *mathematics in the making*: sonar, and aerodynamics. They were so impressive that Émile Picard, in spite of his own patriotism (which non-French cannot help seeing as pure chauvinism), regretted the perspective that young mathematicians might opt in future for applied mathematics only [*Proc. ... 1920*: xxviii]. In general, however, the immediate role of the pure sciences, mathematical and otherwise, was that of providing manpower that could be converted into first-class creative engineers – not restricted to applying a set of standard rules but able to implement theoretical knowledge and make it function in practice; this was also the role of most of the mathematicians that were actually involved in the war effort (if they were not, as was the case in France, sent into the trenches). Nobody will claim that mathematics was in any way decisive for the outcome of the war, nor that WW I applications of mathematics left important traces in the post-war world (civil aviation still belonged to the future).

Picard's worries proved unfounded. Main-stream mathematics soon reverted to the pre-War model, even more swiftly than the precariously erected organization of planned science was dismantled. Aerodynamics of course survived, but only as one current among others.[5]

[3] Since it was a floating-point system with no indication of absolute place, it could only be used for intermediate calculations – just as the slide rule of engineers in quite recent times. Since intermediate calculations have not survived, the exact dating of the implementation can only be determined from indirect arguments. See, for instance, [Høyrup 2002: 314] for further references.

[4] A broad introduction to the whole period is R. Siegmund-Schultze's survey, this volume, pp. 23–82.

[5] The Polish first application of mathematics (group theory) instead of philological methods in code cracking constitutes another isolated current – or, in fact, an omen of what was to come. See E. Rakus-Andersson, this volume, pp. 83–107.

All of this was different in World War II, both quantitatively and qualitatively: the organization of science intended to support the war effort was a major concern of both Axis and Allied powers;[6] mathematically based technologies (radar, sonar, the decipher computer, *the* bomb) can be argued to have been war-decisive; computers, nuclear energy, jet propulsion - all mathematically constructed and computed for the war - have changed our world beyond recognition after 1945. Admittedly, all of these build on pre-war theoretical insights;[7] some of them (computers, jet motors) were indeed not only "in the air" before the war but functioned as prototypes for the devices which reached completion during the war; but in all cases the war, by making available huge means without counting costs and benefits, made it possible to boost a development that otherwise might have taken decades[8] - and perhaps, in cases like the proliferation of DDT and atomic reactors, might have been stopped at an early moment when the problems they create became visible.

During the war, mathematicians in large numbers were recruited, many of them to teach sailors and air-crew members basic trigonometry (etc.), but many also to serve as best-level creative engineers.[9] Afterwards, the latter have often tended to regard what they did dismissively ("I did not write one line that was publishable", as J. Barkley Rosser [1982: 509*f*] summarizes one reaction to his questionnaire on war work in mathematics), perhaps because puzzle-solving with no further theoretical impact did not look important in the mathematician's hindsight; this assessment notwithstanding, what was done depended critically on mathematical ingenuity and training. A striking example is O. R. Frisch and R. Peierls' mathematical formulation of the essential questions surrounding the construction of a uranium bomb in March 1940 and their "back-of-an-envelope" discovery that its critical mass was so small that military use was feasible.[10]

In some cases, of course, the solving of problems defined by the war *did* have important theoretical impact - well-known examples are the emergence of computer science, information theory, Monte Carlo simulation, operations research, and statistical quality control.

6 See, beyond Siegmund-Schultze's article, the contributions in this volume of S. Fukutomi (pp. 153–159), A. N. Shiryaev (pp. 103–107), K. B. Williams (pp. 108–125), T. H. Kjeldsen (pp. 126–152), A. Hodges (pp. 312–325) and T. Makino (pp. 326–335).

7 At times fully detached from every technical application; N. Wiener and E. Hopf had calculated the radiation equilibrium at stellar surfaces, but their theory could be applied to the expanding surface of the exploding bomb [Wiener 1964: 142*f*]. A. A. Markov had investigated his eponymous processes as pure mathematics and illustrated the applicability of the concept on linguistic material [Youschkevitch 1974: 129]; in the Manhattan Project they turned out to be relevant for solving diffusion equations and for describing nuclear branching processes.

8 The parallel to the invention of the place value system in Sumer is striking. In that case, parallel processes not furthered by a military government indeed asked for much longer time: in China the unfolding took more than a millennium, in India it never really took place before the "Indian" system was brought back from abroad.

9 Cf. [Morse 1943].

10 See [Gowing 1964: 40–43, 389–393] and [Dalitz & Peierls 1997: 277–282]. This latter volume presents Peierls as a physicist, but his actual chair was in "applied mathematics"; even a "broad concept" of mathematics does not free us from delimitation problems.

Figure 2. There is no known picture of Turing during the wartime period, but this photograph shows Alan Turing (at left) with his athletic club in 1946. At this point he was engaged in designing a first digital computer at the National Physical Laboratory, London. This design used his wartime knowledge of electronic technology to put his 1936 theory of the universal machine into a practical form. The codebreaking machinery at Bletchley Park, although very advanced, had never actually used Turing's fundamental idea of the universal machine and the stored program, but as soon as the war ended Turing set to work to bring it to reality. [Courtesy: Turing Archive at King's College, Cambridge]

This time, nothing was dismantled after the war (many mathematicians, of course, hurried away from military research) – the Cold War was already on. In the slightly longer run (a decade or so), civil re-application of the new mathematical war techniques caused profound transformation of these and violent acceleration of their development: only the war effort had allowed the creation of the first costly computers, but only commercial use allowed mass production, open competition, intensive development efforts and reduction of costs (actually, stored-program computers like the ENIAC only reached the working stage after the war though at first in military contexts).[11] We may add, on one hand, that only the freeing from the pressure of immediate applicability ("better a fairly satisfactory answer now than the really good answer two years after defeat") gave space for fruitful interaction between theoretical understanding and applications in for instance computer science.[12] On the other hand we may add that the developments born from civil reapplication were then brought back to the military sector with their now immensely increased efficiency (and further, back-and-forth, back-and-forth,...) – a situation that has given rise to the notion of a "military-industrial" or "military-industrial-scientific complex".[13]

When discussing mathematical research for military purposes, both during World War II and in recent decades, we should differentiate several situations and problems.

- Firstly, we must distinguish the application (sometimes creative, sometimes repetitive) of existing mathematical tools (ballistic computation, modelling, ...) from the creation of new mathematical insights and techniques (sequence analy-

[11] See also T. H. Kjeldsen's discussion of the case of operations research, this volume, pp. 126–152.

[12] Cf. Hodges (pp. 312–325) on Turing's post-war work, and Williams (pp. 108–125) on Grace Hopper, both this volume.

[13] Since this concerns scientifically based industry in general and mathematics only in so far as it enters in all scientific technology, this complex is not much discussed in the present volume (cf. note 18). See, however, for a very specific aspect, Davis, in this volume, pp. 174–179). A broad empirical investigation of related questions is [Godement 1978].

Figure 3. When listening to a music CD we enjoy and do not think of the military origin of the coding involved. Similarly, mathematicians going to the wonderful Mathematical Research Institute *Oberwolfach* enjoy the ambience and have no reason to worry about the fact that the institute originated as a military research institution in 1944 - apparently well planned for the purpose though too late to become efficient. [Photo: H. Kastenholz]

sis, Monte Carlo simulation, ...). As a rule, but not consistently, the former type is the chore of mathematicians who are paid by or connected to the military institution itself; new insights directed toward military goals are more likely to come from mathematicians who are less closely bound to the military institution but inspired by problems coming from this source.

- Secondly, we should remember that mathematical military research is a complex activity that cannot be understood exclusively as the mere production of theorems of presumed military use. Several institutions (Süss's original planning of the German Oberwolfach Institute in 1944, the American Mathematics Research Center in Wisconsin) exemplify an efficient model, *a two-way chain* (henceforth the "AMRC chain"), which *grosso modo* works as follows:[14] A core group of highly skilled mathematicians familiar with the direct problems of the military employer (efficiency of bombing, controlled spread of bacteriological agents, better radar detection and avoidance of enemy detection, or whatever it may be) find out which of these can be approached mathematically, undertake an initial translation, and direct the translated problems to other experienced

[14] Concerning the Oberwolfach Institute, this structure follows (conventional whitewashing notwithstanding) from analysis of the material presented by H. Gericke [1984], cf. [Høyrup 1986]; on the same institution, see also [Remmert 1999]. For the Wisconsin Institute, see [*The AMRC Papers*].

mathematicians who are well-informed about and centrally located within the whole mathematical milieu; these parcel out the questions into problems which colleagues may take up as mathematically interesting, perhaps even without knowing that they enter into a network of military relevance; once such questions have been answered, the same chain functions backwards, reassembling the answers and channelling the global solution to the employer.

This is only one among several models. We know that it was planned to function in World War II Germany but was implemented too late to become efficient; we know that it has functioned in the US. We know less about the organization of military mathematical research in the late Soviet Union, but it appears that here, as in production and research in general, the civil and military domains were more sharply separated than in the West.

Rounding off what can be said in "the perspective from mathematics" we may make some general observations.

- Mathematical war research has resulted in certain fundamental theoretical innovations. It is striking, however, that all of these appear to have depended critically on an exceptional mathematician. The names of Turing, von Neumann, Shannon, Wald, and Pontryagin may suffice to make the point.[15]
- However, the utility of mathematics *for the treatment of military problems* does not depend critically on the presence of an exceptional mathematician. Mathematicians in large numbers have proved themselves unexpectedly able to function as creative mathematical engineers, in the sense explained above.
- This ability has largely depended on their capacity to become familiar with methods and approaches of various mathematical disciplines and to synthesize them. The still persistent unity of mathematics is thus demonstrated *ad oculos*, if not in the mathematical journals then in technical application.
- It should not be forgotten that the traditional application of the toolbox of already existing mathematics goes on, now at the level created by recent mathematical research.
- In the wake of World War II and as a consequence of the intertwinement of mathematics with advanced technologies (military as well as civil), mathematics as a subject has changed: discrete mathematics and the "mathematization of complexity" have become increasingly important, in some views to such an extent that they now define the actual essence of the field.[16]

[15] Hodges (this volume, pp. 312–325), as mentioned, analyzes Turing's work. R. V. Gamkrelidze (this volume, pp. 160–173) presents Pontryagin's discovery of the maximum principle in control theory. It may also depend on the extraordinary mathematical competence of all five that their theories or techniques were shaped in so mature form that their applicability outside the military domain resulted almost immediately.

[16] This theme has not been explored in the preceding pages, but see [Booß & Høyrup 1984]. [Bleecker & Booss-Bavnbek 2004], to mention one publication among a thousand possibilities, explores one aspect of the "mathematization of complexity".

Military Perspectives

At the conference, the point was strongly made by Colonel Svend Bergstein that actual war cannot be calculated, no more today than in Clausewitz's times;[17] war and fighting not only involve too many unpredictable external factors but also those aspects of human behaviour which are most atavistic and contrary to reason – in Bergstein's view mostly due to the prevalence of stress and sleep deprivation during combat operations.

Nevertheless, and as a matter of fact, mathematics – that is, mathematical thinking, mathematical methods, and mathematics-based technology – has become an integral and even essential part of modern warfare (though often not recognized as such by a general public which only sees the technology and not the underlying mathematics). This does not mean that mathematics has become the major expense of the military apparatus – mathematics and what goes with it is a cheap way to use costly resources more efficiently, and mathematics is used not least for that reason.

Once more, we may list various aspects of this role and utility of mathematics as discussed at the conference and in other contexts.[18]

- Mathematics serves in managing the institution. Purchases of weapons systems are planned, war-games and logistics are calculated.[19]
- Weapons and weapons systems are optimized and their efficiency during action enhanced. This regards munitions (including missiles and bombs provided with guidance systems); delivery systems (including for instance aeroplanes provided with electronic countermeasure circuitry); the reconnaissance, control and communication interface ("to ensure that the right forces are at the right spot at the right moment, and with the right information about the enemy" – Svend Bergstein); and, across all of these, high-speed cryptography.[20] The improvement of data transmission technologies is of general importance for many of these questions, but the creation of data is not only an evident presupposition for having any data to transmit but in itself something which nowadays often asks for the use of even more sophisticated mathematics than the transmission.[21]

[17] See also S. Bergstein, this volume, pp. 183–215.

[18] Evidently it is difficult to find *any* technology that has been created during the last decades which is not somehow driven by mathematics. The list discusses such facets of the matter as go beyond what holds for any practice that involves computers or microelectronics.

[19] See in this volume Kjeldsen (pp. 126–152), S. Clausen (pp. 216–238), and H. Löfstedt (pp. 239–256). Observe that Clausen's title does not speak of the calculation of *war*, as does that of Bergstein, but of the calculation of *warfare* which includes logistics and other managing matters.

[20] Cf. U. Bernhardt and U. Rehmann (this volume, pp. 257–281).

[21] Interestingly, the analysis of the damages of the intestine of a wounded soldier by magnetic resonance imaging (MRI) and the localization of enemy ground forces by synthetic aperture radar (SAR) build on the same mathematics – both, indeed, by cleverly arranged rapid repetition squeeze out of a "short antenna" as much information as could be gained from an extended antenna without advanced mathematics [Schempp 1998: 44 and *passim*].

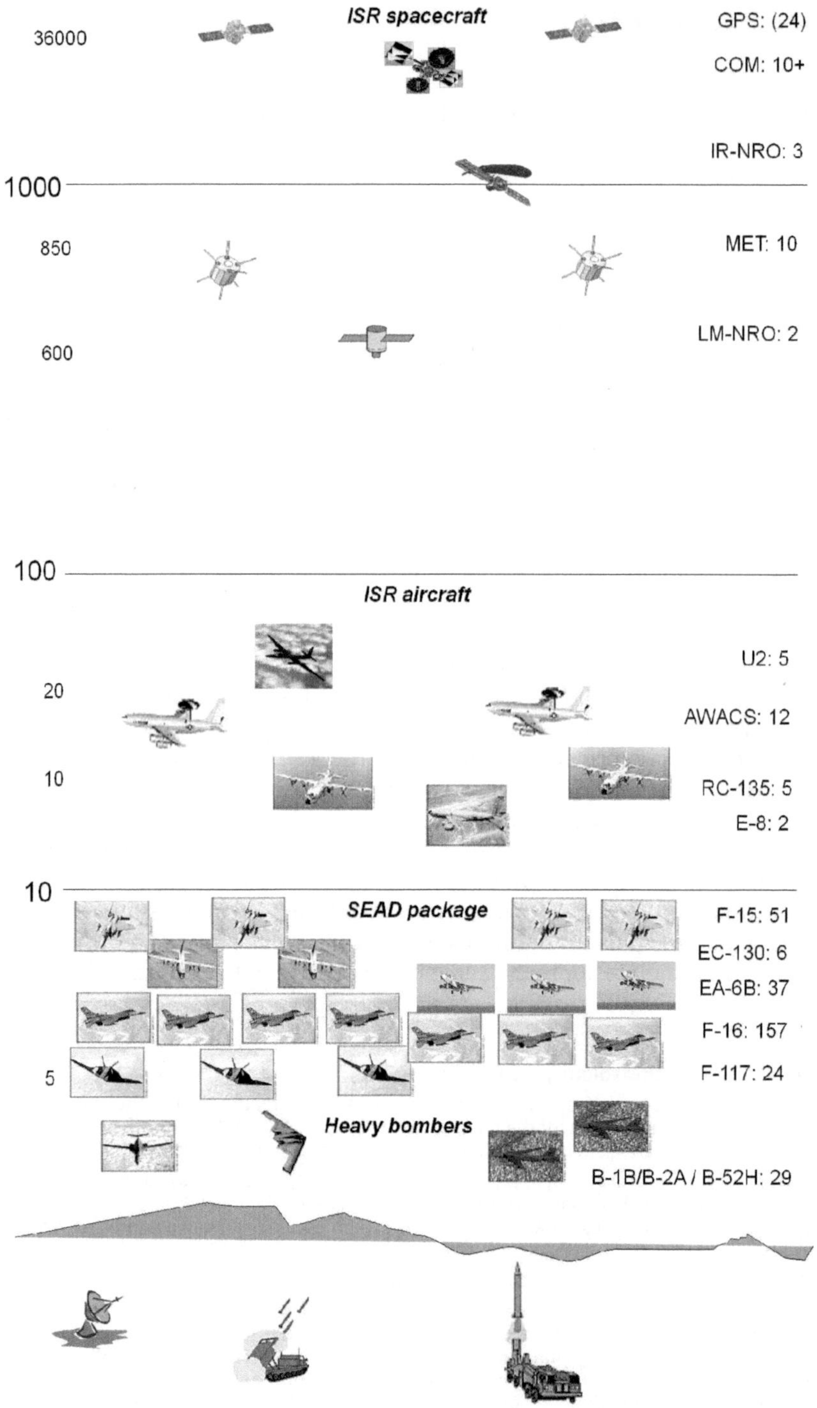
36000
ISR spacecraft
GPS: (24)
COM: 10+
IR-NRO: 3
1000
850
MET: 10
600
LM-NRO: 2
100
ISR aircraft
U2: 5
20
AWACS: 12
10
RC-135: 5
E-8: 2
10
SEAD package
F-15: 51
EC-130: 6
EA-6B: 37
F-16: 157
5
F-117: 24
Heavy bombers
B-1B/B-2A / B-52H: 29

Figure 4 (*opposite*). In World War II, the destruction of a major composite target might ask for the deployment of a thousand bombers. Nowadays a similar task may be effectuated by, say, 29 heavy bombers (lower level of the above schematized front view of an attack - heights are indicated in kilometres). But these have to be supported by another set of 275 fighters and ground attack fighters ("SEAD package") to suppress enemy air defence (bottom). Higher up, 24 "Intelligence-Surveillance-Reconnaissance" (ISR) aircrafts guide the action of the lower levels - and on top, dozens of ISR spacecrafts participate. Much more than informatics is thus involved in the support of the mission itself. So, the dramatic decrease of CEP (see Fig. 5, caption) has to be paid for by dramatic increase of support crafts. However, air raids are still the cheapest way of punitive war (forbidden by international law, but practised), inflicting huge economic losses on the enemy at extremely low operational costs.
[Courtesy: B. Booß-Bavnbek, B. C. Jørgensen, R. O. Rasmussen]

Precision Bombing Acronyms:

AWACS	Airborne warning and control system for air and combat control
B-<>	Long-range bomber with weapon payload of more than 10 tons
COM	Military communication / signals intelligence spacecraft
E-8	Joint surveillance and targeting attack radar system JSTARS
EA-6B	"Prowler" carrier-borne radar jammer
EC-130	"Compass Call" communication jammer
F-<>	Fighter and fighter ground attack aircraft
GPS	Global positioning system navigation satellite
IR-NRO	Infra-red (US) National Reconnaissance Office spacecraft
ISR	Intelligence, surveillance, reconnaissance package
LM-NRO	Imaging radar (US) National Reconnaissance Office spacecraft
MET	Weather satellite
RC-135	"Rivet Joint" signals intelligence gathering aircraft
SEAD	Suppressing enemy air defence package
U-2	Optical spy plane

- Similarly, the strategic planning of the possible use of weapons systems depends on mathematical calculation; even the dismantlement of weapons systems without the risk of destabilizing disequilibrium in the SALT negotiations was analyzed mathematically.[22] Fortunately, nobody implemented the strategy suggested by the naive versions of such planning games – to make a nuclear first strike and promise help to the 80%-annihilated enemy if no counterattack be made, supposing that the enemy would act "rationally" and submit.
- Perhaps unexpected by civilians but emphasized by some military analysts, simple accounting mathematics *performed by mathematically trained independent personnel and not by the active warriors* is mandatory if strategic gains and losses are to be assessed realistically – leading officers, like all of us, are easy victims of self-deceiving optimism and pessimism according to circumstances.[24]
- At the opposite end of the scale, mathematics may also be an indispensable tool. Thus, when the effect of fragmentation bombs on human bodies was to be predicted but humanitarian concerns prohibited testing on pigs, mathematical simulation was put into play.

[22] Cf. J. Scheffran, this volume, pp. 390–412.

[23] This does not disprove the utility of game-theoretical modelling, only the belief that human behaviour is always adequately described by the "rational economic man". Actually, sociobiological models of the same mathematical type indicate that the survival of a species is better guaranteed if egoistic individual sub-optimization is punished. The fear that the enemy might not accept the kind offer but take "irrational" revenge was exactly what made nuclear deterrence work, thus saving our species during the Cold War.

[24] This is the recurrent theme in [Meigs 2002]. Use of the game-theoretical "scenario bundle method" of [Selten 1999] is an advanced analogue.

War	CEP[m]	# bombs
WW II	1100	9140
Korea	330	823
Vietnam	130	128
Gulf	70	38
Kosovo	13	2

Figure 5. The average precision of bombing and firing is commonly characterized by the *Circular Error Probable (CEP)*, that is, the radius of a disc around the goal point within which (on average) 50% of the shots hit while 50% fall outside. (Kolmogorov's approach was more sophisticated). The table shows how CEP has decreased dramatically in aerial bombing over the last 60 years and how the efficiency of a bomber increased correspondingly. The table gives the calculated number of bombs required for "destroying" (i.e., hitting once) a 20 m x 30 m object.

- Ideologically, the waging of war is made more acceptable to the public by the presentation of warfare as precise and hence "more rational and clean". Although that aspect of the matter is not much discussed in the public sphere, this increased precision of weapons (which is real) depends essentially on the application of mathematics. Whereas Hitler preached German invincibility by presenting the Wehrmacht as "Fast as German greyhounds, tough as German *Lederhosen*, hard as Krupp steel", mathematics presents modern warfare as "fast by avionics, precise by GPS, safe by optimized operations planning".
- Similarly, certain mathematical representations of the task to be performed may serve to make the agent see it as a normal manipulation of symbols and thus to eliminate the need for appeals to atavistic instincts – say, seeing a village to be bombed as triangles similar to those of a computer game may facilitate the killing.[25] (Evidently, being at a height of 5 kilometres already has much the same effect).

Utility is one thing, possible backfiring that should be taken into account is another. Firstly, seeing war as "more rational and clean" may deceive (and often appears to deceive) not only the public but also political planners. This is not only devastating for the victims but already pernicious for the planners themselves who may engage recklessly their armed forces in operations and wars that are less easily won than predicted by the machine-rational perception of the character of war.

Less dangerous for planners but just as much for victims is the relative inexpensiveness of present-day mathematically supported asymmetric warfare for the attackers – if the subjugation of Serbia in the Kosovo war cost the subjugator only 7 billion US $, that is, 700 $ per Yugoslav capita, the temptation is great to solve all similar problems in a similar way.[26] (In the moment such a war turns out as things develop to involve the use of ground forces, costs of course explode, and we are brought back to the situation discussed in the previous paragraph).

[25] This issue is touched on by Davis (this volume, pp. 174–179) and by Bernhardt and Rehmann (this volume, pp. 257–281).

Figure 6. The destruction of the bridges across the important international waterway Danube, some of them in the North of Yugoslavia and hence far away from Kosovo where the Yugoslav (here the Varadin Bridge in Novi Sad) military operational capability should be hindered, was unlawful by The Geneva Protocol I. The counter-argument given is that this kind of warfare is, after all, cost-efficient in human lives, even for the target population – as illustrated by the undamaged blocks of flats standing near to the crushed bridge. The as yet mysterious health problems of Nato soldiers who participated in the Gulf and Kosovo wars and the dramatic increase in cancer rates in Iraq tell us that other damages that do not show up on photographs may turn up in medical statistics. An even greater cost of this high-precision warfare supported by mathematics is its very introduction of the concept of justified risk-less punitive wars without bloodshed. This creates invincibility illusions, lowers the barrier against war and talks people into accepting war. [Source: *NATO Crimes in Yugoslavia*, Ministry of Foreign Affairs, Belgrade, 1999]

Another feature of the mathematization of warfare, also contributing to the ongoing militarization of our world but not restricted to the field of easy asymmetric wars and updated "gun boat diplomacy" – actually less important there than in symmetric conflict – is the transformation of the "Krupp model" into an "infinite Krupp model". War and prepared war is always between two (possibly more) parts – Clausewitz would speak of a *Zweikampf*, a duel, which has now become a "duel of systems".[27] In the nineteenth century, Friedrich Krupp would first develop nickel-steel armour that could resist existing shells, then chrome-steel shells that could pierce this armour, then high-carbon armour plate that resisted these, then cap-shot shells that could break this plate – and that was the end of it. In the duel between surface-air missiles and aeroplanes, no physical limit prevents an ongoing sophistication and ensuing arms race. Cap-shot shells were and remained extremely expensive; so are stealth bombers and fighters – but such measures as depend solely on sophistication of computer soft- and hardware have neither budgetary nor intellectual definitive bounds. Processes depending on physics and

[26] As emerges from E. Schmähling's contribution (this volume, pp. 282–296), what is really at stake here is first of all increased pressure on civil objects and the functioning of society. At least in the Kosovo War, mobile military targets were mostly not detected, and *a fortiori* not destroyed. Tactical warfare being unsuccesful in spite of massive electronic support, mathematics showed its serviceability in strategic bombing of civil objects (legally definable as terrorism).

chemistry may have definitive natural boundaries. Those depending solely on mathematics seem to have none. The ensuing virtual absence of limits enhances the stress on both sides, and thus the speed and instability of such a race.

Ethics

Mathematics, according to a familiar view, is a neutral tool. As once formulated by the statistician Jerzy Neyman, "I prove theorems, they are published, and after that I don't know what happens to them".

This is certainly an important feature of the mathematical endeavour, and does not only hold for theorems and theorem production. Also the teaching of mathematics, the production of high-level general mathematical competence in the population, is a precondition not only for the waging of modern war but also for the functioning of our whole technological society (quite apart from the cultural value we suppose it to possess).

But the title "mathematics and war" implies ethical dilemmas. In order to avoid having the ethical discussion end up in non-committal "*I* feel..."/"but *I* feel", we may start by looking at the actual ethical choices of some well-known figures.[28]

- Laurent Schwartz used his high academic prestige to make his resistance to the French and American wars in Algeria and Vietnam more efficacious; he saw no connection between his work in mathematics and his political commitment.[29]
- Niels Bohr, when becoming aware of the German nuclear bomb project, supported the competing Anglo-American project; when discovering the dangers that were to arise from the success of the latter, he issued warnings to responsible politicians (Churchill, Roosevelt) and to the public (the "Open letter") – using his prestige as an originator of the underlying theory and as a collaborator in the project (and arguably overrating the impact his interventions might have).[30]
- Alan Turing, quite sceptical of British society (for political as well as personal reasons), put his outstanding abilities in the service of war with total loyalty when he felt it was needed; unlike Bohr, he did so without ever putting himself into focus.[31]
- Kinnosuke Ogura had been a strong promoter of (Marxist-inspired) democratic modernization of Japan, and had opposed Japanese policies as being parallel to German and Italian Fascism. After the beginning of the aggression against China in 1937, however, patriotism and the prospect of using war as a way to

[27] Cf. Löfstedt, this volume, pp. 239–256.

[28] J. Ryberg (this volume, pp. 352–364) discusses the ethical problem in general terms. The reader may think the following examples into the framework he presents.

[29] See [Schwartz 2001].

[30] See F. Aaserud, this volme, pp. 299–311.

modernization urged him to play a central role in the organization of Japanese mathematics in the service of the military state. After the war he regretted, without specifying too directly what he had done.[32]

- John von Neumann, like Turing, applied his outstanding abilities in war research. Von Neumann did so both during World War II and in the early Cold War; whereas Turing had been a loyal participant about whose personal attitudes in the matter we know nothing, von Neumann made the creation of the H-bomb a personal project which (well served by Stanislaw Ulam and Edward Teller) he did all he could to promote – his aim being to make possible a preemptive first strike.[33]
- Lev S. Pontryagin gave up an extremely fruitful research line in algebraic topology and created control theory. In hindsight this appears to have been caused by a will to serve his socialist country by solving the problems of guiding intercontinental ballistic missiles – thus making *im*possible the same first strike.[34]
- Decades before, G. H. Hardy had tried to avoid that usefulness of his science which consists in "accentuat[ing] the existing inequalities in the distribution of wealth, or more directly promot[ing] the destruction of human life" by concentrating on supposedly useless number theory. Ironically, he repeated this phrase in 1940, when number theory was about to become a cryptographic resource.[35]
- The radical pacifist Lewis Fry Richardson published his path-breaking *Weather Prediction by Numerical Process* in 1922 after having made sure that no less than 64000 "computers" (human beings furnished with desk calculators) would be needed to predict in one day the weather one day ahead. This he saw as a guarantee that numerical weather prediction could not be put to military use.[36]

To what extent can these serve as exemplars and role models? Firstly, they show that two fundamentally different situations must be distinguished. One is that of Schwartz, Hardy and Richardson: deep scepticism towards their own society, or toward aspects of that society as a warring power. The other is that of the remaining examples: they accepted their own society and its warfare or armament policies, either in general or under actual circumstances – certainly with different degrees of identification.

In the second situation, the ethical dilemmas are few. Obviously, one will see no objections to doing his best. Dilemmas, it is true, are not totally absent: one may still, like von Neumann, give an extra push; one may, like Turing, be fully loyal but leave the political decisions to those who are officially entitled to take them (whether politicians, voting citizens in general, or military men); or one may, like Bohr, use one's particular standing and insight and moderate, warn, or point to alternative options.

[31] See Hodges, this volume, pp. 312–325.

[32] See T. Makino, this volume, pp. 327–336.

[33] See, e.g., [Heims 1980].

[34] See Gamkrelidze, this volume, pp. 160–173.

[35] This repetition is found in [Hardy 1967: 120].

The situation of the sceptic is less clear-cut. Very few of us are in a situation (the situation, say, of von Neumann and Pontryagin) where nobody else could do what we are doing; these few may influence matters directly by deciding to cooperate or not to cooperate, but they remain exceptions.

Most mathematicians, if they choose not to cooperate with the military in mathematics research and teaching, will have little effect, and little of what most mathematicians do in research as well as teaching is directed toward a specific application. Deciding to abstain from working with a particular discipline because it seems "corrupt" is mostly futile. Giving up mathematics is giving up not only military applications but anything mathematics can be used for – and whatever cultural value we may ascribe to mathematics.

However, the practice of the mathematician consists in more than the abstract production and dissemination of theorems.[37] Any mathematician is in a *particular* situation, and in any particular situation there are specific conditions and a specific room for decisions.[38] One may, for instance, widen one's own insight and global understanding of the role of mathematics, and try to share it with students, colleagues and the public – or one may choose to remain (and leave others) as blind as comfortable. One may be a teacher in one or the other position, teaching within a highly stratified or a more egalitarian education system; one may organize the research of an institution, one may be a prestigious researcher, or one may be the newly appointed young colleague. One may be in the top of the "AMRC chain" (cf. above, p. 7), one may be in its periphery knowing or not knowing to belong there, or perhaps be wholly outside it. One may belong in a national or institutional context where political choices of consequence are currently accepted by colleagues or in one where they are unwelcome and considered bad taste and lead to social isolation.[39] In each situation, the scope of ethical choices is different, and no general ethical rules or advice can be issued.[40] What can be said in general is that the supposed neutrality of mathematics per se does not entail the neutrality of these ethical choices.

[36] See [Ashford 1985].

[37] How much more is illustrated by P. Davis's probing (this volume, pp. 174–179) of the relations between mathematics, entertainment, and war. Not only mathematical algorithms but also the creation of mentalities and ideologies are at stake here.

[38] W. Göhring (this volume, pp. 336–351) tells the story of an ethically conscious ordinary research mathematician in a particular real-world situation involving such specific down-to-earth conditions as labour legislation, institutional obligations, etc. Fukutomi (this volume, pp. 153–159) relates how a handful of famous university mathematicians from different countries were able to organize a protest movement against the Vietnam War spreading over the globe in a few months of 1966.

[39] As Fukutomi points out, the anti-war movement gained exceptional strength among Japanese mathematicians, both compared to mathematicians elsewhere and to scientists from other disciplines in Japan. In contrast, mathematicians of the US were much more cautious and conservative than US scientists in general during the same years – see [Ladd & Lipset 1972].

An Enlightenment Perspective?

The Enlightenment believed that reason *might* serve general progress; Rousseau and Swift pointed out that too often reason is used in the service of purely technical rationality and for purposes of sub-optimization, with morally and physically disfiguring effects. According to Defoe's Robinson Crusoe, "Reason is the Substance and Original of the Mathematicks". Hugo Grotius the founder of international law, when referring to the method of mathematicians as his inspiration, appears to have had the same identification in mind.[41] Where does that leave mathematics with regard to disfigurement and progress today?

Much of what was said above concerning the utility of mathematics for the military rather points to disfigurement. Most alarming of all are probably not the actual uses but the ideological veil of rationality, cleanness and surgical accuracy which is derived from the mathematization of warfare. By generalization one might claim that this applies not only to the military aspects of our modern technical society but to technically rational society as a whole.

However, one of the ways in which mathematics serves the military points in the opposite direction: that sober-minded elimination of self-deceiving optimism and pessimism which can be provided by mathematical reasoning and calculation. Mathematics-based reason at its best *should* allow us also in larger scale to unlearn conventional wisdom, to undermine facile indoctrination, to distinguish the possible from glib promises. It *might* help us, if not to find any absolutely best way – this is too much to expect from reasoned analysis – then at least to evade the worst. If reason is really "the Substance and Original of the Mathematicks", mathematics might serve to make clear to us that war is fundamentally irrational and unreasonable *not only in commonplace ideological generality but in specific detail*. If mathematics is *not* able to do such things, then its presumed cultural value might be nothing but a convenient excuse for ruthless technical sub-optimization.

Admittedly, technical rationality prevails over reason for the moment, both concerning the general political situation and the uses to which mathematics is put. Mathematical theories are ethically neutral, it has been argued. Mathematics *as a general undertaking*, instead, is ethically ambiguous: responsibility, whether they like to remember it or not, remains with its practitioners, disseminators, and users.

[40] A somewhat less abstract discussion of the matter can be found, however, in [Booß & Høyrup 1984].

[41] See I. M. Jarvad, this volume, pp. 367–389. Grotius goes somewhat further and tries to reshape legal thought as a formal science modelled after mathematics, pointed out by him to deal with its objects in abstraction from their concrete physical character. Similarly, Grotius requires that international relations should be governed by general rules that are independent of sympathy, religion and of the character of states. That principle was the basis for the Peace of Westphalia of 1648 and was mostly respected since then among states that recognized each other as "civilized". The UN Charter proclaimed the universality of the principle; as these lines are being read the reader will know whether the principle still exists or has been reduced to a mere phrase.

References

Ashford, Oliver M. (1985): *Prophet – or Professor: The Life and Work of Lewis Fry Richardson*. Bristol & Boston: Adam Hilger 1985.

Booß, Bernhelm & Høyrup, Jens (1984): *Von Mathematik und Krieg. Über die Bedeutung von Rüstung und militärischen Anforderungen für die Entwicklung der Mathematik in Geschichte und Gegenwart*. (Schriftenreihe Wissenschaft und Frieden, Nr. 1). Marburg: Bund demokratischer Wissenschaftler 1984.
Somewhat updated English translation as pp. 225–278, 343–349; in: Jens Høyrup: *In Measure, Number, and Weight. Studies in Mathematics and Culture*. New York: State University of New York Press 1994.

Bleecker, David & Booss-Bavnbek, Bernhelm (2004): *Index Theory with Applications to Mathematics and Physics*. Boston: International Publishers 2004 (to appear).

Dalitz, Richard H., & Sir Rudolf Ernst Peierls (1997) (eds): *Selected Scientific Papers of Sir Rudolf Peierls. With Commentary*. Singapore and London: World Scientific Publishing, Imperial College Press 1997.

Davis, Chandler (1989): A Hippocratic Oath for Mathematicians?, pp. 44–47; in: Christine Keitel (ed.): *Mathematics, Education, and Society*. Science and Technology Education. Document Series No. 35, Paris: UNESCO 1989.

Epple, Moritz, & Remmert, Volker (2000): 'Eine ungeahnte Synthese zwischen reiner und angewandter Mathematik': Kriegsrelevante mathematische Forschung in Deutschland während des II. Weltkrieges, vol. I, pp. 258–295; in: Doris Kaufmann (ed.): *Geschichte der Kaiser-Wilhelm-Gesellschaft im Nationalsozialismus. Bestandsaufnahme und Perspektiven der Forschung*. Two volumes, Göttingen: Wallstein Verlag 2000.

Gericke, Helmuth (1984): Das Mathematische Forschungsinstitut Oberwolfach, pp. 23–39; in: Willi Jäger, Jürgen Moser, Reinhold Remmert (eds): *Perspectives in Mathematics. Anniversary of Oberwolfach 1984*. Basel: Birkhäuser 1984.

Godement, Roger (1978): Aux sources du modèle scientifique américain I-III. *La Pensée* **201** (Octobre 1978), pp. 33–69, **203** (Février 1979), pp. 95–122, **204** (Avril 1979), pp. 86–110.

Godement, Roger (1994): Science et défense. Une brève histoire du sujet I. *Gazette des Mathématiciens* **61** (1994), pp. 2–60. (Part II has not appeared).

Godement, Roger (2001): Postface. Science, technologie, armament, pp. 377–465; in: idem: *Analyse mathématique*, vol. II. Berlin etc.: Springer 2001.

Gowing, Margaret (1964): *Britain and Atomic Energy 1939–1945*. London: Macmillan 1964.

Gross, Horst-Eckart (1978): Das sich wandelnde Verhältnis von Mathematik und Produktion, pp. 226–269; in: P. Plath & H. J. Sandkühler (eds): *Theorie und Labor. Dialektik als Programm der Naturwissenschaft*. Köln: Pahl-Rugenstein Verlag 1978.

Hardy, Godfrey Harold (1967): *A Mathematician's Apology*. Cambridge: Cambridge University Press. [1]1940, 1967.

Heims, Steve J. (1980): *John von Neumann and Norbert Wiener. From Mathematics to the Technologies of Life and Death*. Cambridge, Mass.: MIT Press 1980.

Høyrup, Jens (1986): [Review of [Gericke 1984]]. *Zentralblatt für Mathematik* **565** (1986), p. 7.

Høyrup, Jens (2002): *Lengths, Widths, Surfaces: A Portrait of Old Babylonian Algebra and Its Kin*. (Studies and Sources in the History of Mathematics and Physical Sciences). New York: Springer 2002.

Ladd, Everett Caril, Jr., & Lipset, Seymour Martin (1972): Politics of Academic Natural Scientists and Engineers. *Science* **176** (1972), pp. 1091–1100.

Meigs, Montgomery C. (2002): *Slide Rules and Submarines. American Scientists and Subsurface Warfare in World War II*. Honolulu, Hawaii: University Press of the Pacific 2002, reprinted from the 1990 edition.

Morse, Marston (1943): Mathematics and the Maximum Scientific Effort in Total War. *Scientific Monthly* **56** (1943), pp. 50–55.

Proc. ... 1920: Comptes Rendus du Congrès International des Mathématiciens (Strasbourg, 22–30 Septembre 1920). Toulouse, 1921.

Remmert, Volker (1999): Vom Umgang mit der Macht. Das Freiburger Mathematische Institut im 'Dritten Reich'. *Zeitschrift für Sozialgeschichte des 20. und 21. Jahrhunderts* **14** (1999), pp. 56–85.

Rosser, J. Barkley (1982): Mathematics and Mathematicians in World War II. *Notices of the American Mathematical Society* 29:6 (1982), pp. 509–515.

Schempp, Walter Johannes (1998): *Magnetic Resonance Imaging. Mathematical Foundations and Applications*. New York etc.: Wiley 1998.

Schwartz, Laurent (2001): *A Mathematician Grappling with his Century*. Boston: Birkhäuser 2001. French ed. 1997.

Selten, Reinhard (1999): The Scenario Bundle Method; in: idem, *Game Theory and Economic Behaviour - Selected Essays*, pp. 291–325. Cheltenham: Edward Elgar 1999.

The AMRC Papers. By Science for the People, Madison Wisconsin Collective. Madison, Wisconsin: Science for the People, 1973.

Tschirner, Martina, & Göbel, Heinz-Werner (1990) (eds): *Wissenschaft im Krieg - Krieg in der Wissenschaft. Ein Symposium an der Phillips-Universität Marburg. 50 Jahre nach Beginn des II. Weltkrieges*. Marburg: Eigenverlag AMW 1990.

Wiener, Norbert (1964): *I am a Mathematician. The Later Life of a Prodigy*. Cambridge, Mass.: M.I.T. Press 1964.

Youschkevitch, Alexander A. (1974): Markov; in: *Dictionary of Scientific Biography*, vol IX, pp. 124–130. New York: Scribner 1974.

I
Perspectives from Mathematics

Military Work in Mathematics 1914–1945: an Attempt at an International Perspective

REINHARD SIEGMUND-SCHULTZE*

After discussing some general methodological points this paper investigates in some detail the use of mathematics (not necessarily confined to academic mathematicians) in military work in the German and American cases, where as of now most knowledge is available. With respect to five other countries (Soviet Union, Great Britain, Italy, France, Japan) the discussion has to be restricted to a rough outline of problems and a collection of the rather scattered sources available. A case study is presented of mathematical war work in ballistics (Wolfgang Haack) which illustrates the complexity of the problems involved and the short distance between basic academic research and applied work for the military. The paper is concluded by appendices (tables) giving condensed information on mathematical war work in the seven countries considered.

1 Introduction and General Discussion

Wars are primarily waged with technologies – be it for the destruction of material assets and human beings or for the control and corruption of minds. Wars are, however, not waged by directly using the sciences or the humanities, or mathematics for that matter. These serve "only" – however crucially and however in themselves responsible for the development of new war technologies – as intellectual techniques behind the technologies. Thus the two World Wars of the 20th century can not simply be declared as the "chemists'" or the "physicists'" wars.[1] This is even more true of an alleged "mathematicians' war". According to Mehrtens,

> "a 'mathematical war' ... [is] hard to imagine. Even with computer science becoming central to military technology, mathematics still remains in the background, being 'just a tool'. The history of the relation of mathematics to the military and to war has therefore to be analysed as a mediated connection."[2]

* Agder University College, Kristiansand, Norway.
Email: Reinhard.Siegmund-Schultze@hia.no

[1] According to chemist J. B. Conant WW II was a "physicists' war" (Gray 1943, p. 25). "By early 1942 OSRD had spent four times as many contract dollars in physics as in chemistry." (Kevles (1987), p. 320). On science in WW II in general see Hartcup (2000).

[2] Mehrtens (1996), p. 87.

That there has been so much talk of "scientific war" in the public and in the historical literature may have *partly* to do with ideological needs to interpret war as something of a higher moral order and to conceal its dirty reality. In this context, mentality conflicts between the military and engineers may have played a role. Vannevar Bush, the head of the American War Research Organisation *Office of Scientific Research and Development* (OSRD), described it in his recollections of 1970:

> Among older military men the engineer was at first regarded as in all probability a thinly disguised salesman, and hence to be kept at arm's length. [(53)...] So all O.S.R.D. personnel promptly became scientists [...}. Even recently when we sent the first astronauts to the moon, the press hailed it as a great scientific achievement. Of course it was nothing of the sort; it was a marvelously skillful engineering job. Now that there is a National Academy of Engineering, perhaps the title of engineer will recover its just recognition.[3]

But this was, of course, only half of the truth, and the other half is going to be at the centre of this paper: Vannevar Bush, the mathematically knowledgeable engineer, author in 1929 of a book on "Operational Circuit Analysis. With an Appendix by N. Wiener" – was certainly aware of the growing importance of basic science in engineering and, in particular, in war technologies. His invention in the early 1930s of the so-called "differential analyzer" – later an important tool in the war – was based on ideas by British "applied mathematicians" Kelvin and Maxwell of the 19th century, even if in itself it was primarily technology.[4] One could go even further and say that the *actual* role of basic science such as mathematics in war-related technologies is concealed rather than exaggerated in the historiography of technology itself because its general nature tends to escape the reader's understanding and – in so far as it is a mere "application" of known knowledge – it is likely to be considered as less "original" than some physical model such as Ludwig Prandtl's boundary layer theory of 1904.[5] Moreover, after the war (!, not during) scientists themselves are often no longer interested in stressing their connection to warfare and they are often rather occupied with an ideological "whitewashing" of their profession.[6]

So let us come back to the quote by Mehrtens on the "mathematical war": Nota bene: Mehrtens does not talk about a "mathematicians' war" but about a "mathematical war" rather.

In fact, we have to introduce another distinction: War technologies require – and are promoted by – intellectual techniques. It is rather irrelevant for the

[3] Bush (1970), p. 54.

[4] So argued in Birkhoff (1980), p. 21. See also Bush (1931).

[5] I found this confirmed when I tried to evaluate Richard von Mises' importance as an "aerodynamicist". He is often mentioned as such by mathematicians, especially in connection with his theory of the "aerodynamic centre" (1917/20) of an airfoil and his "Mises-transformation" (1927) in boundary layer theory, but particularly less so by engineers, even by mathematically inclined ones such as Theodore von Kármán.

[6] As discussed for instance for the British case by Edgerton (1996, p. 20), who also quotes the telling booktitle Beer (1961).

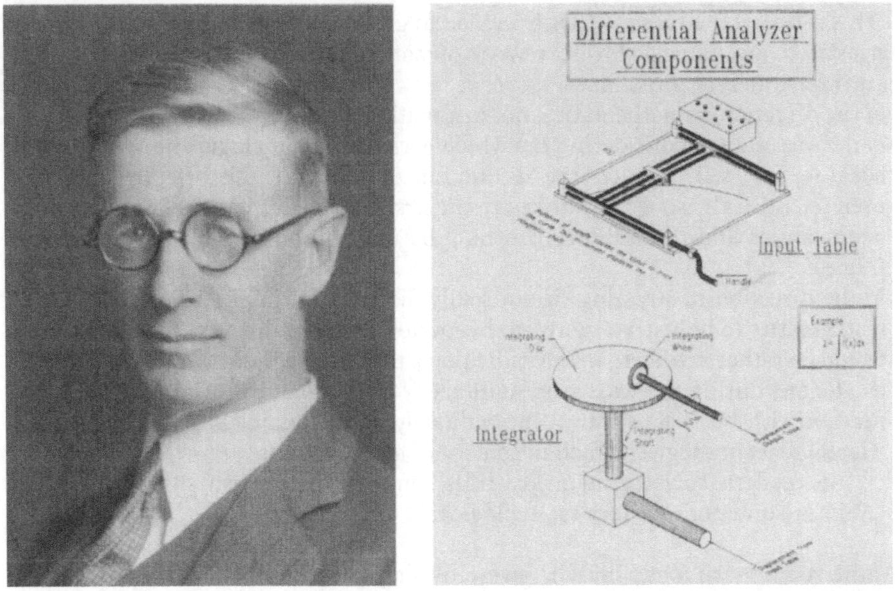

Figure 1. *Left:* Vannevar Bush (1890–1974), head of the American War Research Organisation OSRD, leading in the implementation of military funding of pure research [Source: Masani (1990), p. 166]
Right: A schematic view of his *Differential Analyzer* of 1930, the first analogue computer [Source: Bowles (1996), p. 6]

military who produces and applies these intellectual techniques as long as their existence and application is guaranteed for warfare.

And in fact, we have to ask: Is it really and necessarily the academic mathematicians themselves who are most likely to adapt these intellectual techniques for immediate military use? Are not at least certain social traits required among academic mathematicians or certain social pressure exerted on them before they themselves enter the realm of industrial or military applications? Does a self-serving and self-mobilising[7] "touting" of mathematical theories by mathematicians as "war-relevant" – something which has often occurred in history with or without real basis[8] – already constitute "military work of mathematicians"?

In reality we cannot evade some discussion of the more general type of question such as: "what is mathematics?"; what, in particular, is "applied mathematics?"; who are the people that are doing it?; and what are the historical circumstances necessary for the transfer of that type of intellectual techniques into military technologies?[9]

[7] See the analysis of this notion in Mehrtens (1986/96) and Epple/Remmert (2000).

[8] "It is a fact of history that scientists usually have had to take the initiative in making their gifts known to the government." (Gray 1943, p. 41)

[9] This discussion somewhat duplicates original historical discussion among mathematicians, physicists, and engineers in various countries, for instance on the topic whether "applied mathematics" is just mathematical physics, or whether it includes engineering mathematics, discussions on the mental divide between mathematicans and engineers such as in Kármán (1940/43) etc.

This is one of the reasons that I decided to gradually extend the topic originally suggested to me, namely "Military Work of mathematicians in World War II", in at least three different directions:

- in the direction of mathematics, not just mathematicians;
- in the direction of considering the whole period which includes the World Wars, because, as it will be seen, the second one is politically, historically, mentally, even scientifically so closely linked to the first, and because the relation between mathematics and the military can only be understood in a broader historical frame;
- in the direction of stressing (if not really discussing) the emerging relations of mathematics to industry exactly in this period between the wars. In particular it would be rather artificial, nearly ridiculous, to separate industrial work of mathematicians during the wars from military work. As one historian put it in connection with his discussion of the militarily and economically highly relevant telephone technologies, which need a lot of mathematics:

> In modern societies it makes little sense to distinguish sharply between military and non-military technologies.[10]

Still, some sharp focus shall be given to this paper, and this focus is twofold: it is the role of mathematics in military work and it is an attempt at an international perspective, which has rarely been given in the existing literature.[11]

The article will in particular try to tentatively discuss in the introduction and in several case-studies for the various national environments one central problem: How did the communication channels function between academic mathematics and the application areas, especially the military ones, and to which extent did this depend on the very different social and national preconditions?

If we first look for a moment at the more "epistemic"[12] side of the problem – and not at the equally important social, political and technological sides – there is no doubt that World War II, in particular, required increasingly – on all sides of the battle line – techniques which resembled those that had been developed within the existing *academic discipline mathematics* at universities world-wide decades or centuries before. It is also clear *which* military technology required the most of mathematics. I only mention catchwords for military technologies such as aeronautics in close connection to ballistics, rockets[13] and the atomic bomb, and – as mathematical techniques – computers, numerical analysis, cryptology, operations

[10] Kragh (1996), p. 37. The same point is stressed by mathematician Godement (1978/79), who reserves for mathematics itself only a small part of his study on the American scientific model.

[11] Exception is some historical work with international perspective on single cognitive aspects of applied mathematics in war context. In this category belong some publications on the history of computing e.g. Birkhoff (1980), Todd (1990), and Goldstine (1972), as well as on the history of exterior ballistics (McShane et al. (1953)), and aerodynamics (Kármán 1954).

[12] In allusion to a recent study (to be discussed below) on Germany, Epple/Remmert (2000), esp. pp. 281ff.

[13] For a sketch of the history and mathematics of rockets in various countries (Russia, Germany, U.S.) see Newby (1988).

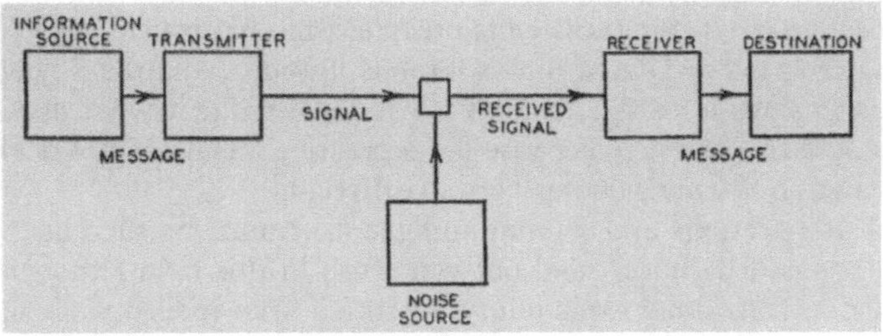

Figure 2. Schematic diagram of a general communication system.
[Source: Shannon/Weaver (1949), p. 5]

research, statistical sequential analysis, prediction theory, control theory.[14] It is also clear that some of those military technologies required *new and very modern mathematical tools* which had to be invented even in pure mathematics and were not available in stock, so to speak.[15]

Hodges, the biographer of Alan Turing, stresses one important aspect of these changes:

> But rapidly developing, and not only at Bletchley Park and Washington, was a new kind of machinery, a new kind of science, in which it was not the physics and chemistry that mattered, but the logical structure of information, communication, and control.[16]

There were, however, both similar and different "epistemic preconditions" for the military application of mathematics in various countries. Internationally minded mathematicians such as American Norbert Wiener from the MIT in Boston realised this. With respect to his work on anti-aircraft fire-control predictors, strongly inspired by pure mathematical research in statistics and what later would become information and communication theories, Wiener wrote retrospectively in his autobiography of 1956:

> At an early stage of my work for the United States military authorities ... the question came up whether anybody abroad was likely to be in possession of ideas similar to mine. I said that they would unquestionably receive no particularly ready reception in Germany; that my own friends Cramer in Sweden, and Lévy, France, might well have been thinking along similar lines, but that if anyone in the world were working on these ideas it would most likely be Kolmogoroff in Russia. This I said because of my knowledge that for twenty or thirty years hardly had either of us ever published a paper on any subject but the other was ready to publish a closely related paper on the same theme.[17]

[14] See the contribution by R. V. Gamkrelidze, one of the creators of the theory as an after-war development, in this volume.

[15] Rosser's (1982) contention that "except in cryptanalysis, hardly any of the mathematics done for the War effort was of a higher level" (509) seems exaggerated.

[16] Hodges (1983), p. 252.

[17] Wiener (1956), p. 261/2. On Wiener see Masani (1990).

Wiener refers in his quote to lacking epistemic preconditions in his field in Germany, and to existing ones in the United States, France, Sweden, and the Soviet Union. But what does this mean for the actual mathematical military work done in the countries considered? Not much, because the respective social and political conditions, not discussed in Wiener's quote, were so different:

Failing theoretical preparations and missing indigenous traditions such as in *Germany* would not necessarily have ruled out war work in this field although those conditions have, in fact, some explanatory potential. That indeed work of this kind was apparently *not* done in Germany in the particular field of prediction theory[18] may also be explained by growing problems of international mathematical communication during the war. *France* under German occupation was no likely place for mathematical war research on a large scale, nor was it neutral *Sweden*. Finally: the actual role and full implication of A. N. Kolmogorov's mathematical work in the *Soviet Union's* war effort is not clear until this day.[19] To have a good feeling for applications among pure mathematicians and to have a man with superior physical insight such as Kolmogorov does not guarantee a solution of the precarious problem of transfer from research to development, from mathematical invention to military innovation.[20]

The other side of the coin is this: during the war(s) a lot of at least potentially applicable theoretical work was done in various countries – whether they were involved in the war effort or not – that escaped attention of men such as Norbert Wiener abroad and was likewise not noticed due to the communication blackout during much of the war(s) and even later in the Cold War. Mathematical work or mathematics-related engineering work that was potentially war-important, such as done in France by E. Borel on game theory and émigrés W. Döblin and F. Pollaczek on Markov chains and queuing problems or in Germany by K. Zuse on digital computers, was not, for various historical reasons, actually applied (= transferred into the war effort) and therefore partly or temporarily ignored in the countries that would write the history of the war and set the norms for the scientific enterprise after 1945, especially the United States.[21] Take the following quote by Garrett Birkhoff, who had turned from abstract algebra to aerodynamics during the American war effort:

> After the defeat of Hitler, the U.S. (along with Canada) was standing triumphant, virtually unscathed in a world in which most advanced countries were prostrate and in ashes. Its scientific preeminence was taken for granted, and most Americans thought of computing machines as almost a national monopoly. They had largely forgotten the tradition of European scientific superiority that had been generally accepted only a decade earlier. What is now known about ENIGMA and the Zuse machines should help to dispel this illusion.[22]

[18] The report by R. Hosemann (1948) on "Verfolgungskurven" (follower curves) is rather pure elementary geometry and does only marginally touch upon probability questions.

[19] But see the remarks by A. N. Shiryaev in this volume.

[20] This problem was discussed for aeronautics in the Soviet Union in Bailes (1978).

[21] See Godement (1978/79) and Lax (1989).

[22] Garrett Birkhoff (1980), p. 29.

Figure 3.
Mechanical computing machine widely used in the pre-computer age, including parts of WW II.
[Source: Meyer zur Capellen (1949), p. 60]

As to the conditions that Wiener himself faced with respect to the application of his mathematical ideas in the United States of the 1920s and 1930s, one should not forget the following: the U.S., so rightly considered the leading country in both pure and applied mathematics after World War II, was rather backward with respect to "academic applied mathematics" in the European sense before the immigration of the 1930s, as will be argued in some more detail below.[23] This is what the same Garrett Birkhoff had to say on a period in the 1920s, when his father George David Birkhoff reigned supreme in American mathematics:

> Wiener was also notable as one of the few Americans of his time who was outstanding in both pure mathematics and his (sic!) applications. How much of this can be attributed to his varied and cosmopolitan early background, and how much to his continuing contacts with non-mathematicians such as G. I. Taylor, it is hard to say.[24]

Birkhoff hints at the important role of international communication as a historical precondition of applied mathematics in academic America. It has been mentioned before that, beginning in 1920s, Wiener collaborated closely with the engineer Vannevar Bush at MIT. Bush's analogue computer *Differential Analyzer* was not only central in applied ballistics during WW II,[25] its emulation in Europe[26] promoted applied research elsewhere in the world. Not least, it served to stimulate crucial developments also with digital computers and in information theory (C. Shannon worked with it). Nevertheless, even Norbert Wiener, a pioneer in international communication[27] and in applications, would have his shortcomings

[23] See Siegmund-Schultze (2003) and some further remarks below.

[24] Birkhoff (1977), p. 63.

[25] On the use of the *differential analyzer* in dealing with the many variables in ballistical calculations, for example in order to take into account the Magnus effect see Gray (1943), p. 90. S. also Godement (1979), vol. 204, p. 94.

[26] Oslo (S. Rosseland), England (D. R. Hartree), partly Germany (W. Cauer, A. Walther).

[27] As a Guggenheim fellow in Göttingen for example.

Figure 4.
Svein Rosseland's analyzer in Oslo (1937), influenced by V. Bush's machine and envied by the Germans. [Source: Willers (1943), p. 234]

Figure 5 (*above left*).
Warren Weaver (1894–1978) – head of the American *Applied Mathematics Panel* (APM), founded 1942.
[Source: Reid (1976), appendix]

Figure 6 (*above right*).
Right: Norbert Wiener (1894–1964) in mathematical war service during World War I.
[Source: Masani (1990), p. 69]

Figure 7 (*left*).
Chief mathematician Thornton Fry (1892–1991) of Bell Labs. [Source: *American Mathematical Monthly* **89** (1982), p. 80]

later on as to his potential use in the American war effort. Not only was he ideologically hardly prepared for an undisputed call to arms, given his pacifist views, which even increased after WW II and the atomic bomb, Wiener was also clumsy both socially and manually, traits he obviously shared with Alan Turing in Great Britain, another important contributor to war research. In fact, Warren Weaver of the American *Applied Mathematics Panel* had considerable problems integrating Wiener into his team-work oriented effort.[28]

So "attitudes" of the mathematicians involved and of other "appliers"[29] of mathematics mattered heavily in the process of application of mathematics, especially in times of war, as one would expect, where militaristic or other emotional positions had their motivational part to play. It is no coincidence that many historians and people involved in the processes themselves have stressed that "applied mathematics" is primarily a social distinction, a "social contract"[30] even. Thornton Fry said in 1953 at a conference in New York City devoted to training in applied mathematics, basing much of his judgement on the involvement of *Bell Labs* in the American war effort:

> The difference between an applied mathematician and a pure mathematician is not the kind of mathematics he knows, it isn't even whether he can create epoch-making new ideas, or like most of us his ability lies principally in interpreting things that are already known. The distinction resides instead in the nature of his interests; in his attitudes, not in his aptitudes. It is almost a social distinction.[31]

But attitudes did *not* influence the process in a very simple or unequivocal way, as Wiener's example shows: the social and political environment matters just as heavily.

As late as 1941 Thornton Fry, then chief mathematician at the *Bell Laboratories*, stated for American industry a "contrast between the ubiquity of mathematics and the fewness of the mathematicians".[32] This he did in a widely-read report for the U.S. government on "Industrial Mathematics". The fact that in the years before World War II, applications of mathematics had not been as a rule the job of academically trained mathematicians, was, however, not restricted just to the United States. Let me give two additional historical examples:

Michael Fortun and Sylvan S. Schweber discussed the origins of *Operations Research* and *Systems Engineering* in Great Britain and the United States and their military use in World War II. They stressed the importance of mathematical techniques from game theory or statistics being used in *OR*[33] but concluded with respect to the people who were really doing the job:

[28] See the discussion in Heims (1980) and Owens (1989).

[29] This English word – herewith "created" – might be more appropriate than the traditional attribute "applied" in connection to scientists or mathematicians.

[30] Davis (1988).

[31] Fry (1953), p. 96.

[32] Fry (1941), p. 255.

[33] Fortun, M. and S. S. Schweber (1993), p. 338.

> Incidentally, mathematicians did not do well as operational analysts during WW II, presumably because they were not used to being parts of 'teams'.[34]

That the problem did not just lie on the subjective side of mathematicians "attitudes" is underlined by Mehrtens, who reports on the job situation for mathematicians coming from universities in the late 1930s in Germany:

> Very few mathematicians found jobs in industry, and the few who did were usually hired in lieu of some other sort of specialist. [Mathematician] Wilhelm Magnus, for example, reports that he was employed as a physicist in electrical industry; to employ a mathematician would have appeared absurd to his boss. But with the preparation for war, industrial work intensified and graduated engineers and physicists were rare; thus mathematicians could slip in.[35]

Now if mathematics could frequently be applied without professional mathematicians being present this was not necessarily always the case, especially not in the more modern fields and under the pressing needs of the war, as the British enterprise at Bletchley Park showed:

> The enigma machine was the central problem that confronted British Intelligence in 1938. But they believed it was unsolvable [...] In particular this department of classicists, [...] did not include a mathematician.[36]

The exigencies of the war finally showed the need for academically trained mathematicians in the last-mentioned case of cryptology as well as in the United States and other places.

More generally the War seems to have triggered in several countries a new step in the professionalisation of mathematics at large, the systematic training of industrial mathematicians being part of this process. This applies also to Germany where the authorities declared the creation during the war of a new type of mathematician, the "Diplommathematiker", which had been much demanded by the German mathematicians themselves.[37] In the United States, however, there seems to have been *missing one more link* in the chain of epistemic-societal preconditions for the application of mathematics in the war.

In fact, the difference between the United States and Germany was that there did not even exist in America around 1940 a considerable supply of academically trained mathematicians with interest or understanding for the needs of the engineers. By comparison the Germans and other European countries could draw on a long tradition of mathematical training of engineers at Technical Universities

[34] Ibid. p. 350. A closer look at ASWORG, the American *Antisubmarine Warfare Operations Research Group* under the lead of physicist Philip Morse, where half of the collaborators were mathematicians, may somewhat modify this picture. See T. Hoff Kjeldsen (2000) and her contribution to this volume. See also McArthur (1990), Meigs (1990), and Rider (1992).

[35] Mehrtens (1996), p. 105.

[36] Hodges (1983), p. 148. As to the mathematical nature of cryptology see also Levinson (1970).

[37] Mehrtens (1996).

– Theodore von Kármán and Richard von Mises bearing witness to that – and even of institutionalised training in applied mathematics at some universities like Göttingen and Berlin.

The deficiencies of the Americans in this respect even around 1940 were clearly stated in articles by Kármán (1943) and by AMS-president Richardson (1943); the latter saw a close connection between the insufficient state of "Applied Mathematics and the Present Crisis" and promoted his famous summer-schools under the title "Program of Advanced Instruction and Research in Mechanics" which was clearly oriented towards training in applied mathematics and took place beginning in 1941.

The role of the war as a catalyst of the development of *academic* applied mathematics in the United States was stressed after the war by Joachim F. Weyl, son of the famous German émigré Hermann Weyl, who organized a Conference on "Training in Applied Mathematics" at Columbia University, New York, in 1953:

> Without the demands, resulting from considerations of national security, applied mathematics in this country might be as dead as a doornail.[39]

Again, what Weyl meant was not just "applications of mathematics in industry" but rather developing germs of institutionalised applied mathematics at American universities – a development which was heavily funded even after the war by state agencies such as the military *Office of Naval Research*. The Second World War, so it seems, had finally led to a transfer of the originally French,[40] later German[41] role model of academically institutionalised applied mathematics to the United States. The pressure of the emergency situation in the U.S. around 1941 was doubtless dominant in this process, but émigrés to the United States such as Richard Courant contributed much to a spread both of this social ideal and of the cognitive mathematical prerequisites[42] for military work of academic mathematicians in WW II.

All examples brought forward so far point to the complicated structure and many-sidedness of epistemic, technical and social prerequisites for the use of

[38] See the publication under the same title as no.8 of the *Bulletin of Brown University* **40** (1943). In 1943 the journal *Quarterly of Applied Mathematics* was launched in close connection with Richardson's program. The maintenance of the name "mechanics" in the program served obviously to stress the area to which the mathematics was being applied.

[39] Weyl (1954), p. 22. Weyl was at that time "Director of Mathematical Sciences Division in the Office of Naval Research". The ONR was the leading state agency in the U.S. after WW II to support both applied and pure mathematics mostly on the grounds that it was defence-related.

[40] Ecole Polytechnique of 1794.

[41] Göttingen with Felix Klein and Carl Runge.

[42] In this latter respect one cannot overestimate the role of mathematical work such as Courant/Friedrichs/Lewy (1928), Courant/Hilbert (1924/37), and Courant/Friedrichs (1948). A detailed historical analysis of this "epistemic side" of the reception of German mathematical work in the U.S. is given in an unpublished work by this author, of which a summary is Siegmund-Schultze (2003). Some analysis also can be found in Goldstine (1972).

mathematics and mathematicians in the war. What was needed for a "successful application" of mathematics in the war was obviously on the *one hand* a bridging of all the mental divides and differences in attitude of the many people involved in a large scale effort; among them were engineers, military men, scientists, government administrators, coming from very different social groups of society.

On the *other hand*, and closely related to the more personal side, military application of mathematics required "continuity" also on the "level of things", be it theories or material artifacts and institutions. This was the continuity (or continuum) of mediation between the "epistemic side", the "heights" of theoretical mathematics, stretching over the technical application, the experiments in wind tunnels[43] and the like all along the way until the actual production of weapons. Of course, a country had to be strong enough economically and militarily to secure the continuity of that mediation or otherwise it had to find appropriate allies.[44] This included the above mentioned "transfer problem" of science and technology, which was important both for economic and military applications but had special traits in times of war. It was not necessarily so that the "transfer problem" was easier to solve in times of war, given the "paradox in technics" as described by sociologist Lewis Mumford, namely that "war stimulates invention, but the army resists it."[45] But at least the necessary money for the applications was not the main concern when they gained such a political priority in the war. In order to consider the similarities and dissimilarities in the realisation of the mentioned epistemic-technical, mental and institutional-material continuities in different social surroundings, case studies for different countries are warranted.

2 Two Case Studies on Germany and the U.S. and Short Remarks on Mathematical War Work in Five Other Countries

If one looks at the different countries and their organisation of military research between 1914 and 1945 one finds – as one would expect – similarities and differences. What was similar was the feeling of growing competition between the nations, in its sharpest and most articulated form, of course, as military threat, and similar measures for research were taken in different countries at comparable points of time. In 1923, the president of the newly founded *International Educa-*

[43] Much on wind tunnel construction in various countries since about 1900 can be found in Anderson (1997), pp. 296–304, and in Lee (1998).

[44] This applies to Poland as an ally to Britain in decoding (s. Hodges (1983) and the contribution by E. Rakus-Andersson in this volume), Japan as an ally to Germany in jet propulsion. Britain (unlike the U.S.) was unable to maintain full computer development on the technical scale, with resulting resentments on the part of the theoreticians such as Turing as described in Hodges (1983).

[45] Quoted from Gray (1943), p. 41.

tion Board of the Rockefeller Philanthropy, Wickliffe Rose, put this very clearly the following way:

> The nations that do not cultivate the sciences cannot hope to hold their own; must take an increasingly subordinate place; [...] and must in the end be dominated by the more progressive states even though these states do not seek to dominate.[46]

This general situation of competition had found its sharpest expression in the First World War. Germany had been – according to leftist British scientist John Desmond Bernal – more successful in the beginning of the war because its industries were based on science. Bernal, in his well-known book of 1939 *The Social Function of Science,* continued that

> the War, and only the War, could bring home to Governments the critical importance of scientific research in the modern economy. This was recognised in Britain by the formation of the Department of Scientific and Industrial Research.[47]

Not surprisingly I will begin my short case studies on the relation of mathematics and the military with the cases of Germany and the United States, the main opponents in the scientific Second World War. Considerably more uncertainty remains especially about the organisation of scientific war work in the Soviet Union, to be tentatively discussed later. In any case the available primary and secondary (historiographic) literature is most extensive with respect to the German and American environments.

Germany[48]

Broad in its approach as to the time frame and problems discussed – and therefore very much in the direction of the present paper – is Mehrtens' contribution of 1996 which is partly based on an earlier German publication by the same author of 1986.

Felix Klein's "small but paradigmatic techno-scientific complex" in Göttingen is discussed, in particular the foundation in 1917 of the *Aerodynamische Versuchsanstalt* (Aerodynamic Proving Ground) AVA in connection with the scientific *Kaiser-Wilhelm-Gesellschaft* but mainly financed by the military. The AVA, together with the more theoretical Institute for Fluid Mechanics (1925) under the leadership of Ludwig Prandtl, became "the most advanced aerodynamic research installation in the world".[49] Aeronautical Research in Germany was "the main

[46] As quoted in Siegmund-Schultze (2001), p. 26.

[47] As quoted in Edgerton (1996), p. 12. For science in WW I see also Hartcup (1988).

[48] See also appendix 5.1 and the case study on Wolfgang Haack's work in ballistics below. The most relevant secondary literature for the German case is Mehrtens (1986/96) and Epple/Remmert (2000).

[49] Mehrtens (1996), p. 97.

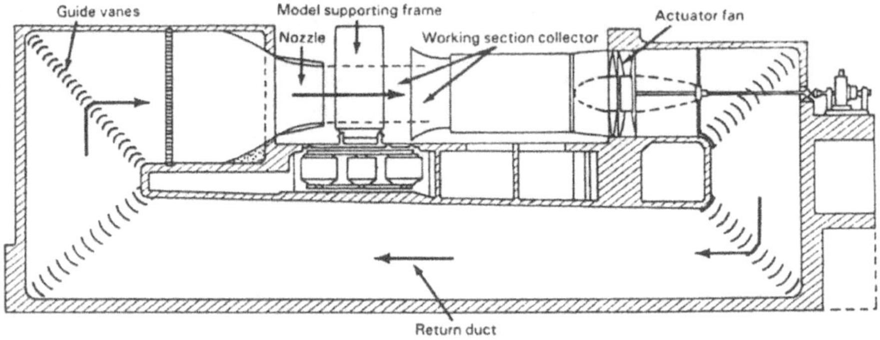

Figure 8. Ludwig Prandtl's wind tunnel at Göttingen, a closed-circuit wind tunnel (1916). [Source: Anderson (1997), p. 300]

consumer of applied mathematics, and especially for complex calculations"[50] and stood for the continuity of military mathematical research from World War I to World War II, especially after the great upswing of the Göttingen, Braunschweig, and Berlin facilities in 1933, when the Nazis came to power. The ideological continuity between the World Wars as to the preparedness of German mathematicians and engineers to engage in military research has also been investigated.[51] In Mehrtens and other publications[52] the German tradition of strong mathematical education of engineers at Technical Universities is stressed as a central prerequisite for the application of mathematics in military and industrial contexts. Little has been published, however, on industrial mathematical research,[53] in particular at the various airplane production plants; some information on respective work at the Siemens & Halske laboratories can be found in (Kragh 1996).

For World War II Mehrtens points out the "poorly coordinated mathematical war-work" in Germany, especially deficiencies – in international comparison – in computers and cryptology, but also, that by 1944 only very few mathematicians were left who were not involved in military work.[54] Things seem to have worked more smoothly in more traditional fields of applied mathematics especially in aerodynamics, which was strongly supported by the regime.[55]

[50] Ibid. p. 105.

[51] For example Siegmund-Schultze (1986) and (1990).

[52] Ludwig (1974), Hensel/Ihmig/Otte (1989).

[53] See Gross (1978).

[54] Mehrtens (1996), pp. 111/12.

[55] In a well-known American report after WW II the following is stated: "In contrast to the Army research it must be said that the German Air Force research under Goering was excellently organized and highly successful." (Goudsmit (1947), p. 147) But note that even Goudsmit acknowledged already the effectiveness of rocket research under army command in Peenemünde (ibid., p. 145/46), which relied heavily on computations from A. Walther's "Institute for Practical Mathematics" in Darmstadt.

A recent German study by Epple and Remmert (2000) presents, for the first time, the structures of the organisation of mathematical research in Germany during World War II. It tries to give a definition of "mathematical war research" and finds two criteria crucial:

1. "Mathematical war research" provides means for the realisation of technical projects which are of immediate or mediate term military interest.
2. There exists an organisational framework which confirms the relevance of that work for the war.

The authors declare it irrelevant for a mathematical work to be called "mathematical war research" whether the mathematician himself is aware of its importance for the war or not.[56]

Epple/Remmert then give historical details on four major streams or strings (Stränge) along which "mathematical war research" was organised in Germany during WW II. They locate first governmental-military institutes,[57] second industrial-military institutes, thirdly research at universities (including technical ones) commissioned by the first two agencies, and, finally, so-called "self-mobilised" mathematical war research at these universities.

Especially in connection with "self-mobilised" research the study confirms the results already found by Mehrtens (1986/96) that the German mathematicians used the pretext of the war for pursuing plans for the erection of a central German mathematical research institute, partly in competition with American (Fry) and Italian (Picone) examples. When the German institute finally came into existence in Oberwolfach in the Black Forest in the last months of the war – the late date being partly a result of the competence struggles in the Nazi hierarchies – it was no longer able to conduct effective war research and served merely as a survival base for German mathematics, including the very pure brand.

Such a listing of war research institutions as in Epple/Remmert is useful, but even more so the theoretical discussion which follows in the article under the title "Epistemic continuities. The example of conformal mappings".[58] In their discussion of the application of conformal mappings in airfoil design the authors show that the code-word "conformal mapping" was used and propagandistically exploited in many "applied" and "pure" mathematical contexts which reached from very simple graphical methods over traditional "pure" complex function theory (L. Bieberbach, P. Koebe) of recently detected importance for applications to sophisticated analytical investigations of the kind which Oswald Teichmüller[59]

[56] Epple/Remmert (2000), p. 266.

[57] Among them the most important aerodynamical research institutions in Göttingen, Berlin and, somewhat later Braunschweig, and Alwin Walther's "Institut für Praktische Mathematik" in Darmstadt. Some institutions, such as the Berlin-Adlershof DVL, seem somewhat undervalued in Epple/Remmert.

[58] Epple/Remmert (2000), pp. 281ff.

[59] The fanatic Nazi Teichmüller could obviously find no fulfillment of his ideological aspirations in his pure mathematical work. He did not see the chain (continuity) till applications, he volunteered for the army and disappeared at the Russian front in 1943. See Schappacher/Scholz (1992).

preferred and performed. This discussion in Epple/Remmert of an example for the "epistemic continuities" is used as another[60] argument for the main thesis which is being brought forward in this study:

> During the war a continuous spectrum of mathematical research activities emerged, which reached from technologically immediately applicable and war-relevant work till domains of so-called "pure mathematics", in which the practical importance of the work on the other side of the spectrum was not easily visible.[61]

USA

On military research in and with mathematics in the US in WW II, there exists a quite extensive literature both of primary and secondary nature[62], where Goldstine (1972), Rees (1980), and Owens (1989) stand out as to importance. Using a "mathematical approach" to war research in a rather broad and general sense, MIT physicist Philip Morse and his *Antisubmarine Warfare Operational Research Group* (ASWORG) since 1942 were crucial in the American war effort:

> In Boston, the operations research scientists showed mathematically that, whatever the tactics employed, the chances of finding a submarine were considerably greater with radar-equipped aircraft than with destroyers.[63]

The cooperation of the U.S. military with scientists in this field of war research has been repeatedly well described.[64] New sophisticated mathematical and statistical methods, such as in linear and non-linear programming, originated over time from this endeavour but were mostly an after-war development.[65]

Aerodynamic research before and during WW II, which had intimate connections to applied mathematics, has been exemplarily documented in publications by the American space agency *NASA*.[66] The lack of academically trained applied mathematicians in the U.S. – unlike Germany at that time – has been mentioned before. Exceptions were Americans N. Wiener, H. Bateman, and F. Murnaghan, the latter two with partly foreign background. Early immigrants, such as Rus-

[60] The other one being the continuous spectrum of institutions for mathematical war research as discussed before by the authors.

[61] Epple/Remmert (2000), p. 265.

[62] See also appendix 5.2, especially for the relevant literature.

[63] Kevles (1987), p. 310.

[64] McArthur (1990), Meigs (1990), Rider (1992), Fortun/Schweber (1993), and Hoff Kjeldsen (2000).

[65] See Hoff Kjeldsen (2000), who shows an impact of the Air Force Programming problem since 1941. See also her contribution to this volume.

[66] See Hansen (1987), Roland (1985), and Bilstein (1989), the latter two, together with many original research reports from the 1930s (e.g., by Th. Theodorsen), also being available on the internet.

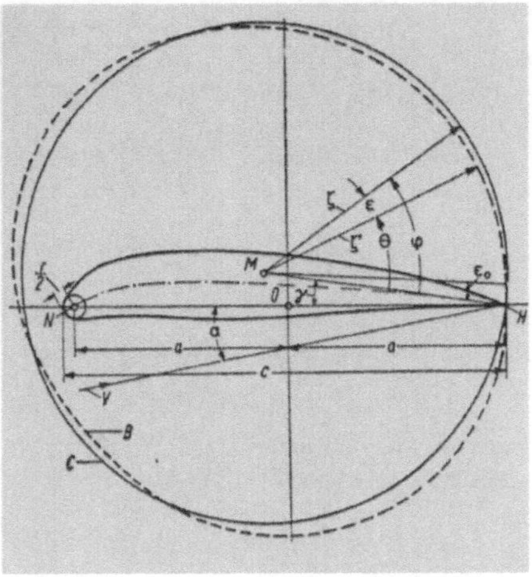

Figure 9.
Using complex function theory in airfoil design.
[Source: Riegels (1961), p. 83]

sian engineers St. Timozenko and S. Lefschetz,[67] aerodynamicists Th. Kármán (Hungary/Germany) and (as already partly American-educated) Th. Theodorsen (Norway) were particularly notable. But the Nazi-caused immigration after 1933 was even more important. Examples are Abraham Wald (statistical sequential analysis), John von Neumann (computers and non-linear hyperbolic equations), Richard Courant and Kurt Friedrichs (work on shock waves).

As to indigenous American traditions the application and creation of mathematics at the *Bell Laboratories* has to be mentioned particularly, since it had clear war relevance especially in statistical prediction and control theory.[69] Bell Labs was "the world's largest and richest institution for industrial research."[69] Mathematically educated engineers such as G. A. Campbell, J. R. Carson, W. A. Shewhart, and Th. Fry,[70] pioneered at Bell Labs in the application of complex functions, operational calculus, probability, and in work on the foundations of communication. A mathematical research division existed at Bell Labs from 1925, led by Thornton Fry and later H. W. Bode.[71]

> Probably the most spectacular development in communications mathematics to take place at Bell Laboratories was the formulation in the 1940s of information theory by C. E. Shannon.[72]

[67] Originally an engineer, Lefschetz – after a long excursion to topology – would only return to more applied fields after WW II.

[68] Millman (1984), pp. 43ff.

[69] Kragh (1996), p. 41.

[70] See Fry (1928) and Shewhart (1931). On early statistical control methods being applied in the U.S. see Bayart/Crépel (1994).

[71] Bode (1953).

[72] Millman (1984), p. 46.

Figure 10.
Claude Shannon (1916–2001) – the founder of information theory in the broader context of war work.
[Source: Masani (1990), p. 155]

The relevance for the military of this research was obvious; cooperation with MIT's department for electrical engineering was tight.[73]

As to state-funded military work in mathematics, the continuities from World War I to World War II are most palpable in the histories both of the *Aberdeen Proving Ground* in Ballistics and of the NACA-related aeronautical research facilities.

During World War I systematic ballistic research had been done in the US in Washington under astronomer Forest Ray Moulton and at the newly founded *Aberdeen Proving Ground* in Maryland under geometer Oswald Veblen. The introduction of finite difference methods in ballistic computation by Moulton and of higher methods of the calculus of variations by G. A. Bliss in Aberdeen, which lay "on the farther boundary of the explored mathematical domain of today"[74] has been described by Goldstine.[75]

The same author describes the continuation of ballistic research on the eve of World War II, when in 1938 a *Ballistic Research Laboratory* was erected at Aberdeen.[76] This laboratory, to which, once again, Veblen and Bliss were attached, would have a decisive catalytic role to play in the development of scientific computing in the U.S. Since 1935 a copy of Bush's differential analyzer was used to

[73] The close contacts between Bell and MIT were personally realized in the collaboration of Wiener and Bush, of Struik and Fry. On the electronic industry in the U.S. and the influence of the War see esp. Godement (1979), vol. 203, pp. 97ff.

[74] Bliss (1927), 313, on his use of "line functions" in ballistics war work.

[75] Goldstine (1972), pp. 75ff. See also Dickson (1918), Moulton (1926), and Bliss (1927).

[76] Goldstine (1972), pp. 123ff.

Figure 11.
Howard Aiken (1900-1973), middle, in front of the Harvard Mark I calculator (circa 1944). [Source: Aspray (1990), p. 212]

compute ballistic firing and bombing tables. Even more important was that the needs of computing at Aberdeen triggered digital computing early in the war, even if the full development – similar to linear and non-linear programming – was basically an after-World War II phenomenon.[77] Goldstine writes that around 1940–41 there was

> worked out an arrangement which resulted in bringing into the Ballistic Research Laboratory a set of standard punch card machines to do ballistic calculations. Also, in 1944, two special multiplying machines were built by IBM and installed at Aberdeen. [...] This marks a renascence of interest in the digital approach, which had proved so important in the days of Bliss, Gronwall, Moulton and Veblen, but which had been temporarily superseded by the acquisition of the differential analyzer in 1935. This idea of doing calculations digitally was to culminate in the ENIAC, the first electronic digital computer, and eventually in the gigantic computer industry of today.[78]

The *National Advisory Committee on Aeronautics* (NACA), the precursor to the later NASA, was founded in 1915 before the entrance of the United States into the war in 1917.[79] In its experimental and research laboratories, particularly the *Langley Laboratory*, built 1917–1920, the Americans tried to catch up with European aeronautical research.[80] Hiring Prandtl's student Max Munk in 1920 gave them access to the modern mathematical approach in aerodynamics. Munk had shown with methods of the calculus of variations that the so-called "induced drag" at the wing tips is minimised in case of the so-called elliptic distribution of lift over the

[77] See Goldstine (1972), esp. his appendix (pp. 349ff.) on the post-war developments in computing in various countries, which were not unaffected by the previous events in the war.

[78] Goldstine (1972), pp. 129/130.

[79] Bilstein (1989).

[80] Roland (1985) and Hansen (1987).

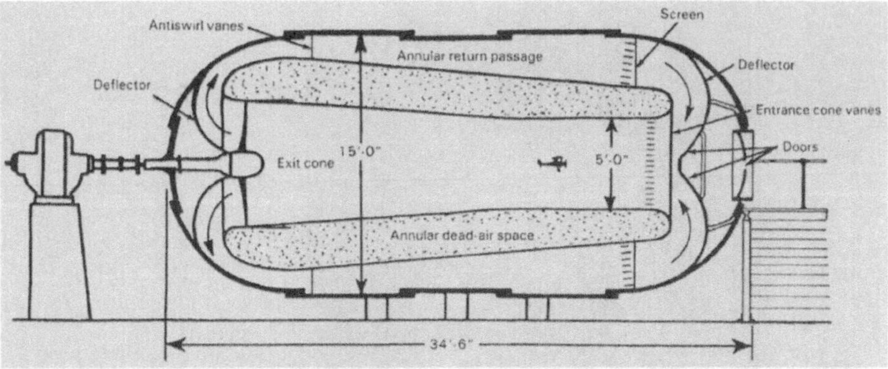

Figure 12.
The NACA Variable Density Tunnel (1922), constructed by German early emigré Max Munk. [Source: Anderson (1997), p. 302]

Figure 13.
Theodore von Kármán (1881–1963) – the leading American aerodynamicist, an early (1929) immigrant from Europe.
[Courtesy: California Institute of Technology Archives]

wing span.[81] Munk introduced a so-called *variable-density wind tunnel* to solve the so-called *Reynold-number scaling problem*. The measurements performed remained partly secret in the 1920s and 1930s and the tunnel would later be seen as an important contribution to the American war effort in WW II.[82] The NACA became a role model for the creation of the OSRD in 1941 and other research committees.[83] But even more important for aerodynamical research in the United States than the NACA facilities proved to be the foundation in 1929 of the *Guggenheim Aeronautics Laboratory at the California Institute of Technology* (GALCIT)

[81] Kármán (1954), p. 65.
[82] Munk (1981), p. 1, and Anderson (1997), pp. 301–304.
[83] Fortun/Schweber (1993), p. 330, and Kevles (1987), p. 292.

in Pasadena under an early immigrant from Germany (earlier Hungary), Theodore von Kármán. This greater importance came particularly from the "multiplication effect" of Kármán's many students, while Munk at NACA did not have students and got into conflict with the prevailing research philosophy there.[84]

Owens (1989) discusses the policies of the *American Mathematical Society* (AMS) during World War II. He argues that M. Morse and M. H. Stone, as spokesmen for the AMS, were working with all the assumptions of the status quo. They believed that, as leading researchers in pure mathematics, they alone were best suited to organise the utilisation of mathematical skill on behalf of the war. They had only little understanding of or respect for the Washington bureaucracies that were in charge of the general mobilisation of scientific workers.

But when the war came, American mathematicians who sat in a joint committee of the *National Academy of Sciences* (NAS) and the *National Research Council* (NRC) since spring 1942, simply did not know what to do next and Warren Weaver had to step in with his *Applied Mathematics Panel* (APM), founded in November 1942 within the *Office of Research and Development* (OSRD). On the work of the AMP much has been written especially by Rees (1980) and Owens (1989).

The discussion in Owens (1989) shows once again the complexity of the problems of mediation between the various participants in war work to enable mathematics really to be applied in the war. On the more cognitive side problems in mutual understanding between engineers and mathematicians were at the same time discussed by key figures like Kármán (1940/43), who had experiences with both kinds of participants.

Briefly on mathematical war work in the Soviet Union[85]

Much has been published in recent years on the development of the leading Moscow and Leningrad schools of mathematics before WW II, on ideological intrusion of Stalinism into mathematics and the role of people like E. Kolman, on the political behaviour of great Russian mathematicians such as A. N. Kolmogorov in those years, on the sufferings of mathematicians during the war.[86] Most of the discussion has been about pure research and basic science, with only few passing remarks on applications.[87] Frequently the point is being made that mathematics came out of the various crises relatively undamaged just because it was so pure and apolitical. Not much, however, is known about applied mathematics in general

[84] Hanle (1982), p. 90/91. In Munk's case the mentality problems between the theoretician (Munk) and engineers weighed heavily and contributed to his early resignation from NACA in 1926. See Roland (1985), vol. I, pp. 87–98.

[85] See also appendix 5.3.

[86] Two examples in English are Zdrakovska/Duren (1993) and Lorentz (2002).

[87] Lorentz (2002, p. 177) mentions that during WW I pure mathematician and mathematical physicist A. A. Friedmann (Fridman) became interested in plane navigation and construction and served as a pilot for the Russian army. Lorentz reports also on a political trial during the blockade of Leningrad 1941/42 against the "group Rose-Koshlyakov", of which N. V. Rose was dean of the Mathematical-Mechanical Faculty of Leningrad University, counter-admiral

and mathematical war work in the Soviet Union in particular. The work available is mostly in Russian.[88]

Some literature in English[89] on general science policies in the Soviet Union which does not address mathematics complements the picture. It is generally assumed that there existed problems of "transfer" from invention to innovation in some sciences in the SU partly due to a certain divide between theoretical science and engineering.[90] Some Western observers seem to even principally doubt the relevance of Soviet basic science for the war effort.[91] Much of the historical perspective on and interest in Soviet science and mathematics is, of course, inspired by the later successes of this country in the space age. In fact, the foundations of Soviet rocketry and jet propulsion were laid between the two World Wars. Those technologies were clearly relevant for the military; at the same time their smooth development was apparently exceptional in the Soviet research environment, comparable to other exceptions such as results in Soviet synthetic rubber research.[92]

In 1918, Moscow saw the foundation of the *Central aero-hydrodynamic institute* (CAGI or TsAGI) under the famous aerodynamicist N. E. Zukovskij (Joukowski).[93]

> Defense-related institutes, such as the Central Aviation Institute (TsAGI) were among the best equipped, and were held up as models.[94]

The Soviet Union could rely on a strong tradition[95] of applied mathematics and theoretical mechanics in Russia (P. L. Cebysev, A. M. Ljapunov, V. A. Steklov).[96] America, partly Germany/France, profited from this tradition and from Russian influence, especially in more applied fields like differential equations and mechan-

of the Soviet Navy and author of many textbooks. Lorentz says that Rose died in jail while N. S. Koshlyakov "survived in special camp(s) (sharashkas) for working on war-related problems." (Ibid., p. 212)

88 Golubev (1955) and *Moscow University* (1975) are most relevant for mathematical war work. Bogoljubov/Urbanskij (1985), contains material on N. M. Krylov and his Ukranian school in Kiev. Ogibalov/Kudrjasova (1980) and Vogt (1998) give some information on institutions but do not go into the mathematics produced.

89 In particular Bailes (1978).

90 There was "a strong tradition of pure research in Russian science and (a) high status accorded to such work by comparison with technical innovation." (Bailes 1978, 342). A similar divide existed probably also in other countries such as in the U.K. see Bush (1970), 54, and earlier also in Germany and the US.

91 "While the eastern front was turned by sheer brute force, the western powers had space and time for developments in which force was not the only element." (Hodges 1983, p. 251)

92 Bailes (1978), p. 340.

93 Golubev (1955), p. 98. Lenin called Zukovskij the "father of Russian aviation."

94 Bailes (1978), p. 346.

95 So Russians could connect to an uninterrupted tradition of "applied mathematics" in some sense in close neighborhood to academic mathematics from the 19th to the 20th century, unlike the Germans, as is known from Felix Klein's reform around 1900.

96 To Lorentz (2002), p. 176, Steklov was more of an organizer than a scholar, but as such important especially due to his influence with Lenin. But this judgement may more reflect on Lorentz as a pure mathematician.

Figure 14.
V. V. Golubev (1884–1954) – one of the main organizers of mathematical war work in the SU. [Source: Golubev (1955), p. 28–29]

ics. Among the emigrants from Russia/Soviet Union to America were Timozenko, Tamarkin, Lefschetz, Besicovich, etc.

Golubev (1955) is the most critical of the publications available on mathematical war work, although it was written in the immediate aftermath of war and Stalinism and does not give sources or bibliographic information. Being both a mathematician and a military officer, Vladimir Vasil'evic Golubev (1884–1954) was obviously himself a key figure in the organisation of mathematical war work. He began as a student of "pure" mathematician D. F. Egorov. He started with analytic functions and changed to applied aerodynamics around 1930. In 1932 he became chair for higher mathematics at the Airforce Academy "N. E. Zukovskij" in Moscow and at the same time dean at University faculty math and mechanics. His monograph in Russian of 1948 "Lectures on the theory of wings" gave a systematical outline of Zukovskij's ideas.[97]

There seems to have existed close collaboration of pure and applied mathematicians and common responsibility for the training of students at the *Scientific Research Institute for Mathematics and Mechanics* (NIIMM, founded in Moscow 1922) at least til 1935.[98] Golubev maintains:

[97] Ogibalov/Kudrjasova (1980), p. 11. Golubev (1955), p. 121.

[98] There existed for instance practical courses, introduced in 1923, for mathematics and mechanics students who had to work either at TsAGI or in airplane factories (Ogibalov/Kudrjasova 1980, p. 9). In 1935 the NIIMM was separated, but according to Golubev (1955), p. 110, without hampering the cooperation of pure mathematicians and mechanists.

Figure 15.
N. E. Zukovskij (1847–1921) – engineer, mathematician and "father of Soviet aviation" (Lenin). [Source: Golubev (1955), p. 82–83]

> The period from 1917 till 1930 was a revolution in the development of mechanics at Moscow University. The mutual approachment of University theoretical mechanics and technics ... did not have any support in the ministry before 1917.[99]

Golubev argues that before 1917 mechanics was considered as "applied mathematics" (in the sense of Lagrange), while it now was a "science rather" and was in much closer relation to experimental facilities and applications.[100] Obviously there was a new notion of "applied mathematics" created also in the Soviet Union in the 1920s and 1930s which pointed – similarly to other countries – in the direction of "engineering mathematics." Insofar as the situation in the SU reflected an international process of "modernization" of mathematics, even if Golubev recognizes some "exaggerations such as a 'laboratory brigade method'"[101] under the special ideological conditions in the SU which were apparently removed in a new reform 1930–1933.

Venerable traditions and increased cooperation between mechanicists and mathematicians alone could not guarantee successful war research, and details of the "transfer" from invention to innovation in the case of mathematics (for instance in the production of the well-known "Katjusa"-missiles, where apparently

[99] Golubev (1955), p. 107. My translation from Russian.
[100] Golubev (1955), p. 115.
[101] Ibid., p. 108.

mathematical work by A. A. Kosmodemjanskij played a role) under the particular political conditions of the SU have still to be investigated. Little is known, for instance, about the concrete collaboration between mathematicians and the military during the war.[102] For example: To what extent was A. N. Kolmogorov aware of the military applicability of his work on statistical prediction theory or was the latter even commissioned by the military?[103]

The War caused not only the destruction of the aerodynamical laboratory in Moscow by bombardment in February 1942, and with it of Zukovskij's models and wind tunnel of 1904.[104] The evacuation of mathematicians and mechanists from Moscow in the fall 1941, the reduced possibilities for experiments and somewhat lower teaching duties led – according to Golubev – also to some increase in theoretical mathematical reflections among mechanists and to the preparation of after-war publications.[105]

Briefly on mathematical war work in Great Britain[106]

Britain could, in her mathematical war work, draw on a rather broad and versatile mathematical background ranging from numerical methods (E. T. Whittaker, L. F. Richardson) over substantial aerodynamic research (G. I. Taylor)[107] to applications of pure research such as logic and number theory in cryptology and computer theory (A. Turing). The origins of operational research in Britain in connection with radar development are also well known.[108] Fox (1990) reports on his war work in the Oxford group on "relaxation methods" under R. V. Southwell beginning in 1939 and on his subsequent (1943) engagement in the "new Admiralty Computing Service at Bath."[109]

There existed huge differences as to the institutionalisation and even the notion of "applied mathematics" as compared to Germany and later the U.S.: that notion

[102] More specific questions see in Section 3 below.

[103] According to one American observer there was "a very real connection between the success of the superb Russian offensive and the fact that a Stalin prize is given for a paper entitled 'On Dispersion, Probability of Hits and Mathematical Expectations of the Number of Hits'" Gray (1943), p. 149. Shiryaev says (this volume) that the publication was urged by the military, while Kolmogorov saw "primarily methodological interest" in his work. N. Wiener assumes the existence of more detailed work by Kolmogorov, adapted to applications: "I am by no means convinced that Kolmogoroff was not independently aware of the possibility of the applications I made. If that was so, he must have had to keep them out of general publication because of their importance for the military-scientific work." (Wiener 1956, p. 261)

[104] Golubev (1955), p. 116/117.

[105] Golubev (1955), p. 118.

[106] See also appendix 5.4, especially for selected literature.

[107] Batchelor (1996). He worked on the statistical theory of turbulence parallel with Kolmogorov.

[108] Fortun/Schweber (1993).

[109] Fox (1990), p. 292.

did not, as a rule, include engineering mathematics.[110] Some sources report a particularly strong social divide between engineers and scientists in Britain.[111]

There were differences between British and American science policies in the war too, especially with respect to team work and secrecy regulations[112] and with respect to the highly personalised style of directing research, particularly in Britain.[113] Connected to the different social and political environments there remained a certain intellectual superiority complex among some British scientists, especially over Americans, which surfaced for instance in the discussion of ideas underlying computers.[114] However, the lack of economic preconditions commensurate with the Americans was clearly palpable and effective in the war effort, particularly in technical computer development.

There seems to have existed a rather broad tradition in Britain of both leftist and (not necessarily overlapping) pacifist views among some pure mathematicians as revealed in Hardy's book (1940) which triggered some discussion on the morality of war research.[115]

Briefly on mathematical war work in Italy[116]

Pure mathematical research had a dominant position as to the number of chairs at universities in Italy until the first World War in comparison to other sciences, even to physics.[117] Thus the shift towards applications which was initiated by Italian Fascism under Mussolini, especially after 1927, and which was enhanced due to economic and political autarky policies after 1935, was not just "detrimental" but had some "healthy" aspects for mathematics as well.

However, the concrete forms of change symbolised in mathematics by the substitution of V. Volterra by G. Marconi (later by marshal P. Badoglio) in leading positions (CNR, Academy) had strong political connotations, particularly so when those measures later were supplemented by antisemitic campaigns (racial laws of 1938) as well.[118]

[110] McVittie (1953), p. 45.

[111] Bush (1970), p. 54, reflecting on the war years: "It was then true that to the British – and they have not quite recovered from it yet – the engineer was a kind of second-class citizen compared to the scientist." According to Ashford (1985, p. 117) was the inventor of the relaxation method, R. Southwell, "a rare example of an engineer with outstanding mathematical abilities."

[112] See Bush (1970), p. 281/2, and Hodges (1983), p. 222, for the British doubts in American secrecy policies because of "lacking the habits of deference, secretiveness and deviousness."

[113] Think of the well-known Tizard/Lindemann dispute as for instance discussed in Edgerton (1996), p. 21.

[114] In 1946, Turing spoke of an "American tradition of solving one's difficulties by means of much equipment rather than by thought." (Hodges 1983, p. 352)

[115] Turing himself said at one point about his work in cryptology that he was "rather doubtful about the morality of such things." (Hodges 1983, 120) See also Levinson (1970). As to L. F. Richardson's pacifist views see Ashford (1985).

[116] See also appendix 5.5, esp. for selected literature.

[117] Orlando (1998), p. 149.

The Volta-Conference[119] of 1935 in Rome on "High Velocities in Aviation", where "the treatment of the foreign guests [...] was almost like that of royalty"[120] revealed a rather developed state of Italian aerodynamic research. One participant, the German applied mathematician A. Busemann (1971) reported on "General Arturo Crocco (the father of Luigi Crocco, who worked in Princeton later), a very able chairman and aeronautical scientist in Italy, in both research and teaching."[121] He also visited "the new Italian aerodynamical research center in Guidonia near Rome, equipped with a variety of high-velocity wind tunnels of Ackeret's [Jacob Ackeret, Swiss] design."[122]

However, the central event as to applications of mathematics in industry and military was the creation in 1927/31 of the "National Institute for the Application of the Calculus" (INAC) of the Italian Research Council under Mauro Picone in Rome.

Picone's biography, on which a collaborator at his institute, L. Amerio (1987), gives detailed information, is interesting because it tells something about the ideological background and motivations for some applied mathematicians to pursue their field.[123] At the same time Picone's biography is indicative of the role of ballistics as a general field of application of mathematics in many different countries (Vahlen, Veblen, Bliss, Moulton, Charbonnier etc.). Trained as a pure mathematician and specialising in partial differential equations around 1910, Picone became involved in ballistic work for the military during World War I. He had to calculate ballistic tables and soon recognised the need for numerical methods.[124] In the 1920s he worked on the famous Ritz method for the application of the calculus of variations in mathematical physics. In June 1923, after Mussolini's "march on Rome" of 1922, Picone wrote a letter to then minister of education Giovanni Gentile, who at that time launched a Fascist educational reform, congratulating him on joining the Fascist party and indicating that he himself was a member.[125] Still in Naples and under continued impression of his experience of WW I, Picone created a privately funded "Institute for Calculus" in 1927, which was later on taken over by the Italian Research Council and renamed in "National Institute for the Application of the Calculus" (INAC) in 1931/32, after Picone had been called to

118 On the general political background of Italian mathematics see above all Israel/Nurzia (1989).

119 In honour of Alessandro Volta (1745–1827).

120 Busemann (1971), p. 7. More general on the Volta Conference: Anderson (1997), pp. 400–403.

121 Ibid. Crocco constructed a wind tunnel near Rome already in 1903. Anderson (1997), p. 299.

122 Ibid., p. 10. See also Kármán (1954, p. 133) reporting on the Volta conference and on Busemann's talk there on aerodynamic properties of sweptback wings.

123 This background he shared with Germans such as the Nazi and ballistician Theodor Vahlen. See Siegmund-Schultze (1984).

124 More details on Picone's ballistic work in WW I are possibly in Picone (1934) which I have not seen.

125 Nastasi (1990), p. 31.

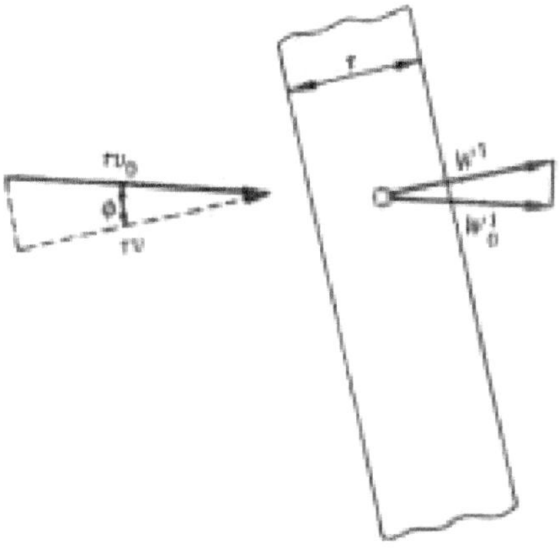

Figure 16.
Adolf Busemann (1901–1986) and his diagram for a swept wing of 1935.
[Source: Anderson (1997), p. 402]

a professorship at the University of Rome. In his report of 1938 on the activities of his institute during the quadriennium years XII til XV (in Roman numerals) of Mussolini's rule, Picone first stressed the theoretical work done by collaborators such as Fabio Conforto and Carlo Miranda on problems of the calculus of variations and eigenvalue problems. In fact, one of the most important pure mathematical after-effects of the INAC was Miranda's book (1955) on elliptical differential equations. But these more general theoretical results notwithstanding, Picone assured industry of its right to an exclusive exploitation of the concrete calculations and algorithms performed and developed for it at the Institute. Focussing more on economic than military applications Picone, nevertheless, concluded his report with the following remarks:

> Much of the calculative work done at the Institute was commissioned by the Ministry of National Defence: the stability of aeronautical structures, direction of air bombardment, stabilisation of submarine torpedoes, ballistic tables for long distance artillery, military radio communication have provided material for long and complicated research, which fortunately has been finished in time with the help of the mature mathematicians and numerical procedures which belong to the Institute.

> And the Institute for the Application of the Calculus, today under the command of his excellency marshal of Italy Pietro Badoglio [1871–1956, apparently chief of the CNR after Marconi's death in 1937], is decided to maintain and increase its efficacy always ready to respond to any request to secure the power of the Italian army.[126]

Already before World War II there was much collaboration with German applied mathematicians and industry/military; the Germans saw the INAC as somewhat of a role model for their own developments in applied mathematics.[127] There is not much said, in the available reports on the INAC, on the mechanical devices used in the numerical work. Because the reports tell about a great number of "human calculators and computists" one might speculate that the amount of instruments was not very considerable and insufficient even for the time period considered. After WW II (1951) the Picone Institute gained recognition by UNESCO and got the title of a "Centro Internazionale di Calcolo" after a positive review by American Herman Goldstine.[128]

In 1940 the Italian "Royal National Institute of Higher Mathematics" was founded under mathematician Francesco Severi in Rome, and it partly compensated for losses in pure mathematics.

Briefly on mathematical war work in France[129]

As to the general background of applied mathematics in France in the inter-war years one is often referred in the literature to an overly bureaucratic and Paris-centred science and education system, to a lack of contact between engineering schools and universities quite contrary to the earlier great traditions of the Ecole Polytechnique as a research institution.[130] The leading *Ecole Normale Supérieure* was rather distant to applications;[131] the first modern French applied mathematician became known only after WW II: J.-L. Lions (1928–2001). Misguided developments in French aeronautics seem to have prevented extensive mathematical research in that field and seem to have reversed promising and exemplary developments as the early one at the Eiffel laboratories.[132]

[126] Picone (1938), p. 6. Translation from Italian by the author.

[127] See documents (letters) reproduced in Picone (1959). On the INAC see also Salvadori (1938).

[128] Picone (1959), pp. 160/61, Amerio (1987), p. 23.

[129] See also appendix 5.6.

[130] On both points mentioned see Germain (1953).

[131] Andler (1994).

[132] According to Chadeau there was a focus on strong engines instead of aerodynamics: "From 1925 on, when international developments favored innovations with strong scientific components, French products fell behind those of foreign manufacturers. [...] Not until the mid-fifties, then, did French technology catch up with international standards. In the field of jet propulsion a first step was taken in 1950 [...]" (Chadeau 1988, pp. 28/29)

The poor French reception of international scientific results[133] seems to have improved somewhat in the 1920s, testified to by both developments in ballistics[134] and the founding in 1926 of the *Institut Henri Poincaré* (IHP). However, the developments in stochastics at that institute were theoretical rather than applied.[135] The founding in 1939 of the French CNRS led to the creation of "mathematical laboratories" at IHP.[136] As to their importance for mathematical war research, however, the outcome was doubtful, not least due to the political situation of France under German occupation since June 1940.[137] French deficiencies in computation and communication technologies became clear after WW II.[138] The results in mathematical war research in France given in appendix 5.6 are probably very incomplete and due to the insufficient historical work available. One wonders, for instance, whether the promising developments of M. d'Ocagne's "nomographie" of around 1900 did not have followers in France as well.

Briefly on mathematical war work in Japan[139]

The main problem of the historiography of mathematical war research in Japan is that almost no primary historical sources are available.[140] The secondary literature on Japan contains almost no information on mathematics in the War. Indirect conclusions from the existing fields of application and general historical works are therefore needed.

According to Sasaki (2002, p. 231) Japanese mathematics began to emancipate itself from the traditional Chinese mathematics called *wasan* before the Meji Restoration which started in 1868. This emancipation was originally "Learning Western Mathematics as Military Science," where navy officers played a leading role. It was followed by "Germanization of the Political System and of Learning, 1881–1945", and at the same time "the intellectual and political mood in Japan became authoritarian and militaristic."[141] On this period Sasaki writes:

> At all institutions where higher mathematics was taught prior to the outbreak of World War II – the imperial universities as well as the teacher-training schools – mathematical research was an intellectual activity aimed at

[133] Pyenson (1996).

[134] See Charbonnier (1924).

[135] Siegmund-Schultze (2001), pp. 156–177.

[136] These were the labs for "mathematics" (under É. Cartan), "theoretical physics" (L. de Broglie), statistics (M. Fréchet), ballistics (G. Valiron), astrophysics (M. Mineur), and numerical calculus (L. Couffignal). See Siegmund-Schultze (2001), p. 176.

[137] Picard (1990).

[138] See the reports by L.Couffignal, from his after-war mission in Germany, ordered by the CNRS (Picard 1990, p. 97). See also Kragh (1996).

[139] See also appendix 5.7.

[140] According to (Yamazaki 1995, 167) were "the important materials relating to the matter burnt on orders of the military at the end of the War."

[141] Sasaki (2002), p. 238.

turning Japan into a first-class world power. This is one of the reasons why the German cultural ethos was so influential in modern Japan and why so many young Japanese mathematicians studied in Germany. [...] It is therefore not surprising that, with a few exceptions, mathematicians were directly or indirectly mobilized for the war without any serious resistance. After all, some were already engaged in research that had direct military application.[142]

The "German ethos" in mathematics did, however, not necessarily imply a drive for applications and it has been argued that "the distinction between pure and applied mathematics may have been sharper in Japan than in other countries."[143] There existed social and epistemic barriers between mathematicians and engineers, more generally there was a lack of interdisciplinarity in research.[144]

The tradition of Japanese colonialism, militarism[145] and racism was reinforced in 1931 with the Japanese invasion of Manchuria. The period considered between the two World Wars largely coincides with the so-called Showa Period (1926–1945), another period of modernisation[146] of Japanese science, of increase in the number of graduate students etc.[147] In 1932 the Japanese broke the American code (two-digit encipherment system) but they were apparently unable to read the new one of 1939. Drea (1991) says that Japan's communication intelligence was more successful than commonly assumed. However, no indication of the use of sophisticated mathematical methods in decoding is given in Drea's publications.[148] As to scientific computing, Suekane (1980) argues that the Japanese were unable to produce punched-card machines of their own early in the 20th century and had to import them as late as 1941. Some developments with binary circuits in Japanese industry since about 1938 notwithstanding, the Japanese tradition of the ubiquitous abacus (soroban) contributed to a delay in computer development until well after World War II.

[142] Sasaki (2002), p. 246.

[143] Sasaki (2002), p. 248. The author adds that "there have been few Japanese mathematicians in the mold of Norbert Wiener." (Ibid., p. 248) "It is certain that modern Japan has produced as many excellent physicists or theoretical engineers as talented mathematicians. Unfortunately, however, as a general rule Japanese mathematicians do not seem to be good at fields connected with other sciences. [...] In general, pure mathematics may have been too autonomous in Japan." (Sasaki 1999, p. 43)

[144] As stated in Grunden's dissertation of 1998, of which only an abstract was available to this author.

[145] According to Makino (in this volume), the French mathematician Emile Borel, visiting Japan in the 1920s, found the country "too nationalist", although he himself was used to sharp utterances of nationalism by his French compatriots.

[146] See Makino (in this volume) on the Japanese mathematician Kinnosuke Ogura (1885–1962), who first protested against Ludwig Bieberbach's racist theories in mathematics (1935), only to support the policies of the Japanese military under the slogan of an "anti-feudalistic modernization".

[147] Shimao (1989), p. 86.

[148] Drea (1991), 188. S. Fukutomi's contribution in this volume gives the impression that mathematicians turned to military cryptography rather late in WW II and did not apply very sophisticated methods.

The Japanese Technology Board (founded in 1941 with strong focus on aeronautics) was the "true starting point of Japanese science and technology policy."[149]

During WW II Japan had a strong ally, Germany,[150] where several Japanese mathematicians had studied before the war. According to one historian, however, the mobilisation of science in Japan compared to the US was a "downright failure".[151]

3 Some Open Historiographical Problems

A listing of some open historiographical problems of different kinds (internal mathematical, sociological, moral etc.), which have been partly touched upon before, may conclude the more general discussion in this article:[152]

- Where exactly is the boundary between industrial and military research in war times, if there exists one?
- How did secrecy regulations during the war influence mathematical research and what connection existed to secrecy regulations in other fields, such as engineering?[153]
- How did mental barriers and ideological prejudices between the different social groups involved (engineers, military, mathematicians, politicians) influence war research in different national surroundings? Can Meigs's observation that "American scientists [...] measured and analyzed, in a way far more unbiased than their counterparts in uniform"[154] be applied to other national contexts?
- Does totalitarian political rule promote or hinder mathematical applications for the military?
- What does the historical material discussed say on the spin-off problem?[155] Is, in particular, Lewis Mumford's "paradox in technics" confirmed, according to which "war stimulates invention, but the army resists it"?[156]

149 Yamazaki (1995), p. 168.

150 For the rather one-sided technology transfer from Germany to Japan see Braun (1987).

151 "Nothing was produced by the mobilisation of science comparable with the atomic bomb, radar, rockets, operations research, etc., which were achieved by the mobilization of science in the allied nations. However the Japan programme exerted a great influence upon the postwar order of research." (Low (1990), p. 350) The reference to rockets is somewhat misleading, because Germany led the field here.

152 More and more detailed examples suggested by the special national environments are given below in the appendices.

153 It seems as though mathematicians were still able to partly communicate mathematical information both nationally and internationally due to the esoteric character of their language. Thus secrecy policies applied to them in quite a different way than to engineers, which made also for mental differences between the two social groups, the engineers and mathematicians. But at the same time there existed national differences in secrecy policies, which were not restricted to the military and not to times of war.

154 Meigs (1990), pp. 219/20.

155 As discussed for instance in Booss/Høyrup (1984).

156 On this one finds much in Meigs (1990), in particular in connection with the discussion of Admiral King's "dogmatic authoritarian personality and an almost bitter determination to protect the Navy from any outside meddling" (Meigs (1990), p. 218).

- Did the War put a halt to much of fundamental (mathematical) research, or has the notion of fundamental research changed rather, for instance in the direction of applied mathematics as an academic discipline?[157]

And more internal:
- What kind of historical connection existed between the development of numerical methods and mathematical instruments: was the correlation necessarily positive or did sometimes a lack of instruments promote the development of more sophisticated numerical algorithms?[158]
- Is it justified to say that geometrical mathematical methods were relatively less important in war research than analytical, as suggested by some authors, and could this be connected to the upswing in numerical analysis and computer development?[159]

4 Wolfgang Haack's *Geschoßformen kleinsten Wellenwiderstands* (1941) as a Typical Piece of Mathematical War Research[160]

In cursory form and without going into mathematical details I shall discuss a paper of 1941 by Wolfgang Haack (1902–1994), who started as a "pure" mathematician before the war, but who would later, not least due to his war experiences, become one of the leading applied mathematicians with strong theoretical leanings[161] in the Federal Republic of Germany. Unlike many of his other results,

[157] Note the following remark by Bell Lab president and president of the National Academy of Sciences Frank B. Jewett, which may however have been coloured by his desire to lift restrictions on scientific communication still in place after WW II: "It wasn't long before practically all fundamental research in the physical sciences ceased, as did the advanced training of men." (Jewett 1947/71, p. 405).

[158] As for instance suggested by L. Collatz in Walther (1948–1953), volume 3, p. 34, and in Weissinger (1985), p. 117.

[159] For industrial research: "It is not easy to say why advanced geometry plays no larger part in industrial research; however, the fact remains that it does not."(Fry (1941), p. 269)

[160] The title reads in its English translation of 1948 "Projectile Shapes for Smallest Wave Drag" (Haack (1948)). I have not seen this translation which has apparently only limited circulation in American libraries. All translations from German are mine.

[161] Bernhelm Booss-Bavnbek considers Haack as one of his mathematical ancestors (personal communication). With his joint work with Günter Hellwig in the early 1950s on boundary value problems for elliptic partial differential equations, not satisfying the *Fredholm Alternative*, Haack contributed to what later was to be called the index theory of Atiyah and Singer. Haack reports on this work in his recollections Haack (1989), p. 225ff. See also Haack (1987) and (1998).

Figure 17.
Wolfgang Haack (1902–1994) in his retrospection on 85 years.
[Source: Haack (1987), p. 23]

Haack's paper, which I am going to discuss, is apparently not, theoretically, of a far reaching importance for "applied mathematics" as a whole or even for "aerodynamics", "ballistics" or "gas dynamics" as scientific fields of their own standing.[162] But lying on the borderlines of the four fields mentioned, the circumstances of the creation of Haack's paper in 1940/41 are revealing with respect to the practices of war research in mathematics at that time and exactly with respect to the "continuity of epistemic, technical, mental and institutional prerequisites for war research," as discussed before. At the same time, work by Busemann of 1942, inspired by Haack's contribution, proved to be of a more lasting, theoretical kind.[163]

In 1939 Haack was professor for mathematics and geometry at the Technical University Karlsruhe. The work on the shape of projectiles was commissioned by the "Aeronautical Proving Ground Hermann Göring" in Braunschweig. Haack did the work as an act of self-mobilisation: he phoned the head of the research department of Göring's aviation ministry, A. Baeumker, immediately after the outbreak of the war in 1939. Haack was a differential geometer and he says in his recollections:

> This special field was obviously not very much tuned to war purposes. But nevertheless, it was mathematics, and mathematics is being needed everywhere.[165]

Haack's research, which he did in collaboration with his wife, a theoretical physicist,[166] resulted in a talk in October 1941 in Peenemünde (the well known

[162] In a letter to this author, dated 29 February 1992, Haack stressed that he used in his report "but very elementary mathematical methods".

[163] Work on conical flow as quoted by Kármán (1954), pp. 117, 141. See also letter by Busemann to Kármán, 6 April 1946 (Kármán Papers 4.27, Caltech) according to which he had meanwhile extended his "criticism of Haack's paper", which he obviously found theoretically insufficient.

[164] Haack (1941), p. 15.

[165] Haack (1989), p. 2. Later on, in an interview with this author on 26 June 1991, Haack added that he was afraid to be called to the army as an ordinary soldier, because he had been classified as a driver only in examination.

[166] Haack interview 26.6.1991, Haack (1941, p. 15): "lively collaboration", also Haack (1989, p. 5).

rocket proving ground, where Wernher von Braun was working) before the *Lilienthal-Gesellschaft für Luftfahrtforschung*. The talk was published by that society as a secret report with limited circulation (it is therefore still difficult to obtain today)[167] but it was translated after the war in a report by Brown University.[168]

Supersonic speed of projectiles in compressible fluids had been known for long as a phenomenon of ballistics. The new problem of applied mathematics here was to precisely calculate the air resistance to shapes of projectiles (profiles, especially the angles of the noses), which could not be considered as mere pointmasses. Connected to this were phenomena known from fluid dynamics, such as the well-known *boundary layer* and its possible separation from the body moving through the flow.[169] Transonic and supersonic flow problems emerged at the same time also in aviation where the resulting consequences for so-called sweptback wings etc. were discussed by applied mathematicians such as Busemann.[170] To interconnect the results of the ballisticians and the aerodynamicists as done in the work by Kármán, Busemann and others was no easy task, not least because of the low level of international cooperation and standardization on the part of the ballisticians, which Kármán would criticise for instance at the Volta Conference in 1935.[171] In the same talk Kármán stated with respect to the experimental material available concerning the drag of different bodies at supersonic speeds that there were "very few wind tunnel experiments and [...] very numerous ballistic tests." (p. 49)

The governing partial differential equations of this type of problems were non-linear and thus mostly beyond the reach of analytical methods. Simplification of the assumptions concerning the flow, in particular linearisation of the problem, was required, which resulted in a Volterra-type integral equation which still was only solvable after further simplification.

This had already been done by J. Ackeret in 1928 and Kármán and N. Moore in 1932, that means in times of peace, and again by Kármán at the Volta Conference in Rome on "High Velocities in Aviation" in 1935, stressing the analogy "between the wave resistance of slender projectile and the induced drag of a wing."[172] On a more general theoretical level this became later part of a chapter on "Flow in Three Dimensions" of the classic monograph published by the German émigrés to

[167] My copy is from the library of the *Deutsche Forschungs- und Versuchsanstalt für Luft- und Raumfahrt e.V.* in Göttingen, the successor institution to Ludwig Prandtl's AVA.

[168] Haack (1948). See remarks above on this translation.

[169] Haack (1941), p. 28. It was Busemann himself, who had helped Haack in the interpretation of his results in the sense of aerodynamical wing theory (Haack (1941), p. 15 and p. 20, fn.). Busemann was head of the institute for gas dynamics at the Braunschweig aeronautical proving ground.

[170] See Volta-conference of 1935, and Kármán (1947), esp. 1954, pp. 103–141 "Supersonic Aerodynamics", where Kármán is also shortly alluding to the work of the American task force after the war in 1945 "collecting German papers and documents" (117). Kármán also reported that the Americans were much impressed in 1945 by the German wind tunnel experiments with sweptback wings in Völkenroda near Braunschweig and that this "led to the birth of the present B-47 airplane, the first bomber with sweptback wings in this country." (Kármán 1954, p. 134)

[171] Kármán (1936), p. 50, Fn.1.

[172] Sears (1947), p. 361.

the U.S. Richard Courant and Kurt Friedrichs in 1948 under the title "Supersonic Flow and Shock Waves".

For the immediate requirements of artillery in the Second World War the results reached by Kármán were, however, not sufficient:

Haack stressed in his publication in the secret reports of the *Lilienthal-Gesellschaft für Luftfahrtforschung* in 1941 that the practical demands of ballistics (especially anti-aircraft artillery) during warfare required the consideration of calibre and volume (mass) of the projectile in the first place and not, as Kármán had done, calibre and length.

This led Haack to a generalisation of the problem of the optimal cross section (line of profile) of the projectile for the three parameters calibre (K), volume (M), and length (L). He considered the variational problem for the optimal cross section (especially the shape of the nose) in the three different cases where two of the three parameters are fixed. The variational problem and its solution by Haack were closely related to one which the Prandtl student Max Munk had solved in his influential dissertation in wing theory in 1919.[173]

Kármán in 1932 and 1935 had considered the problem of calibre and length fixed ("K-L-projectiles", according to Haack) and had theoretically found a singularity of the solution which hinted at a blunt nose. This, he admitted, was "somewhat contradictory to our assumptions",[174] and only additional assumptions for slender projectiles enabled him to make the optimality (with respect to air resistance) of a *pointed* nose plausible. But Kármán could not derive the pointed nose theoretically because he "ignored of all properties of the solution just a differential-geometric one,"[175] as Haack stated it with some pride in his recollections of 1989, alluding to his field of special mathematical competence. But there were other reasons as well for the fact that Kármán did not reach a definite solution at that time in 1935, reasons which Haack commented on in a letter to the author on February, 29, 1992 with the following words:

> Kármán [...] found no solution. The problem was not urgent at that time and remained unsolved ("blieb liegen"). 1940, when supersonic aircraft was built, the interest in the problem had increased. I generalised the problem and succeeded [...] understandable that the aerodynamicists got angry not to have hit on it.

Haack, however, who basically used Ackeret/Kármán's methods of linearisation and subsequent simplification of the occurring Volterra equation, arrived at a different conclusion in the case of fixed calibre and volume (mass):

> This solution is the only one [of the three considered] which is in accordance with the method of approximation [the Ackeret/Kármán method], because the projectile is absolutely sharp (acute).[176]

[173] Kármán (1954), pp. 65/66.

[174] Kármán (1936), p. 45.

[175] "Ich hatte beobachtet, dass Herr von Kármán ausgerechnet eine differentialgeometrische Eigenschaft der Lösung übersehen hatte und dadurch zu diesem Widerspruch gekommen war." (Haack 1989, p. 4)

[176] Haack (1941), p. 20.

By "absolutely sharp" Haack meant a cross section with two symmetrical inflection points with respect to the axis (of rotation). So in this case of the K-M-projectiles – unlike Kármán's case of the "K-L-projectiles" – no further theoretical adaption of the solution for slender projectiles seemed necessary. What, however, was necessary was the numerical or graphical calculation of concrete cross sections of K-M-projectiles and an experimental verification of their properties – after all several steps of approximation (and therefore possible sources of error) had been used also in this case – not least due to Ackeret/Kármán's general method of linearization of the original differential equation.

The first confirmation for Haack's theory came soon. As Haack writes in his paper:

> Following the Western campaign several French experimental projectiles were captured (erbeutet) which had a curiously sharp shape. They agree almost completely with the optimal projectiles for given volume and calibre.[177]

Further practical and experimental confirmation was needed.

Haack gives in his paper many hints to bolster up the plausibility of his results and to estimate the boundaries of its applicability.[178] He refers to graphical methods by Busemann and numerical methods by G. I. Taylor in the solution of the original differential equations which secure the validity of the linearised theory at least for conic shapes of the projectiles.[179] He reproduces photographic pictures from wind tunnel experiments where the Mach lines (Schlierenaufnahmen, "Machsche Linien") of conic projectiles for supersonic speed are visible.[180]

As to M-L-projectiles, for which Haack also had found a theoretical solution, he had the shapes checked in the Göttingen wind tunnel for Mach-number $U/c = 2$ – following a proposal and financing by the Braunschweig institute[181] – and found excellent coincidence with the theoretical values for the air resistance (Widerstandsbeiwerte c_w).

Finally, in order to get full vindication for his optimal shapes, he let the governing differential equation integrate graphically in two special cases and thanked a "Miss Kistner for the laborious computations".[182] At one point in his research report Haack also hinted at lack of time for further wind tunnel experiments.[183]

[177] Haack (1941), p. 14. Photographs of the French projectiles are reproduced in Haack's paper, p. 25. Haack wrote in a letter to this author (20.1.92) that he saw the French projectiles only after he had submitted an earlier version of his manuscript in Braunschweig.

[178] "Grenzen für die Brauchbarkeit der folgenden Untersuchungen festzulegen"(Haack 1941, p. 17).

[179] Haack (1941), p. 16.

[180] Ibid., p. 17.

[181] Ibid., p. 15.

[182] Ibid., p. 26.

[183] Haack (1941), p. 21: "Ist aus Zeitmangel unterblieben."

So there is no doubt that the job he did was urgent and that his results were really used to produce anti-aircraft projectiles. With undisguised pride[184] Haack reported on the "success" of his theory of projectiles which allowed for a considerable extension of reach of German anti-aircraft artillery. Haack, who generally[185] was not given to very critical or self-critical attitudes with respect to the role of mathematics or of himself in the Third Reich, but nevertheless or for that reason was influential in after-war West German mathematics (for example president of the DMV in 1961/62), reports in his recollections of 1989:

> In the first months of 1944 mass production [of the projectiles] began in Enzisfeld near [...] Wien. But this could not remain hidden to the enemy (Gegner). He destroyed each large production site of the shells within short time such as their number was just sufficient to protect important military objects with the new ammunition.[186]

Haack says in his recollections of 1989 that Kármán himself acknowledged the value of his war work, when he was in Germany immediately after the war in 1945 to collect German manuscripts:

> I am happy to meet you. Yes I did not realise in Rome in 1935 [the Volta Conference mentioned] that these optimal projectiles can be computed with our method, and your paper interested me greatly.[187]

And, as was seen, the baton was passed over to the Americans and they found - in addition to their own results - mathematical "intellectual techniques" in occupied Germany to go on in artillery or in supersonic bomber warfare or wherever else.

What makes this small piece of mathematical war research so interesting and - I believe - typical, is the close interconnection it reveals, the "continuity" as we called it, of a great many levels of intellectual and practical activities which led to the "success" of the mathematical war work. Particular mathematical activities of very different kinds and theoretical levels (pure analysis, numerical algorithms, graphical methods) were involved. There was a gradual approach of two different engineering disciplines, ballistics and aerodynamics, that were about to recognise what they had in common mathematically. There was a "pure mathematician" (Haack) who made use of his special expertise (differential geometry) at one point, where even the mathematically well-educated engineer Kármán had failed. There were engineering techniques and facilities (wind tunnel experiments, calculation devices) needed. There was a lot of oral communication and collaboration between the pure mathematician Haack and the applied mathematician Busemann, and of a mathematician and a physicist (Haack's wife).

184 Haack would however stress in a letter to this author (20.1.1992) the importance of his work "in an age of civil supersonic flight" trying to somewhat diminish the author's insistence on war research as the topic of Haack's report.

185 See also Haack (1998).

186 Haack (1989), p. 6.

187 Haack (1989), p. 35.

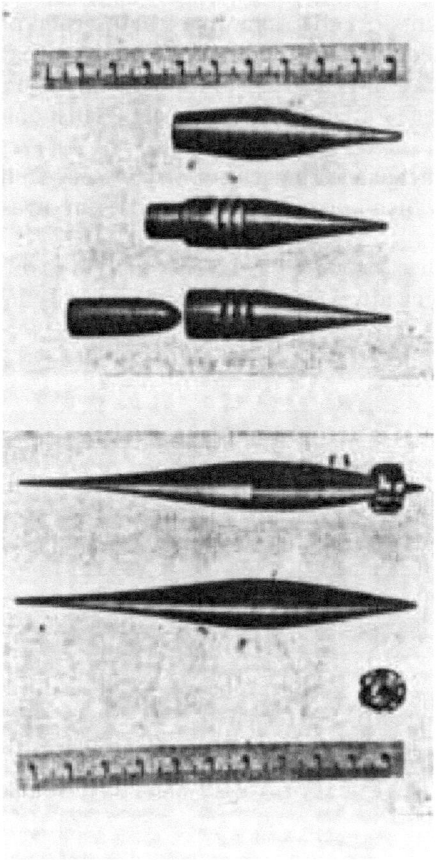

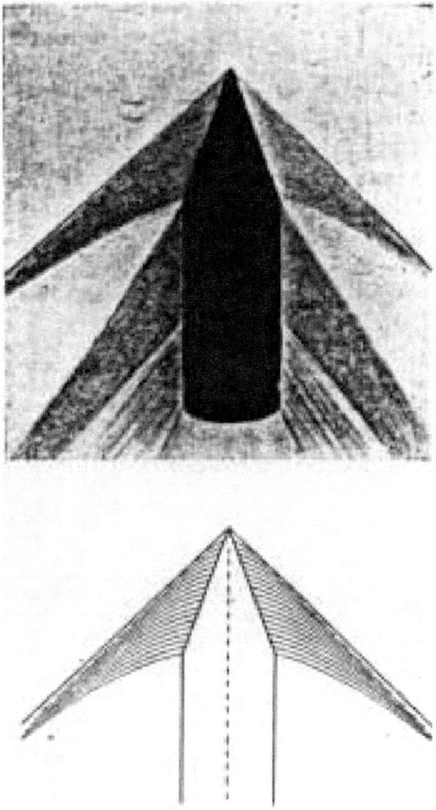

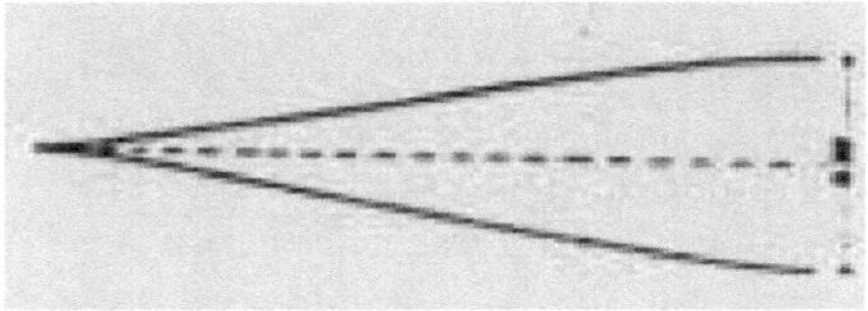

Figure 18 (*above left*).
Captured French experimental projectiles.
[Source: Haack (1941), p. 25]

Figure 19 (*above right*).
Schlieren picture of a cone projectile from the wind tunnel at $U = 1{,}54$ c (Taylor); *below*: Mach lines at velocity $U = 1{,}38$ c. [Source: Haack (1941), p. 17]

Figure 20. K-M-projectile. Optimal projectile (with two inflection points) for given calibre and mass. [Source: Haack (1941), p. 20]

Not just in the background of this story but rather dominant is, of course, the peculiar political situation of the war. One sees the indifference of an allegedly apolitical mathematician[188] to the destructive and deadly aspect of the military use of his science. One notes the abundance of money from the military for math-

[188] The attitude of the mathematician implied to separate his political positions from the scientific part of his work is expressed in Haack's remark, in a letter, dated 20.1.1992, to this

ematical work and wind tunnel experiments. At the same time the urgency of the job becomes clear which had the consequence of stopping the level of logical and mathematical sophistication when the result was needed, while Kármán had broken up his work in 1935 in times of peace for lack of a theoretically satisfying result.

Finally, the case study enables also some general judgement on the epistemic status of fluid dynamics at that point, on which one of its leading representatives said in 1948:

> For modern fluid dynamics d'Alembert's saying has been typical in many instances: 'Just go ahead, belief will come afterwards'![189]

author: "The letter to Min.Dir. Bäumker has in my opinion nothing to do with the evaluation of my work and is a pure matter of privacy." Note, however, that Haack himself made this episode public in his recollections of 1989.

[189] Görtler (1948), p. 11. My translation from German.

5 Seven Appendices (Tables) on Mathematical War Work in Different Countries[190]

5.1 Military work in mathematics (1914–1945): selected notes on the case of Germany

5.1.1 Main institutions and events connected to military application of mathematics

1917:	Aerodynamische Versuchsanstalt (AVA) Göttingen, financed by military
1925:	Institute for Fluid Mechanics Göttingen under Ludwig Prandtl
1928:	Darmstadt: *Institut für Praktische Mathematik* under Alwin Walther: during WW II calculations for rocket project in Peenemünde
1933:	upswing with rearmament, e.g., 1934 DFL (aviation) in Braunschweig
1937:	upswing of mathematics at DVL (aviation) Berlin-Adlershof: political "oasis"[191]
1942:	"Diplom" in mathematics introduced: "creation" of industrial mathematician
1944/45:	Reichsinstitut in Oberwolfach founded, originally for war-relevant research

5.1.2 Some fields of application, methods, mathematicians/appliers involved[192]

field	method	mathematicians/appl.	source
ballistics, supersonic flight	diff. and integr. equations, calculus of variations	Busemann, Haack	Busemann (1971) Haack (1941)
airfoil design	complex functions	Schmieden, Lagally	Epple/Remmert
rocket-project	calculation of	Darmstadt institute	Walther (1948ff.)
	numerical analysis	Collatz	Walther (1948ff.)

5.1.3 Social/epistemic environment for applications of mathematics

- traditions at Technical Universities of strong mathematical education for engineers, and growing understanding of applications among mathematicians (Felix Klein in Göttingen)
- loss of famous applied mathematicians in 1933 (Courant, Mises etc.)

[190] In so far as the following tables contain – for the sake of completeness – some overlap in the facts with the previous text, at least the sources will not be indicated again.

[191] Weissinger (1985), p. 129.

[192] A particularly important source is Walther (ed., 1948–1953), which contains much German research that had remained secret during the war.

5.1.4 Historical primary and secondary literature/sources

Busemann (1971), Epple/Remmert (2000), Görtler (1948), Gross (1978), Haack (1941/1989), Ludwig (1974), Mehrtens (1986/96), Petzold (1992), Siegmund-Schultze (1986/89/90), Trischler (1992), Walther (1948–53), Weissinger (1985)

5.1.5 Some specific problems for historiography

- To what extent did deficiencies in research (statistics, differential equations, celestial mechanics) from before the war influence German war research?
- How did the German material support for mathematics compare to the conditions in the U.S.?

5.2 Military work in mathematics (1914–1945): selected notes on the case of United States

5.2.1 Main institutions and events connected to military application of mathematics

1915: foundation of *National Advisory Committee on Aeronautics* (NACA)[193]

1918: Aberdeen Proving Ground: ballistics under O. Veblen and G. A. Bliss

1920: NACA's *Langley Laboratory* with German immigrant Max Munk

1925: mathematical division in Bell Labs

1929: immigration of Th. v. Kármán and foundation of GALCIT at Caltech

1931: Bush's differential analyzer: since 1935 and during WW II in Aberdeen ballistics

1938: *Ballistic Research Laboratory* at Aberdeen Proving Ground

1941: Brown summer school founded (R. G. D. Richardson)

1942: *Applied Mathematics Panel* (APM) under W. Weaver in Bush's OSRD

1942: First OR groups founded in US

1943: Foundating of *Quarterly of Applied Mathematics*

1943: Los Alamos: Atom bomb project with strong need for calculations

1944: Mark I by H.Aiken, Courant/Friedrichs "Shockwaves"

[193] See as an early NACA-report Wilson (1915).

5.2.2 Some fields of application, methods, mathematicians/appliers involved

field	method	mathematicians/appl.	source
air foil design	calculus of variations	Max Munk at NACA	Munk (1981)
ballistics in WW I	finite differences, calculus of variations	F. R. Moulton, G. A. Bliss	Moulton (1926), Bliss (1927)
war industry	statistical sequential	A. Wald	Rees (1980)
ballistics, supersonic flight	diff./int. equations	Kármán	Kármán (1936)
computers	logic, electrotechniques	von Neumann, Bush	Goldstine (1972)
gas dynamics, atom bomb	nonlinear hyperbolic equations	von Neumann, Courant	Goldstine (1972), Courant/Friedrichs (1948)
operations	game theory, statistics	von Neumann, Ph. Morse	Fortun/Schweber (1993)
anti-aircraft	statistical prediction	N. Wiener	Wiener (1956)
information	stochastics, computing	Cl. Shannon	Millman (1984)

5.2.3 Social/epistemic environment for applications of mathematics

- lack of academic applied mathematics with exception of MIT (Boston) and Caltech (Pasadena)
- strong traditions in industrial research at Bell labs (statistics and electrical engineering)
- important influence of British statistics (estimation tests, J. Neyman) in the 1930s
- transfer of German/European ideals to America with immigrations after 1933
- superior material conditions of American mathematics compared with Europe (instruments)

5.2.4 Historical primary and secondary literature/sources

Birkhoff (1977/80), Bilstein (1989), Bliss (1927), Bode (1953), Bush (1929, 1931, 1970), Dickson (1918), Fortun/Schweber (1993), Fry (1928, 1941, 1953), Goldstine (1972), Gray (1943), Hanle (1982), Hansen (1987), Heims (1980), Jewett (1947/71), Kármán (1936ff.), Masani (1990), McShane et al (1953), Meigs (1990), Millman (1984), Morawetz (1992), Morse/Hart (1941), Moulton (1926), Munk (1981), Owens (1989), Rees (1980), Richardson (1943), Roland (1985), Rosser (1982), Rothrock (1919), Shewhart (1931), Todd (1990), Vivian (1919), Wallis (1980), Weyl (1953), Wilson (1915), Zant (1943/44).

5.2.5 Some specific problems for historiography

- How differently would applied mathematics look in the U.S. today without WW II, and what would have been the consequence for pure mathematics, which seems to have profited greatly from military support even after the war?

5.3 Military work in mathematics (1914–1945): selected notes on the case of the Soviet Union

5.3.1 Main institutions and events connected to military application of mathematics

1891: wind tunnel in Moscow of 2 feet diameter built by Zukovskij (Joukowski)[194]

WW I mathematical physicist A. A. Friedmann (Fridman) serves as a pilot.

Training of pilots for the war at Moscow University under Zukovskij[195]

1918: Central aero-hydrodynamic institute (CAGI or TsAGI) in Moscow, director Zukovskij, afterwards S. A. Caplygin, later M. V. Keldys, who became Academy president 1961–1975

1921: Academy institute in Petrograd/Leningrad under Steklov, named after him and shifted to Moscow following his death in 1926

1922: *Scientific Research Institute for Mathematics and Mechanics* (NIIMM), founded at Moscow University, led for example by A. Ja. Khinchin since 1932 until split in 1935

1928: new wind tunnel at CAGI[196]

1929: journal “Vestnik prikladnoj matematiki i mechaniki” (“Applied mathematics and Mechanics”)

1932: V. V. Golubev has chair for higher mathematics at the Airforce Academy “N. E. Zukovski”, at the same time dean of University faculty math and mechanics

1941: (June) mathematicians of University of Moscow gradually evacuated to Aschabad, Turkmenistan, later to Sverdlovsk (Ukraine); in September 1943 returned to Moscow[198]

5.3.2 Some fields of application, methods, mathematicians/appliers involved[199]

field	method	mathematicians/appl.	source
aerodynamics: flutter	part. diff. equations	M. V. Keldys	MU, p. 81
naval anti-aircraft	nomography	N. A. Glagolev	MU, p. 78
airplane design	geometry, gas dynamics	S. A. Christianovic	MU, p. 80, Gol., p. 120
Katjusa rockets	gas dynamics, diff. equations	A. A. Kosmodemjanskij	MU, p. 83
stability of projectile	Ljapunov functions	N. G. Cetaev	MU, p. 82
shrapnel	plasticity	A. A. Iljusin (not S.V.)	MU, p. 85, Gol., p. 117
anti-aircraft	statistical prediction	A. N. Kolmogorov	Shiryaev (this vol.)

[194] According to Kármán (1954) p. 14. Zukovskij’s wind tunnel is also mentioned by Lee (1998), p. 11. Lee and Anderson (1997, p. 299) refer to another one built by Zukovskij with support by the wealthy patron D. P. Rjabucinskij at Koutchino near Moscow in 1904. There was obviously another one in Moscow, mentioned by Golubev (1955), p. 117.

5.3.3 Social/epistemic environment for applications of mathematics

- strong tradition of applied mathematics and theoretical mechanics in Russia (N. G. Cebysev, A. M. Ljapunov, V. A. Steklov)
- particularly strong tradition in mathematical foundations of aerodynamics around 1900: N. E. Zukovskij, S. A. Caplygin
- influence of Russian/Soviet emigrants above all in America
- language problems and political problems of mathematical communication with the West, reinforced 1933
- secret German-Soviet collaboration in aeronautics until 1933
- close collaboration of pure and applied mathematicians at Moscow University
- foundations of Soviet rocketry and jet propulsion laid between the Two World Wars
- mathematician Keldys "theoretician of Soviet space program" after World War II
- possibly problems of "transfer" from invention to innovation in SU
- propagandistic role of aeronautics between the World Wars, also intentionally with respect to defence,[200] involved mathematician and geophysicist O. Yu. Shmidt (1891–1956)
- superiority of German military aviation over the Soviets became clear already during the Spanish Civil War 1937
- speed of war planes belatedly (1939) declared as the first priority over distance (bombers)[201]

5.3.4 Historical primary and secondary literature/sources

Bailes (1978), Golubev (1955), Moscow University (MU, 1975), Lorentz (2002), Ogibalov, P.M. and L.V.Kudrjasova (1980), Palmer (2000), Tobies (1985), Vogt (1998)

5.3.5 Some specific open problems for historiography

- Was there a specific influence of Stalinism, in particular of the Moscow trials (1936/37), on mathematics, possibly in favour of applied as opposed to "pure" mathematics?[202]

[195] Golubev (1955), p. 97.

[196] This apparently even influenced Kármán's tunnel in Aachen (Tobies 1985, p. 73).

[197] Vogt (1998), p. 181.

[198] Ogibalov/Kudrjasova (1980), p. 14.

[199] Most of the details from Golubev (1955) [Gol] or *Moscow University* [MU]... (1975). All mathematicians mentioned are from Moscow.

[200] Bailes (1978), Palmer (2000), Vogt (1998). However according to Bailes (p. 394) "the achievement of world aviation records cost [the Soviet Union] dearly in military unpreparedness."

[201] Bailes (1978), pp. 394/95.

[202] Golubev (1955), p. 108. In spite of the well-known attacks against pure mathematicians such as N. N. Lusin, which had some utilitaristic overtones (see Lorentz (2002), p. 206, Lusin subsequently did some more applied work), there was obviously no automatism in this respect. Erickson (1972), p. 256, argues that the military purge in June 1937 damaged the "radio-technical programme" severely.

- Which foreign support as to scientific information received the Soviets during the War, for instance under the American Lend-Lease Act?[203]
- Which were the meeting places and channels of communication between Soviet mathematicians and the military?
- How were secrecy regulations enforced or did mathematicians follow self-imposed rules?
- Which developments in mathematics during the war proved crucial in the later Soviet space program?

5.4 Military work in mathematics (1914–1945): selected notes on the case of Great Britain

5.4.1 Main institutions and events connected to military application of mathematics

1909:	*Advisory Committee on Aeronautics* (later *Aeronautical Research Committee or Council*)
1913:	Mathematical Laboratory at University of Edinburgh[204]
	Royal Aircraft Establishment, one of the principal research establishments in inter-war[205]
1939:	Bletchley Park project started
1939:	Operational Analysis starts at Royal Air Force Command
1935–1945:	Results on relaxation methods at Imperial College London during war only published in "confidential version of the Royal Society Proceedings"[206]
1943:	Admiralty Computing Service at Bath[207]

5.4.2 Some fields of application, methods, mathematicians/appliers involved

field	method	mathematicians/appl.	source
cryptology	logic, coding theory, statistics	A. Turing	Hodges (1983), Hinsley/Stripp (1993)
meteorology	finite difference methods	L. F. Richardson	Ashford (1985)
elasticity	relaxation method	R. Southwell	Zienkiewicz (2000), Fox (1990)
aeronautics	statistic turbulence	G. I. Taylor	Kármán (1954)
aeronautics	complex functions	H. Glauert, S. Goldstein	Goldstein (1969)
operations research	various, including statistics	physicists, mathematicians at Bomber Command	Edgerton, Fortun/Schweber (1993)

[203] Erickson (1972, p. 243) argues that the scientific support for the Soviet Union due to Lend-Lease has been overestimated and domestic development undervalued at least with respect to radar.

[204] Whittaker/Robinson (1924).

5.4.3 Social/epistemic environment for applications of mathematics

- particularly strong social divide between engineers and scientists?
- different notion of "applied mathematics" as compared to Germany and later U.S.: not including engineering mathematics
- Kelvin/Maxwell as mathematical background for Bush's "differential analyzer"
- Air Ministry in inter-war years largest R&D spending institution in Britain[208]
- certain superiority complex over Americans maintained, in connection with ideas underlying computers, but lack of economic preconditions commensurate with the Americans
- technical-economic superiority of U.K. over the Poles in connection with cryptology[209]
- differences between British and American science policies in the war, especially with respect to team work and secrecy regulations (Hodges, Bush)
- highly personalised style of leading research: Tizard/Lindemann dispute

5.4.4 Historical primary and secondary literature/sources

Ashford (1985), Batchelor (1996), Birkhoff (1980), Bush (1970), Edgerton (1996), Fortun/Schweber (1993), Fox (1990), Hardy (1940), Hinsley/Stripp (1993), Hodges (1983), Levinson (1970), Whittaker/Robinson (1924), Zienkiewicz (2000)

5.4.5 Specific problems for historiography

- Why did OR not fare well in England after the war (Fortun/Schweber (1993), p. 343), although the origins were in Britain?

5.5 Military work in mathematics (1914–1945): selected notes on the case of Italy

5.5.1 Main institutions and events connected to military application of mathematics

1903: first wind tunnel constructed near Rome by Arturo Crocco

WW I ballistic work by Mauro Picone

1923: Consiglio Nazionale delle Ricerche (CNR), restructured 1927

1931: "National Institute for the Application of the Calculus" (INAC) under CNR and Mauro Picone in Rome

1935: International Volta-Conference in Rome on "High Velocities in Aviation"

about 1935: Italian aerodynamic research centre in Guidonia near Rome

[205] Edgerton (1996), p. 4.

[206] Zienkiewicz (2000), p. 9.

[207] Fox (1990), p. 292.

[208] Edgerton (1996), p. 3.

[209] British cooperated with the Poles who made early use (1932) of mathematicians in cryptology (Hodges 1983, p. 170). The Polish had gradually "to turn to the technically superior West" (Hodges 1983, p. 235), clearly so, of course after 1939. See also Rakus-Andersson, this volume.

5.5.2 Some fields of application, methods, mathematicians/appliers involved

field	method	mathematicians/appl.	source
ballistics	numerical methods	Picone	Amerio (1987)
aerodynamics		Arturo Crocco	Anderson (1994)
aerodynamics: stability, flutter	eigenvalue problems, calculus of variations, numerical methods	Picone-Institute INAC	Picone (1938), (1959)
pure spin-off	elliptic differential equations	C. Miranda at INAC	Picone (1938)

5.5.3 Social/epistemic environment for applications of mathematics

- pure mathematics had a dominant position in Italy until WWI in comparison to other sciences: thus partly healthy reorientation towards applications under Mussolini, especially after 1927
- change symbolized in mathematics by substitution of Volterra by Marconi (later by marshal P. Badoglio) in leading positions (CNR, Academy) with strong political connotation
- CNR's committee for mathematics later suppressed in favor of a sub-committee for applied mathematics of the committee of physics[210]
- 1923 Picone congratulates minister of education Gentile on membership in Fascist Party
- 1935: Ethiopian War, followed by boycott of Italy, growing autarky policies and increased demands for Picone's institute
- missing steel industry created problems for airplane production especially in isolation[211]
- 1938 racial laws in Italy: expulsion of Jewish mathematicians (Orlando)
- 1940 Italian "Royal National Institute of Higher Mathematics" under Severi in Rome: partly compensated for losses in pure mathematics
- collaboration with German applied mathematicians and industry/military: Italians as a role model for Germans (Picone 1959)
- after WW II (1951) recognition of Picone Institute by UNESCO under the title "Centro Internazionale di Calcolo" after a positive review by American Herman Goldstine

5.5.4 Historical primary and secondary literature/sources

Amerio (1987), Busemann (1971), Guerragio (1987), Israel/Nurzia (1989), Nastasi (1990), Ogburn (1949), Orlando (1998), Picone (1934-1959), Russo (1986), Salvadori (1939), Todd (1990)

5.5.5 Specific problems for historiography

- Unclear level of computing facilities at INAC: large number of "human calculators and computists" mentioned by Salvadori and others indicates rather low level

[210] See Russo (1986), p. 307.

[211] Ogburn (1949), p. 90.

5.6 Military work in mathematics (1914–1945): selected notes on the case of France

5.6.1 Main institutions and events connected to military application of mathematics

1909: *Ecole Nationale Supérieure d'Aeronautique*[212]
G. Eiffel's laboratories with aerodynamic experiments[213]

1922: *Le Mémorial de l'Artillerie Française*: internationally oriented

1926: *Institut Henri Poincaré* (IHP): theoretical rather than applied stochastics

1939: CNRS with laboratories at IHP: doubtful outcome

5.6.2 Some fields of application, mathematical methods, mathematicians/ appliers involved

field	method	mathematicians/appl.	source
ballistics in WW I	analytic functions	A. Denjoy, G. Valiron, P. Lévy	Charbonnier (1924)

5.6.3 Social/epistemic environment for applications of mathematics

- overly bureaucratic and Paris centred science and education system (Germain)
- lack of contact between engineering schools and universities (Germain)
- indirect shot since WW I urged new mathematics in French ext. ballistics (Charbonnier)
- poor French reception of international results (Pyenson), seems to improve in 1920s
- historic influence of metric system in French artillery on British and American artillery[214]
- leading *Ecole Normale Supérieure* distant to applications
- criticism of the state of applied mathematics and mathematical physics in France in the context of the foundation of the IHP 1926[215]
- misguided developments in French aeronautics: focus on engines instead of aerodynamics
- 1938 disbandment of French Strategic Air Force
- no defence research under German occupation (1940) possible without collaboration

[212] This school was nationalized after the creation of the French Air Ministry in September 1928, but mainly produced bureaucrats who as students originally had often done brilliant theoretical work (Chadeau 1988, p. 39).

[213] Hashimoto (1994), Both Prandtl and von Mises referred to the experiments there. However as a private institution the Eiffel laboratories were rather isolated in France itself (Chadeau 1988, p. 40). For the influence on Japanese aerodynamics see Braun (1987, p. 13).

[214] Gray (1943), p. 77.

[215] Siegmund-Schultze (2001).

- attempts to involve French prisoners of war in German war research[216]
- French deficiencies in computation and communication technologies clear after WW II

5.6.4 Historical primary and secondary literature/sources

Charbonnier (1924), Chadeau (1988), Germain (1953), Hashimoto (1994), Pestre (1994), Picard (1990), Pyenson (1996), Andler (1994), Orlando (1998), Kragh (1996), Siegmund-Schultze (1986/2001)

5.6.5 Some specific problems for historiography

- According to Charbonnier (1924): superior French mathematical methods in WW I ballistics
- How did the inner logic of an applied science such as ballistics stimulate the interest of mathematicians in new applications, including applications in wars?[217]
- "Phony war" (drôle de guerre) also in French science 1939, i.e., defence research not taken seriously?

5.7 Military work in mathematics (1914–1945): selected notes on the case of Japan

5.7.1 Main institutions and events connected to military application of mathematics

1931: Japanese invasion of Manchuria, new scientific institutions, partly in occupied territories

1932: Japanese broke American code (two-digit encipherment system)[218] but unable to read the new one of 1939

1933: Ultra-nationalists oppose introduction of metrical system which is delayed till 1958[219]

Wind tunnel erected by W. Margoulis, former director of the Eiffel laboratories in Paris.[220] "Extensive wind tunnel tests" for the *Aeronautical Research Institution of Tokyo Imperial University*[221]

1937: Japanese-Chinese War, Japan becomes the first country outside the United States to build and successfully operate a cyclotron[222]

[216] Siegmund-Schultze (1986).

[217] Charbonnier (1924), p. 577, is talking about tendencies in ballistics to create a coherent logical structure of the field. According to him there were no applications in the moment: but hopefully, as he maintained, in the future.

[218] Drea (1991), p. 188.

[219] Shimao (1989), p. 58.

[220] Braun (1987), p. 13. See also Hashimoto (1994) for the general importance for aerodynamics of the Eiffel experiments and tables, not just in Japan.

[221] Yamazaki (1995), p. 171.

[222] Low (1990), p. 350.

1941: Technology Board (focus on aeronautics): "true starting point of Japanese science and technology policy"[223]

Pearl Harbor: "with few exceptions mathematicians were directly or indirectly mobilised for the war"[224]

1943: S. Tomonaga (mathematical physicist)[225] driven into applied electrical engineering (Naval Institute of Technology).

1944: Institute of Mathematical Statistics established in 1944[226]

1946: late foundation of "Japan Mathematico-Physical Society": sign of belated modernisation[227]

5.7.2 Some fields of application, methods, mathematicians/appliers involved

field	**method**	**mathematicians/appl.**	**source**
electrical engineering		S. Tomonaga	Low (1990)
circuit design	lattice theory, logic	Nakajima at NEC	Suekane (1980)
war production	statististics	statistical Institute	Itagaki (2002)
airplane design		aeronautical institute Tokyo	Braun (1997)

5.7.3 Social/epistemic environment for applications of mathematics

- tradition of Japanese colonialism, militarism and racism since Chinese-Japanese War of 1895, reinforced 1931
- backwardness in scientific computation, partly due to strong abacus tradition (Suekane (1980))
- 1926–1945 Showa Period: period of modernisation of Japanese science
- pioneering results in theoretical physics: H. Yukawa: nobel prize theoretical physics 1949 for prediction of meson
- Japan had a mathematically strong ally, Germany, where several Japanese had studied
- social and epistemic barrier between mathematicians and engineers felt: pure mathematics too autonomous, more generally, lack of inter-disciplinarity in research

[223] Yamazaki (1995), p. 168.

[224] Sasaki (1999), p. 37.

[225] S. Tomonaga had worked on the mathematical foundations of meson theory before, esp. on "renormalization theory, which made quantum electrodynamics consistent with the special theory of relativity." (Low (1990), p. 356)

[226] According to a personal communication by Prof. Itagaki Ryoichi (2002), based on "The hundred-years history of mathematics in Japan" of 1983/84, this institute was erected by the Ministry of Education on the ground that statistical research will help to increase the production of goods in wartime.

[227] Interpreted this way by Sasaki (1999).

- 1940–42 Japanese military showed interest in atomic weapons, but found prospects discouraging during the war, therefore concentration on radar[228]
- mobilisation of science in Japan compared to the US a "downright failure"[229]

5.7.4 Historical primary and secondary literature/sources

Braun (1987), Drea (1991/95), Hashimoto (1994), Low (1990), Sasaki (1999), Shimao (1989), Suekane (1980), Yamazaki (1995), Itagaki Ryoichi (personal inf. 2002), *The hundred-years history of mathematics in Japan* (Japanese), 2 vols. Tokyo: Iwanami-Shoten 1983–1984

5.7.5 Specific open problems for historiography

- Controversial statements in the literature with respect to dependence on Germany[230]
- Unclear whether Japan's communication intelligence used sophisticated mathematical methods

[228] Low (1990), p. 350.

[229] Low (1990), p. 350.

[230] According to Braun (1987), p. 15, the Japanese were largely independent in jet technology, while Shimao (1989), p. 171, maintains that they did import it from Germany.

References

Amerio, L.: Mauro Picone e l'Istituto per le Applicazioni del Calcolo; in: A. Guerragio (ed., 1987), pp. 15–23.

Anderson, J. D.: *A History of Aerodynamics*, Cambridge University Press 1997.

Andler, M.: Les mathématiques à l'École normale supérieure au XXe siècle: une esquisse; in: Sirinelli (1994), pp. 351–404.

Ashford, O. M: *Prophet or Professor. The Life and Work of Lewis Fry Richardson*; Bristol and Boston: Adam Hilger 1985.

Aspray, William: *John von Neumann and the Origins of Modern Computing*; Cambridge, Mass.: MIT Press 1990.

Bailes, Kendall E.: *Technology and Society under Lenin and Stalin*; Princeton: Princeton University Press 1978.

Batchelor, G.: *The Life and Legacy of G. I. Taylor*; Cambridge: Cambridge University Press 1996.

Bayart, Denis & Crépel, Pierre: Statistical control of manufacture; in: Grattan-Guinness (ed.) (1994), volume II, pp. 1386–1391.

Beer, G. de: *The Sciences Were Never at War*; Cambridge: Cambridge University Press 1961.

Bilstein, R. E.: *Orders of Magnitude. A History of the NACA and the NASA*, 1915–1990; Washington, D.C.: NASA 1989 (also under http://www.hq.nasa.gov.)

Birkhoff, Garrett: Some leaders in American mathematics 1891–1941; in: Tarwater (ed. 1977), pp. 25–78.

Birkhoff, Garrett: Computer developments 1935–1955, as seen from Cambridge, U.S.A.; in: Metropolis, N. et al. (1980), pp. 21–30.

Bliss, Gilbert A.: Mathematics; *The Scientific Monthly* **24** (1927), pp. 308–319.

Bode, H. W.: On the operation of a mathematics group in an industrial environment; in: Weyl, F. J. (ed.) 1953, pp. 84–87.

Bogoljubov, A. N. & Urbanskij, V. M.: Organisation and evolution of mathematical institutes at the Ukranian Academy of Sciences (Russian); *Istoriko-matematiceskie issledovanija* **28** (1985), pp. 160–187.

Booß, B. & Høyrup, J.: *Von Mathematik und Krieg*; Marburg: Bund Demokratischer Wissenschaftler 1984.

Bowles, M. D.: U.S. technological enthusiasm and british technological scepticism in the age of the analog brain; *IEEE Annals of the History of Computing* **18** (1996), No. 4, pp. 5–15.

Braun, H.-J.: Technology transfer under conditions of war: German aero-technology in Japan during the Second World War; *History of Technology* **11** (1987), pp. 1–23.

Burnham, J. C. (ed.): *Science in America. Historical Selections*; New York: Holt, Rinehart and Winston 1971.

Busemann, A.: Compressible fluids in the thirties; *Annual Review of Fluid Mechanics* **3** (1971), pp. 1–12.

Bush, V.: *Operational Circuit Analysis. With an Appendix by N. Wiener*; New York 1929.

Bush, V.: The differential analyzer. A new machine for solving differential equations; *Journal of the Franklin Institute* **212** (1931), pp. 447–448.

Bush, V.: *Pieces of Action*; New York: William Morrow 1970.

Chadeau, E.: Government, industry, and the nation: the growth of aeronautical technology in France (1900–1950); *Aerospace Historian* **35** (1988), March, pp. 26–44.

Charbonnier, P.: Sur l'état actuel de la balistique extérieure théorique (1924); in: *Proceedings of the International Mathematical Congress Toronto 1924*, ed. J. C. Fields, Toronto University Press 1928, volume II, pp. 571–593.

Courant, R. & Friedrichs, K.: *Supersonic Flow and Shock Waves*; New York: Interscience 1948.

Courant, R.; Friedrichs, K., & Lewy, H.: Über die partiellen Differenzengleichungen der mathematischen Physik; *Mathematische Annalen* **100** (1928), pp. 32–74. [english in: *IBM Journal of Research and Development* **11**(1967), pp. 214–234]

Courant, R. & Hilbert, D.: *Methoden der mathematischen Physik*; 2 volumes, Berlin: Springer 1924 and 1937.

Davis, Ph.: Applied mathematics as social contract; *Mathematics Magazine* **61** (1988), pp. 139–147.

Dickson, L. E.: Mathematics in war perspective; *Bulletin American Mathematical Society* **25** (1918/19), pp. 289–311.

Drea, Edward J.: Reading each other's mail: Japanese communications intelligence, 1920–1941; *The Journal of Military History* **55** (April 1991), pp. 185–205.

Drea, Edward J.: Were the Japanese codes secure" *Cryptologia: a quarterly journal devoted to cryptology* **19** (1995), pp. 113–136.

Edgerton, D. E. H.: British scientific intellectuals and the relations of science, technology, and war; in: Forman, P. & Sánchez-Ron, J. M. (1996), pp. 1–35.

Epple, M. & Remmert, V.: "Eine ungeahnte Synthese zwischen reiner und angewandter Mathematik." Kriegsrelevante mathematische Forschung in Deutschland während des II. Weltkrieges; in: Doris Kaufmann (ed.): *Geschichte der Kaiser-Wilhelm-Gesellschaft im Nationalsozialismus. Bestandsaufnahme und Perspektiven der Forschung*; 2 volumes, Göttingen: Wallstein 2000, Band I, pp. 258–295.

Erickson, J.: Radio-location and the air-defence problem: The design and development of Soviet radar 1934–40; *Science Studies* **2** (1972), pp. 241–268.

Forman, P. & Sánchez-Ron, J. M. (eds.): *National Military Establishments and the Advancement of Science and Technology.* Studies in the 20th Century History; Dordrecht, Boston, London: Kluwer 1996.

Fortun, M. & Schweber, S. S.: Scientists and the state: the legacy of World War II; in: Gavroglu, K., Christianidis, J. & Nicolaidis, E. (eds.), *Trends in the Historiography of Science*, Dordrecht etc: Kluwer 1993, pp. 327–354.

Fox, L.: Early numerical analysis in the United Kingdom; in: Nash (1990), pp. 280–300.

Fry, Th.: *Probability and Its Engineering Uses*; New York 1928.

Fry, Th.: Industrial mathematics; *The Bell System Technical Journal* **20** (1941), pp. 255–292 [report for U.S. National Research Council 1940, also as supplement to *American Mathematical Monthly* **48** (1941), June/July, pp. 1–38.]

Fry, Th.: Applied mathematics as a responsibility of the mathematical profession; in: Weyl, F. J. (ed.) 1953, pp. 89–97.

Germain, P.: Applied mathematics in France; in: Weyl, F. J.(ed.) 1953, pp. 48–53.

Godement, R.: Aux sources du modèle scientifique americain; *La pensée* **201** (1978), pp. 33–35; **203** (1979), pp. 95–122; **204** (1979), pp. 86–110.

Goldstein, S.: Fluid Mechanics in the first half of this century; *Annual Review of Fluid Mechanics* **1** (1969), pp. 1–28.

Goldstine, H. H.: *The Computer from Pascal to von Neumann*; Princeton: Princeton University Press 1972.

Golubev, V. V.: Mechanics at Moscow University before the Great Socialist October Revolution and in the Soviet period (Russian); *Istoriko-matematiceskie issledovanija* **8** (1955), pp. 77–126.

Görtler, H.: Übersicht; in: Walther, A. (ed.) 1948–1953, volume 5, *Mathematische Grundlagen der Strömungsmechanik* (1948), pp. 1–12.

Goudsmit, S. A.: *Alsos*; New York: Henry Schuman Inc. 1947.

Grattan-Guinness, I. (ed.): *Companion Enyclopedia of the History and Philosophy of the Mathematical Sciences*, 2 Volumes; London/New York: Routledge 1994.

Gray, G. W.: *Science at War*; New York/London: Harper 1943.

Gross, H.-E.: Das sich wandelnde Verhältnis von Mathematik und Produktion; in: Plath, P. & Sandkühler, H. J.: *Theorie und Labor. Dialektik als Programm der Naturwissenschaft*; Köln: Pahl-Rugenstein 1978, pp. 226–269.

Grunden, W.E.: *Science under the Rising Sun: Weapon's Development and the Organization of Scientific Research in World War II Japan*; Santa Barbara: PhD-Thesis University of California 1998, 385 pp.

Guerragio, A. (ed.): *La matematica italiana tra le due guerre mondiali*; Bologna: Pitagora Editrice 1987.

Haack, W.: Geschossformen kleinsten Wellenwiderstandes; *Bericht der Lilienthal-Gesellschaft für Luftfahrtforschung* **139** (1941), part 1, pp. 14–28.

Haack, W.: *Projectile Shapes for Smallest Wave Drag*; Providence: Brown University, Report A9-T-3, 53 leaves, 1948.

Haack, W.: Rückschau über 85 Jahre, *Wissenschaftsmagazin* (TU Berlin), 1987, no. 10, pp. 20–23.

Haack, W.: *Erinnerungen* (Auf Kassetten gesprochen von 1984–1989); Berlin 1989 (mimeographed ms., 292 pp.).

Haack, W.: 1935 bis 1937 Dozent an der Technischen Hochschule Berlin – Erinnerungen; in: Begehr, H. (ed.): *Mathematik in Berlin. Geschichte und Dokumentation*; 2 volumes, Aachen: Shaker 1998, volume 2, pp. 205–217.

Hanle, P.: *Bringing Aerodynamics to America*; Cambridge, MA: MIT Press 1982.

Hansen, J. R.: *Engineer in Charge: A History of the Langley Aeronautical Laboratory*, 1917–1958; Washington, D.C.: NASA 1987.

Hardy, G. H.: *A Mathematician's Apology*; Cambridge: Cambridge University Press 1940.

Hartcup, G.: *The War of Invention, Scientific Developments, 1914–1914*; London etc.: Brassey's Defence Publishers 1988.

Hartcup, G.: *The Effect of Science on the Second World War*; London etc.: MacMillan 2000.

Hashimoto, Takehiko: Graphical calculation and early aeronautical engineers; *Historia Scientiarum* (2) **3** (1994), pp. 159–183.

Heims, Steve J., *John von Neumann and Norbert Wiener. From Mathematics to the Technologies of Life and Death*, Cambridge, Mass.: MIT Press, 1980.

Hensel, S., Ihmig, K.-N., & Otte, M.: *Mathematik und Technik im 19. Jahrhundert in Deutschland*; Göttingen: Vandenhoeck & Ruprecht 1989.

Hinsley, F. H., & Stripp, A. (eds.): *Code Breakers. The Inside Story of Bletchley Park*; Oxford: Oxford University Press 1993.

Hodges, A.: *Alan Turing, the Enigma*; New York: Simon & Schuster 1983.

Hoff Kjeldsen, T.: A contextualized historical analysis of the Kuhn-Tucker Theorem in nonlinear programming. The impact of World War II; *Historia Mathematica* **27** (2000), pp. 331–361.

Hosemann, R.: Verfolgungskurven; in: Walther (ed., 1948–53), volume 3 (1948), pp. 269–307.

Israel, G., & Nurzia, L.: Fundamental trends and conflicts in Italian mathematics between the two World Wars; *Archives Internationales d'Histoire des Sciences* **39** (1989), pp. 112–143.

Jewett, F. B.: The future of scientific research in the postwar world (1947); in: Burnham (1971), pp. 398–413.

Kármán, Th. v.: The problem of resistance in compressible fluids; in: *Atti del V Convegno della "Fondazione Alessandro Volta", Roma 1935*; Rome 1936, pp. 5–59.

Kármán, Th. v.: The Engineer grapples with nonlinear problems; *Bulletin American Mathematical Society* **46** (1940), pp. 615–683.

Kármán,Th. v.: Tooling up mathematics for engineering; *Quarterly of Applied Mathematics* **1** (1943), pp. 2–6.

Kármán,Th. v.: The role of fluid mechanics in modern warfare (1943); in: *Collected Works* IV (1956), pp. 193–205.

Kármán, Th. v.: Supersonic aerodynamics – principles and applications; *Journal of the Aeronautical Sciences* **14** (1947), pp. 373–409.

Kármán, Th. v.: *Aerodynamics. Selected Topics in the Light of Their Historical Development*; Ithaca: Cornell University Press 1954.

Kármán, Th. v.: *Collected Works of Theodore von Kármán*; London: Butterworths 1956 (volumes I–IV).

Kevles, D. J.: *The Physicists. The History of a Scientific Community in Modern America* (1977); Cambridge, Mass.: Harvard University Press 1987.

Kragh, H.: Telephone technology and its interaction with science and the military; in: Forman, P., & Sánchez-Ron, J. M. (1996), pp. 37–67.

Lax, P.: The flowering of applied mathematics in America; *SIAM Review* **31**, no. 4 (1989), pp. 533–541.

Lee, L. J.: The origin of the wind tunnel in Europe 1871–1890; *Air Power History* **45**, summer 1998, pp. 6–15.

Levinson, N.: Coding theory: a counter example to G. H. Hardy's conception of applied mathematics; *American Mathematical Monthly* **77** (1970), pp. 249–258.

Lorentz, G. G.: Mathematics and politics in the Soviet Union from 1928 to 1953; *Journal of Approximation Theory* **116** (2002), pp. 169–223.

Low, M. F.: Japan's secret war? 'Instant' scientific manpower and Japan's World War II atomic bomb project; *Annals of Science* **47** (1990), pp. 347–360.

Ludwig, K.-H.: *Technik und Ingenieure im Dritten Reich*; Düsseldorf: Athenäum 1974.

Masani, P. R.: *Norbert Wiener 1894–1964*; Basel etc.: Birkhäuser 1990.

McArthur, C. W.: *Operations Analysis in the U.S. Army Eighth Air Force in World War II*; Providence, R. I.: American Mathematical Society, and London: London Mathematical Society 1990.

McShane, E. J., Kelley, J. L. & Reno, F. V.: *Exterior Ballistics*; Denver: University of Denver Press 1953.

McVittie, G. C.: The training of applied mathematicians in Britain; in: Weyl, J. F. (ed., 1953), pp. 45–48.

Mehrtens, H.: Angewandte Mathematik und Anwendungen der Mathematik im nationalsozialistischen Deutschland, *Geschichte und Gesellschaft* **12** (1986), pp. 317–347.

Mehrtens, H.: Mathematics and war: Germany, 1900–1945; in Forman, P. & Sánchez-Ron, J. M. (eds., 1996), pp. 87–134.

Meigs, M.C.: *Slide Rules and Submarines. American Scientists and Subsurface Warfare in World War II*; Honolulu, Hawaii: University Press of the Pacific 1990.

Metropolis, N., Howlett, J., & Rota, G.-C. (eds.): *A History of Computing in the Twentieth Century. A Collection of Essays*; New York, London etc.: Academic Press 1980.

Meyer zur Capellen, W.: *Mathematische Instrumente*, 3. ed.; Leipzig: Akademische Verlagsgesellschaft 1949.

Millman, S. (ed.): *A History of Engineering and Science in the Bell System. Communication Sciences (1925–1980)*; AT&T Bell Laboratories 1984.

Miranda, C.: *Equazioni alle derivate parziali di tipo ellittico*; Berlin: Springer 1955.

Morawetz, C.S.: Giants; *American Mathematical Monthly* **99** (1992), pp. 819–828.

Morse, M., & Hart, W. L.: Mathematics in the defense program; *American Mathematical Monthly* **48** (1941), pp. 293–302.

Moscow University during the Great Patriotic War (Russian, anon.); Moscow: The University of Moscow 1975.

Moulton, F. R.: *New Methods in Exterior Ballistics*; Chicago: University of Chicago Press 1926.

Munk, M.: My early aerodynamic research – thoughts and memories; *Annual Review of Fluid Mechanics* **13** (1981), pp. 1–7.

Nash, St. G. (ed.): *A History of Scientific Computing*; New York: ACM Press 1990.

Nastasi, P. (ed.): *Per l'archivio della corrispondenza dei matematici italiani: il fondo 'Giovanni Gentile'*; Progetto Ricerche Storiche e Metodologiche Università Commerciale Luigi Bocconi, Milano 1990, 99 pp.

Newby, J. C.: The history and mathematics of rockets; *Physics in Technology* **19** (1988), pp. 172–180.

Nihon no sūgaku 100 nenchi [A Hundred Years History of Mathematics in Japan] (Japanese), 2 vols. Tokyo: Iwanami-Shoten 1983–1984.

Ogibalov, P. M., & Kudrjasova, L. V.: On the establishment and development of Moscow University faculty of mechanics and mathematics (Russian); *Istorija i metodologija estestvennych nauk (Matematika/Mechanika)* **25** (1980), pp. 3–22.

Ogburn, W. F.: Aviation and international relations; in: Ogburn, W. F. (ed.), *Technology and International Relations*; Chicago: University of Chicago Press 1949, pp. 86–101.

Orlando, L.: Physics in the 1930s: Jewish physicists' contribution to the realization of the 'new tasks' of physics in Italy; *Historical Studies in the Physical and Biological Sciences* **29** (1998), 1, pp. 141–181.

Owens, L.: Mathematics and war. Warren Weaver and the Applied Mathematics Panel 1942–1945; in Rowe/McCleary (1989), Vol.II, pp. 287–305.

Palmer, S. W.: Peasants into pilots. Soviet air-mindedness as an ideology of dominance; *Technology & Culture* **41** (2000), pp. 1–26.

Parshall, K. H., & Rice, A. C. (eds.): *Mathematics Unbound: The Evolution of an International Mathematical Research Community, 1800–1945*; History of Mathematics, Volume 23; Providence: American Mathematical Society 2002.

Pestre, D.: *Physique et physiciens en France 1918–1940*; Paris: éditions des archives contemporaines 1984.

Petzold, H.: *Moderne Rechenkünstler. Die Industrialisierung der Rechentechnik in Deutschland*; München: Beck 1992.

Picard, J.-F.: *La république des savants. La recherche française et le C.N.R.S.*; Paris: Flammarion 1990.

Picone, M. (1934): *L'artiglieria italiana nella guerra mondiale*; Roma: Tip.del Senato 1934.

Picone, M.: L'attività dell'Istituto per la Applicazioni des Calcolo nel quadriennio XII–XV E. F.; *Società Italiana per il progresso delle scienze* **16** (1938), pp. 1–6.

Picone, M. (ed.): *Giudizi sull'opera trentennale dell'istituto nazionale per le applicazioni del calcolo*; Roma: tipografia Pio X, 1959.

Pyenson, L.: On the military and the exact sciences in France; in: Forman, P., & Sánchez-Ron, J. M. (1996), pp. 135–152.

Rees, M.: The mathematical sciences and World War II; *American Mathematical Monthly* **87** (1980), pp. 607–621.

Reid, C.: *Courant in Göttingen and New York*; New York etc.: Springer 1976.

Richardson, R. G. D.: Applied mathematics and the present crisis; *American Mathematical Monthly* **50** (1943), pp. 415–423.

Rider, R.: Operational research; in: Grattan-Guinness, I. (ed.) 1994, volume I, pp. 837–842.

Riegels, F. W.: *Aerofoil Sections. Results from Wind-tunnel Investigations. Theoretical Foundations*; London: Butterworths 1961.

Roland, A.: *Model Research*; Volumes 1 & 2, The NASA History Series, Washington, D.C.: NASA 1985 (also on http://history.nasa.gov.).

Rosser, J. B.: Mathematics and mathematicians in World War II; *Notices American Mathematical Society* **29** (1982), pp. 509–515.

Rothrock, D. A.: American mathematicians in war service, *American Mathematical Monthly* **26** (1919), pp. 40–44.

Rowe, D., & McCleary, J. (eds.): *The History of Modern Mathematics. Volume II: Institutions and Applications*; Boston etc: Academic Press 1989.

Russo, A.: Science and Industry in Italy between the two world wars; *Historical Studies in the Physical Sciences* **16** (1986), pp. 281–320.

Salvadori, M. G.: The activity of the Italian National Institute for the Application of Calculus; in: *Proceedings of the Fifth International Congress for Applied Mechanics held at Harvard University* ... 1938, New York: Wiley 1939, pp. 260–264.

Sasaki, Chikara: Science and the Japanese Empire 1868–1945: an overview; in: P. Petitjean et al. (eds.), *Science and Empires*, Dordrecht etc.: Kluwer 1992, pp. 243–246.

Sasaki, Chikara: The development of mathematical research in twentieth-century Japan, in: Sasaki, C.: Preprint "Collected Papers", University of Tokyo, 1999, pp. 27–44.

Sasaki, Chikara: The emergence of the Japanese mathematical community in the modern western style, 1855–1945; in Parshall/Rice (2002), pp. 229–252.

Schappacher, N., & Scholz, E. (eds.): Oswald Teichmüller – Leben und Werk; *Jahresbericht Deutsche Mathematiker-Vereinigung* **94** (1992), pp. 1–39.

Sears, W. R.: On projectiles of minimum wave drag; *Quarterly of Applied Mathematics* **4** (1947), pp. 361–366.

Shannon, C. E. and Weaver, W.: *The Mathematical Theory of Communication*; Urbana: University of Illinois Press 1949.

Shewhart, W. A.: *Economic Control of Quality of Manufactured Products*; New York 1931.

Shimao, Eikoh: Some aspects of Japanese science; *Annals of Science* **46** (1989), pp. 69–91.

Siegmund-Schultze, R.: Theodor Vahlen – zum Schuldanteil eines deutschen Mathematikers am faschistischen Mißbrauch der Wissenschaft; *NTM-Schriftenreihe für Geschichte der Naturwissenschaften, Technik und Medizin* **21** (1984), 1, pp. 17–32.

Siegmund-Schultze, R.: Faschistische Pläne zur 'Neuordnung' der europäischen Wissenschaft. Das Beispiel Mathematik; *NTM-Schriftenreihe* **23** (1986), 2, pp. 1–17.

Siegmund-Schultze, R.: Zur Sozialgeschichte der Mathematik an der Berliner Universität; *NTM-Schriftenreihe* **26** (1989), 1, pp. 49–68.

Siegmund-Schultze, R.: Über die Haltung deutscher Mathematiker zur faschistischen Expansions- und Okkupationspolitik in Europa; in: Tschirner, M., & Göbel, H.-W. (eds.): *Wissenschaft im Krieg – Krieg in der Wissenschaft*, Marburg 1990, pp. 189–195.

Siegmund-Schultze, R.: *Rockefeller and the Internationalization of Mathematics Between the Two World Wars. Documents and Studies for the Social History of Mathematics in the 20th Century*; Basel etc.: Birkhäuser 2001.

Siegmund-Schultze, R.: The late arrival of academic applied mathematics in the United States: a paradox, theses, and literature, *NTM (N.S.)* **11** (2003), pp. 116–127.

Sirinelli, J.-F. (ed.): *École normale supérieure. Le livre du bicentenaire*; Paris: Presses Universitaires de France 1994.

Suekane, R.: Early history of computing in Japan; in: Metropolis et al. (eds. 1980), pp. 575–578.

Tarwater; J.D. (ed.): *The Bicentennial Tribute to American Mathematics 1776–1976*; Washington: MAA, 1977.

Tobies, R.: Zu den Beziehungen deutscher und sowjetischer Mathematiker während der Zeit der Weimarer Republik; *Mitteilungen der Mathematischen Gesellschaft der DDR* 1985, No.1, pp. 66–80.

Todd, J. L.: The prehistory and early history of computation at the U.S. National Bureau of Standards; in: Nash, St. G. (ed.) 1990, pp. 251–268.

Trischler, H.: *Luft- und Raumfahrtforschung in Deutschland 1900–1970*; Frankfurt/New York: Campus 1992.

Vivian, R.: Statistics in relation to the war; *American Mathematical Monthly* **26** (1919), pp. 32–35.

Vogt, A.: Von Petersburg nach Moskau: Zur Geschichte der russisch-sowjetischen Mathematik zwischen 1875 und 1975; in: Henning, A. (ed.): *Wissenschaftsgeschichte in Osteuropa*; Wiesbaden: Harrassowitz 1998, pp. 165–183.

Wallis, W. A.: The Statistical Research Group, 1942–1945; *Journal of the American Statistical Association* **75** (1980), pp. 320–330.

Walther, A. (ed.): *Angewandte Mathematik*; in: *Naturforschung und Medizin in Deutschland 1939–1946*. Für Deutschland bestimmte Ausgabe der FIAT Review of German Science; volumes 3–7 (1948–1953), Weinheim: Chemie; Wiesbaden: Dieterich 1953.

Weissinger, J.: Erinnerungen an meine Zeit in der DVL 1937–1945; *Jahrbuch Überblicke Mathematik* (1985), pp. 105–129.

Weyl, F. J. (ed.): *Proceedings of a Conference on Training in Applied Mathematics*; AMS and NRC, Columbia University, New York City, 22–24 October 1953, manuscript, 97 pp. (Available in Mathematische Bibliothek Oberwolfach/Schwarzwald, Inv. Nr. 196).

Whittaker, E. T., & Robinson, J.: *The Calculus of Observations*; London: Blackie 1924.

Wiener, N.: *I Am a Mathematician*; London: Victor Gollancz 1956.

Willers, A.: *Mathematische Instrumente*; München, Berlin 1943.

Wilson, E. B.: *Theory of Airplane Encountering Gusts*; First Annual Report of the National Advisory Committee for Aeronautics; U.S. Government Printing Office 1915.

Yamazaki, Masakatsu: The mobilization of science and technology during the Second World War in Japan; *Historia Scientiarum* (2) **5** (1995), No.1, pp. 167–181.

Zant, J. H.: Mathematics in the Army Specialized Training Program; *Journal of Engineering Education* **34** (1943/44), pp. 703–706.

Zienkiewicz, O. C.: Achievements and some unsolved problems of the finite element method; *International Journal for Numerical Methods in Engineering* **47** (2000), pp. 9–28.

Zdravkovska, S., & Duren, P. L. (eds.): *Golden Years of Moscow Mathematics*; History of Mathematics, vol. 6, Providence: American Mathematical Society 1993.

The Brains behind the Enigma Code Breaking before the Second World War

ELISABETH RAKUS-ANDERSSON*

The German Enigma encoding machine and the contributions of famous cryptologists who broke it are still topics that fascinate both scientists and the general public. After the monarchy of Kaiser Wilhelm II fell, the Weimar republic came into being, and the idea of equipping the armed forces with machine ciphers already found realization in 1926. The German cipher machine, called Enigma, alarmed the general staffs of neighbouring countries, especially Poland and France. This work describes the efforts of cryptanalysts who solved the mystery of Enigma during the 1930s before the beginning of the war.

1 Introduction

In the late 1920s, everything indicated that the small Reichswehr would be converted into a modern million-man army. All the German political parties that came to power voted on the same program, which had as the main assumption to take away from Poland some districts regarded by the German government as "lost territories". The military build-up program of the German army forces had to involve the development of secret intelligence operations [Kozaczuk 1984, Garliński 1999]. These operations always belonged to two categories, namely, cryptography ("secret writing") and cryptology (the study of secret writing, especially for purposes of decryptment – "the breaking" or "reading" of secret correspondence by the third party).

Cryptology deals basically with ciphers, which either transpose (shift) or substitute letters for the original letters in a message, and with codes, which replace entire words and phrases by arbitrary symbols commonly consisting of letters and numbers [Kahn 1967, Kozaczuk 1984].

Rather early the German army commanders realized that they should introduce a cryptography device that was both secure and could satisfy the requirements of speed and convenience. In spring of 1918, the German Navy contemplated the use of cipher machines. The inventor Hugo Koch designed the "Enigma" machine in Holland in 1919 and sold the patent later to Dr. Arthur Scherbius. Scherbius could improve Koch's design and hoped to win the market for this machine in the busi-

* Blekinge Institute of Technology, Department of Health, Science and Mathematics, S-37179 Karlskrona, Sweden. E-mail: Elisabeth.Andersson@bth.se

ness world, but his best customer would be the German government, especially the armed forces [Kozaczuk 1984, Garliński 1999]. In 1926 the navy, and in 1928 the army, introduced cipher machines that at first were modified versions of the civilian model "Enigma" [Kozaczuk 1984, Gaj 1989]. Two years later, in 1930, a military version of the device was constructed. In 1933–34, the Germans adopted Enigma as a basic, unitary cipher system for the armed forces as well as military intelligence (the Abwehr), S.S. formations, the Nazi Party security and Political Intelligence Service (S.D.).

Cryptologists could easily recognize an Enigma cipher by its perfect spread of letters. There were no correlates with natural languages, and the statistical calculations of frequencies of the letters were completely useless.

It was not strange that the Polish authorities were the most engaged of all countries in solving the German coding system. The other German neighbours, such as France or England, were not threatened by German actions to the same degree as Poland. The Polish cryptologists found the solution of the Enigma secret before the war began, and this story shows the historical background of progress made by Polish mathematicians.

2 The Cryptology Course in Poznań

The general staffs of the neighbouring countries were alarmed by the German machine cipher, and cryptologists, who received some messages from the monitoring stations (placed in Poland in Warsaw, Poznań, Starogard and Krzesławice) set to work on breaking the code [Kozaczuk 1967].

There were few persons skilful at cryptology in Poland at this time. At the General Staff's Cipher Bureau in Warsaw, specialists were trained and distinguished from the clerks who enciphered and deciphered messages. The German ciphers and codes monitored by the Poles were only partially intercepted, which gave a reason for educating the cryptologists and equipping them with all necessary knowledge about the ciphers.

The need for organizing the course in cryptology was justified by the actual political situation. In the first days of January 1929, the students at Poznań University's Mathematics Institute were preparing for their final examinations in mathematics, where the examiner was Prof. Zdzisław Krygowski. He had already prepared a list of those third- and fourth-year mathematics students who knew German and had marks of at least "good" in their course work. Afterwards the selected students were asked to assemble at the Institute, where two officers from the Polish General Staff in Warsaw, Major Franciszek Pokorny and Lieutenant Maksymilian Ciężki informed them that a cryptology course was organized and invited them to participate in it [Kozaczuk 1984, Gaj 1989, Sebag-Montefiore 2000]. The students who chose the course were pledged to secrecy concerning both the existence of the course and their participation in it. The lessons were held twice a week in the evenings and conducted by Cipher Bureau cryptologists commuting from Warsaw.

After several weeks of the course, the lecturer presented authentic Reichswehr ciphergrams for the students to solve. This system had already been broken and the Germans called it Double Dice [Kozaczuk 1984, Garliński 1999]. A couple of hours later, three students: Marian Rejewski, Henryk Zygalski and Jerzy Różycki (Fig. 1, 2, 3) presented their solutions. These three students, as the best ones, continued working on cryptology in the underground basements of the Command Post in Poznań.

Figure 1.
Marian Rejewsk in 1944.[1]

Figure 2.
Jerzy Różycki in 1933.[1]

Figure 3.
Henryk Zygalski in 1941.[1]

This work was a kind of laboratory that gave a broad opportunity for experimentation. The mentioned team of skilled students, engaged in cryptology work, did not read long dispatches but tried to work out methods of breaking German cipher keys, which were changed periodically. Radio intercepts were delivered by courier from Warsaw and from the nearest monitoring station in Poznań while the solutions were sent to Warsaw by airplane.

Solutions of some non-mechanical military German ciphers became a matter of routine. The cryptologists learned to exploit the mistakes made by the German cipher clerks as well as certain regularities they discovered. One of these was a rule that a cipher text must contain at least fifty letters. Thus every message sent by the Germans had the letter X added to enlarge the text to fifty signs.

The discovery of such rules made it easier to solve the cipher problems. But strange things began to happen, namely, the enciphered information was sent without any stable rules at all, and all the doubts disappeared – this was a complicated machine cipher inaccessible for standard methods of breaking.

[1] From [Garliński 1999]

3 The Enigma

The Poznań cipher office was closed in the summer of 1932. The main purpose for the course in cryptology and for the primary work with reading German messages, which were monitored by the Polish radio network, was to find talented students and train them in working on the new German machine cipher [Hodges 1983, Kozaczuk 1984].

The fact that the Polish authorities turned to mathematicians could have been partially explained by the great development of the Polish mathematical school at that time. Marian Rejewski, who had spent a year in Göttingen for studies, felt that the famous German school of mathematics belonged to the past. Among those professors he had met in Germany were no such outstanding persons as Poland's Stefan Banach and Wacław Sierpiński whose work had such great importance for twentieth-century mathematics [Kuratowski 1980]. The achievements made in mathematics in Poland at the beginning of the twentieth-century instilled a strong belief in the power of the subject. The members of the Polish government, who represented Poland as a new state with rather poor resources after World War I, realized that all intellectual possibilities should have been utilized in the service of the country.

On 1 September 1932, Marian Rejewski, Jerzy Różycki and Henryk Zygalski began working as regular employees at the Cipher Bureau in the general staff building (the "Saxon Palace") in Warsaw [Kozaczuk 1976, 1984, Bauer 2002]. The efforts undertaken during the years 1928–1930 to solve the new machine cryptosystem led nowhere. However, the three mathematicians got a different problem to solve during their first weeks in the General Cipher Bureau, namely, to break the four-letter German naval code [Kozaczuk 1984].

From a dozen short messages, one was selected for closer study. It consisted of only six groups, each of four letters. After thorough analysis the mathematicians noticed that the letter Y occurred at the beginning of a large number of code groups. In German, many question expressions (Wer?, Wo?, Wohin?, Wann?, Welcher?) begin with the same letter, and this regularity could have been present in the code. Next, they noticed that, following this six-group message, another station sent on the same wavelength a short signal consisting of only four signs. By assuming that the first message was a question, they guessed that the second might be an answer. Such a short reply could be a number, maybe a year. The question-answer system was recognized with a possibility that the answer could be a date, e.g., a year. The solution of this six-word signal led to the gradual reconstruction of the entire German naval code used in the second half of 1932. Even after fifty years, Marian Rejewski remembered that YOPY meant "when", YWIN – "where", BAUG – "and", and KEZL – "cancel the final letter".

The Polish mathematicians could read most of the messages in navy code even if the Germans tried to make them difficult. They transposed the alphabetical order, omitted certain letters of the alphabet, or from time to time remitted false code groups.

Even if the effects concerning the navy code were very apparent, the Enigma remained unsolved. The attempts were made to solve the mystery included the mathematicians' efforts as well as the predictions of clairvoyants. One of the preliminary findings was formulated as: "If we write two cipher texts with identical beginnings one below the other, identical letters in the same places will occur in the average twice as often as when we place texts with different beginnings in the same manner".

The work on Enigma required great concentration and at least eighty intercepts per day. Marian Rejewski obtained a commercial Enigma machine used by business firms. The machine resembled a typewriter, with an additional panel built into the lid. The panel contained twenty-six little circular glass windows having, like the keyboard, the letters of the alphabet. A number of glow lamps were built in the panel underside. Inside the machine was a set of three rotors, or rotating drums, and a reversing drum, all sitting on the common axle. The machine had also a stationary drum, called the entry ring. With every stroke of a key one or more rotors rotated, and at the same time the corresponding glow lamp lit up and illuminated the letter in the window above it. The machine was designed in a way that allowed finding the association between the plain text and the cipher. If one struck the key with the letter coming from the clear text, then a corresponding cipher letter would appear in the window. Conversely, when another person tapped out a cipher text, the letters illuminated in turn would spell the plain text. In order to conduct a secret dialog, both parties had to possess the same device set [Kozaczuk 1984].

The commercial Enigma only provided the general insight into the construction of the machine. It was easy to guess that the military version of Enigma would probably have a different wiring system and additional components. The cryptologists had to continue studying the system from the mathematical side. They needed group theory [Kozaczuk 1984, Freedman 2000], especially the properties of permutation groups, which were very useful when working on the military Enigma.

4 International Cooperation

France and Czechoslovakia, like Poland, were threatened by German expansionism. They also were natural allies for Poland in collecting knowledge about German devices and war plans. In 1932, a man in France initiated contacts with the Polish General Staff. Captain Gustave Bertrand, chief of French radio intelligence established a direct cooperation with the Poles, especially for work resulting in solutions of the Enigma problem [Bertrand 1973, Kozaczuk 1984].

In October 1932, French military intelligence made a great contribution to solving the Enigma mystery, thanks to a special opportunity. A French intelligence officer Captain Henri Navarre reported that a man came to him and introduced himself as an employee of the Reichswehr cryptography agency [Bertrand 1973, Garliński 1999, Sebag-Montefiore 2000, Bauer 2002]. Moreover, he offered his services in return for money. Captain Bertrand, who was responsible for technical and scientific intelligence and ciphers, checked carefully all the information collected about the German agent and decided to investigate the first documents, which the man delivered. The samples were recognized as authentic. The newly recruited agent Hans-Thilo Schmidt received the pseudonym “Asche”. The documents delivered by “Asche” were both original and of great importance. During the long cooperation with Captain Bertrand, who met the German collaborator regularly, “Asche” left the following reports [Kozaczuk 1977, 1979, 1984]:

- Materials of the organization of the Reichswehr Cryptographic Agency;
- Various codes used in the German armed forces: A, B, C, D, E and code “Black”;
- Documents concerning keys to manual ciphers used by civil staff and army signals service for quick contact between civil and military authorities;
- Documents on machine ciphers: operating instructions for Enigma, keying instructions and monthly tables of army keys for December 1931, 1932, 1933 and the first half of 1934;
- Materials on an earlier Enigma model from 1930 and a document including one cipher text and a corresponding plain text.

Unfortunately, “Asche” never had an opportunity to get into his hands the most important materials such as the dossier of Enigma, containing the scheme of the machine’s wiring.

After receiving the first materials, which threw a new light on the Enigma mystery, Captain Bertrand contacted the Polish Cipher Bureau and arranged to visit Warsaw [Bertrand 1973, Kozaczuk 1984]. The materials he came with awoke a great interest among the cryptologists, because they constituted the first written papers referring to Enigma. Even if the Polish cryptologists had worked on the Enigma cipher since 1927, they had only intercepts as the base of their investigations.

Bertrand described his first meeting with the Poles on 7–11 December 1932 in Warsaw as “historic”, and he did not exaggerate when he used this word [Bertrand 1973]. The meeting gave rise to a long and friendly cooperation between Polish and French intelligence specialists. During the Warsaw meeting in December 1932, a number of tasks were decided between Bertrand and his Polish colleagues. The head of the Polish Cipher Bureau, which had been reorganized a year earlier, was Major Gwido Langer. The French were to concentrate on delivering intelligence reports from Germany to help in the code breaking, while the Poles were responsible for theoretical studies of Enigma intercepts. It was also decided to establish closer connections with the intelligence unit in Czechoslovakia in order to create a triple entente of cryptological services. Captain Bertrand (Fig. 4) was to use the pseudonym “Bolek”, Major Langer – “Luc”, and the Czechoslovak officer – “Raoul”. In the later 1930s, the B-L-R triangle was active on the Bolek–Luc line, due to the deteriorated Polish–Czech relations prevailing at that time [Kozaczuk 1976].

It is worth emphasizing that the principle of very strict secrecy was introduced into the Polish–French contacts, and even the three Polish cryptologists – Rejewski, Zygalski and Różycki – did not know anything about the origin of delivered

Figure 4.
General Gustave Bertrand in 1940.[2]

materials. These played an important role in the studies on Enigma in conjunction with the mathematical analysis, which had already been carried out on the intercepts.

The instructions brought by Bertrand gave a general idea of the military Enigma's appearance and operating principles, but said nothing about its inner structure. The electrical connections within the rotors, the variable contacts and other components were still unknown. If one of the last mentioned elements had been missing, then the entire effort would have given no results.

5 Breaking the Enigma System

The precise mathematical analysis was combined with intuitive reconstruction of individual parts. As the following example shows, even knowledge about the German mentality helped in solving the posed problem. Rejewski had found that in the commercial Enigma the letters of the alphabet were represented on the circumference of the entry ring in the same order in which they appeared on the German typewriter keyboard. It was assumed that the military model had its entry ring organized in the same way as the commercial model, but that assertion was wrong. Further, in January 1933 he came to the conclusion that the wiring on the entry ring in the military Enigma was in alphabetical order. The hypothesis was proved to be correct and helped to designate the connections in one of the rotors. The belief in German Ordnung ("order") made the work faster and simpler [Rejewski 1980-2].

[2] From [Kozaczuk 1984].

Solution of the Enigma system involved two distinct matters:
The theoretical reconstruction of the cipher machine was done. The most important task was to determine the Enigma's wiring. The cryptologists first discovered functions of the reflector, or "reversing drum". Afterwards, they reconstructed all the connections in the machine, which had a system of rotors as essential components. Even a special commutator was built into the system. The Poles were able to construct exact replicas of Enigma.

Secondly, the cryptologists developed methods for reconstructing the Enigma keys on the basis of intercepts that were supplied daily by monitoring stations.

The main code break came in the last days of December 1932. The practical reading of messages began during the second ten days of January 1933.

At the beginning of February 1933, the Cipher Bureau ordered the AVA Radio Manufacturing Company in Warsaw to build fifteen replicas of the military Enigma [Kozaczuk 1984]. The machines had to be made of the same components and with identical wiring, to work in the same way as the original military machine known to the German cipher staff as E-Eins [Sebag-Montefiore 2000]. The AVA copy model had typing keys instead of caps, and the upper part of the machine was changed as well. The illuminated windows were covered with cellophane on which the letters were written in the appropriate order.

The first attempt to produce a replica of Enigma was a failure. The cipher text going through the machine resembled some exotic language, but not German. The mathematicians discovered soon that the producers had forgotten about the capital letters, which slipped over the keys only for reading signals. This altered the order of internal connections. The mistake was corrected soon and AVA kept building more copies [Kozaczuk 1984].

A few weeks after the first copies of Enigma were constructed, the cryptologists received a series of German military signals, which indicated a correspondence between the district number 1 in Königsberg and number 2 in Stettin. The messages were unreadable and all the methods to translate them became useless. This cipher was sent on another Enigma machine called Enigma II. Later, it was discovered that it was the ordinary Enigma equipped with eight rotors and an automatic writing device. It had been used only for sending the highest military commands and had been evaluated as unreliable. After several weeks this version of Enigma was withdrawn from use.

Recovery of the settings (starting positions of rotors) and keys to the messages in the various Enigma nets happened by the method of elimination. The connections in the commutator were found by using the lattice method. During the first months after the Enigma solution, further elements of the key were obtained manually, by turning the metal rotors as many as 17576 ways. For top secrecy, the mathematicians did the job by themselves without any help from their assistants. One must add that the Germans changed the connections of the commutator regularly and more and more often, which forced an additional effort in finding the new conditions. The situation was improved when Rejewski invented a cyclometer, which had two sets of Enigma rotors linked together electrically. The cyclometer enabled the cryptologists to create a catalogue of possible settings of the rotors. After that, the comparison of intercepts with the catalogue made it possible to recover the keys faster [Hodges 1983, Kozaczuk 1984].

Other inventions included a clock, devised by Jerzy Różycki, that determined the right position of one of the rotors on a certain day in a given Enigma net.

In late June 1934, the three mathematicians experienced the exciting decryptment of a transmission they could read as "To all commandants of the airfields throughout Germany". The signal ordered "the transportation to Berlin, alive or dead, of Karl Ernst, adjutant to S.A. chief".

Thus 1934 was the year when the cryptology team of the Polish Cipher Bureau broke the ciphers of the German Army (Heer) and the codes of S.D. as well as codes and ciphers of the German Navy. The Kriegsmarine used three kinds of Enigma keys: operational, staff and admirals. The last key was resistant to breaking for a long time [Kozaczuk 1984].

6 Devices as a Reaction for Changes of Enigma Settings

The struggle against the German machine cipher had not ended in 1933, with the solution of Enigma and the building of its copy. In order to read the messages of the German Army, Air Force, and Navy, it was not enough to break the system once. The changes in it had to be detected and the reaction had to follow as well. Since the numbers of intercepts were growing proportionally to the Wehrmacht's expansion, several Polish lieutenants and captains were sent to the Cipher Bureau for training.

By the beginning of 1936, the Germans were using six kinds of keys to Enigma machines which were intended use by the supreme civil authorities: the staffs of the Armed Forces, the Army, S.S. staffs, S.S. operational units, special situations (the key "A"). The precautions and the secrecy concerning Enigma were growing in Germany as the war approached. The Germans carefully kept records of machine starting positions to prevent repetition of the same combinations. From 1 October 1936, they changed the settings every day. The responses to these changes from the Polish side resembled a duel between Poland's Cipher Bureau and Nazi Chi-Dienst (the German Cryptological Service). Three periods could be recognized in this battle [Kozaczuk 1984].

In the first period, 1933–35, little "changes" were made by the Germans, and the apparatus that had been developed by the Poles was sufficient for continuous decryptment.

In the second period from 1936 to November 1938, each change came very fast and to keep pace with them, the Polish cryptologists had to use all their knowledge and experience while they worked with the same tools and resources.

In the third period from the late 1938 to September 1939, a new wartime generation of Enigmas appeared with further complications.

Once the luck was on the Polish side, when German cipher clerks had committed great errors. They had often selected message keys (the first six letters at the beginnings of messages) in a stereotypic manner. For example, they could strike the same letter three times (AAA) or they could strike letters in alphabetical order (ABC). There was another possibility that they could use letters that lay next to each other down or diagonally across the keyboard, which was against regulations.

The cycle principle discovered by Marian Rejewski let him distinguish the proper regular message keys from the chaotic ones, which were introduced by mistake [Rejewski 1980-1, 1980-2, Hodges 1983, Kozaczuk 1984, Gaj 1999, Bauer 2002].

In 1937, important changes were made in the Polish Cipher Bureau. Its German section, B.S.-4, was separated from headquarters and moved out of the city. In the specially constructed new buildings hidden in woods not far from Pyry in the south of Warsaw, working conditions were better than at the cramped quarters in Warsaw. Another purpose in moving the German section with its cryptologists was to better protect the secrecy of their operations. The Abwehr carefully tested all people who were suspected of being traitors, and the secret German agents were present everywhere.

At the Polish B.S.-4, a strict prohibition was introduced against talking to anyone, even to fellows from the Cipher Bureau, about Enigma [Woytak 1979].

The Polish General Staff ordered an experiment to be carried out in January 1938. The test was to show how many of the intercepted Wehrmacht ciphers the cryptologists could read. The results of tests, which were conducted during two weeks, showed that about 75 percent of the messages had been decrypted.

7 French and English Efforts in Breaking Enigma

During these years Captain Bertrand visited the Polish Cipher Bureau many times. Expecting the approach of war it was important to ensure communication in the Warsaw-Paris-Prague triangle. After visiting the new nest of the Polish cryptologists in Pyry, which was called Wicher (Wind), Bertrand went to Prague to speak to representatives of the Czechoslovak General Staff, which even in May 1938 still looked to the future with hope.

In France, Bertrand's services were occupied with non-machine ciphers, leaving Enigma to the Poles [Kozaczuk 1979]. The French cryptologists were able to read the secret radio correspondence from Germany and Italy as well as from other countries. The French intelligence services also spread false information about French codes and ciphers, and in this way the Germans got "the mobilization code" of France's military intelligence while the Italians got the French B.D.G. naval code. During the war, the false information that was transmitted in these codes by the allies caused many damages and defeats for the Fascist countries. In Rotterdam one could buy various codes and ciphers, risking spending a lot of money for nothing, and even Captain Bertrand visited this exchange using the pseudonym Victor Hugo.

After the annexation of Austria by Germany in 1938, the British began to show more interest in intelligence contacts with their future allies. Bertrand was invited to London, where for the first time he met the British cipher experts. He came with Asche's papers that he gave to the British [Bertrand 1973]. The British cryptological service in the 1930s was part of the Foreign Office and contained some military sections. Officially known as the Government Code and Cipher School, or G.C.C.S., it was also called Room 47 of the Foreign Office until 1939 [Calvocoressi

1977, Hodges 1983, Kozaczuk 1984, Gaj 1989, Freedman 2000]. Afterwards it was called Station X or Bletchley Park. The chief of the G.C.C.S., Commander Alastair Denniston, was a professional naval intelligence officer. He successfully worked at breaking German codes and ciphers in the famous Room 40 at the Admiralty during the years 1914–18. The chief cryptologist at G.C.C.S., Alfred Dillwyn Knox, had been a worker of Room 40 during World War I. In the middle of the thirties, G.C.C.S. worked hard on breaking the German machine cipher, but failed to make progress. Knox managed, likely in 1938, to solve the cipher of General Franco's Army based on the commercial version of Enigma, but the military Enigma still was a riddle. The lack of success with the military Enigma in England could be attributed to the shorter time of investigation when compared with the Poles, and less mathematical effort involved in their analysis.

The British were rather reserved towards Bertrand's proposals to join forces with the French and Polish intelligence sections.

8 The Bombe as a Response for Further Changes in the Enigma System

The international situation became severe and nobody had doubts anymore that Germany would prepare more and more aggressive plans directed towards their neighbours. Suddenly, on 15 September 1938, two weeks before the Munich conference, the Germans altered the rules for enciphering message keys used by the twenty thousand Enigma machines. Now, the Enigma operator himself could select the basic position, a different one each time when he sent a message.

The Polish mathematicians, who met the difficulties in their work every day, thought of constructing a device that was more efficient than the cyclometer and could take over the long calculations. In October 1938, Marian Rejewski invented the mathematical model of an aggregate, which was left to designers in AVA.

The bombe, as the device for recovering Enigma's daily keys was christened, was a true invention [Hodges 1983, Kozaczuk 1977, 1979, 1984, Sebag-Montefiore 2000, Bauer 2002]. This was an electro-mechanical aggregate based on six Polish Enigmas combined with additional devices and transmissions. An electrically driven system of rotors turned round automatically, creating in each bombe 17567 different combinations of letters within two hours. When the rotors were placed in the sought-for position, a light appeared, the motors stopped automatically, and the cryptologist read the indications. By setting the bombes in action (in November 1938), the daily keys could be recovered within two hours. Almost at the same time, the B.S.-4 worked out a method for breaking the doubly enciphered individual message keys, which were formed according to the changes of the Enigma system introduced in September 1938. The new Polish method was based on using a special series of perforated paper sheets with a capacity of fifty-one holes by fifty-one. Each series consisted of twenty-six sheets. Theoretically, the method was based on so-called females, that is, on manipulating the sheets to match the coincident places in the pre-programmed system. Designed mainly by Henryk Zygalski, the system was quite independent of the number of connections in the German Enigma's commutator [Hodges 1983, Sebag-Montefiore 2000].

The Germans were cautious and once again changed the Enigma ciphers on 15 December 1938. This time the change involved not operating procedures but components. The Germans introduced two additional rotors per device, raising the number from three to five [Hodges 1983, Kozaczuk 1984, Freedman 2000]. This innovation, combined with the new keying procedure, made the process of decryptment almost impossible to continue with. The costs of further operations with further Enigma breaking seemed to be too high for the Polish government.

9 A Gift for the Allies

At the beginning of 1939 the Polish General Staff decided to broaden its exchange of information on Enigma with possible allies. The constant connection was already established with the French staff, but the French still did not know that Enigma had been solved in 1932 and that the replicas had been created. General Bertrand wrote in his book, published in 1973, that Enigma had been broken in Poland in 1939 [Bertrand 1973]. Since the Polish concept of the solution was kept as a close mystery, the French became impatient and weighed the possibility of arranging a common meeting with French, Polish and British representatives. Bertrand hoped that such a meeting would improve the Polish-British contacts that were rather cool [Bertrand 1973]. At the end of 1938, Bertrand succeeded in organizing the meeting in Paris, where top officers of the cryptological services were present. Before, Bertrand had to make a trip to London in order to persuade Commander Denniston of the need of taking part in the meeting. Denniston did not believe that anything new could be expected in Paris. He had never been in touch with the Polish cryptologists and did not imagine that they could have any achievements with Enigma.

This first meeting with British representatives took place on 9–10 January 1939 in Paris, at the French Military Intelligence offices. Major Bertrand, Captain Henri Braquenié from the air force staff and an army staff officer represented the French. Three British experts and the Poles – Colonel Gwido Langer and Major Maksymilian Ciężki – were the other members of the meeting.

At the Paris meeting it was agreed that the next conferences would be held in Warsaw and London if something new would come to light.

The principle of the highest security about the development of work on Enigma was observed in contacts between Poland and its French and British allies. The Poles did not leave any decrypted messages to their foreign partners up to the summer of 1939. Only summaries about the German armed forces were exchanged. Close contacts in radio intelligence between Poland and France did not mean sharing the secrets of cryptological methods.

But the great moment came. In July 1939, Bertrand received a telegraphed invitation from Gwido Langer (Luc) with words "there is something new..." [Bertrand 1973]. During the evening on 24 July, an international meeting was held in Warsaw and Langer informed his French and British colleagues that not only Enigma's secret had been penetrated but also the machine itself had been reconstructed. The French would receive one copy, the British another. The next morning all the participants of the Warsaw meeting drove to the new B.S.-4 place in the Kabackie Woods. A working meeting took place and the reconstructed Enigma was shown. Marian Rejewski said about himself and his colleagues: "We showed and told them everything what we knew about Enigma". Besides Denniston and Knox, who participated in the meeting, there was another person coming from England. It was possible that the Deputy head of the British military intelligence, Colonel Steward Menzies, appeared incognito as a "Professor Sandwich", a mathematician from Oxford [Rejewski 1980-2].

After seeing the Polish Enigma replica, Denniston and Knox wanted to contact London at once to order engineers and electricians to come down. It was unnecessary because Langer told them that it would be a machine each for Paris and London. The two Enigma machines soon arrived in Paris in diplomatic luggage. The perforated sheets with instructions for using were also enclosed. On August 16 Bertrand, accompanied by a British diplomatic courier, took one of the machines to London and at Victoria Station personally handed it to Colonel Menzies [Bertrand 1973, Kozaczuk 1984, Gaj 1989, Freedman 2000, Sebag-Montefiore 2000, Bauer 2002]. A few days later Knox sent greetings to the three Polish mathematicians, which included the words in both Polish and English: "My sincere thanks for your cooperation and patience. A.D. Knox".

10 The Mathematical Solution of Enigma

The applications of mathematics to cryptology expanded rapidly with the introduction of cipher machines. The use of permutation theory, combined with other methods of cryptological analysis, contributed to the breaking in Poland in 1932/33 of the German machine cipher, Enigma. The description of some examples, giving insight into Enigma decrypment, is based on two available reports written by Marian Rejewski, which can be found in the Sikorski Historical Institute in London and the military Historical Institute in Warsaw respectively. Other sources of information are Rejewski's publications which appeared in 1980 [Rejewski 1980-1, 1980-2, Kozaczuk 1984].

Description of the machine
Enigma was a device that was used for the mechanical encipherment of plain texts.

It had a twenty-six-letter keyboard and, behind it, a panel with twenty-six letters illuminated by glow lamps, which were placed under them. The main ciphering components were the three cipher drums or rotors and a fourth stationary reflector or reversing drum. All the rotors sat on the common axle. The reversing drum could be moved towards or away from the rotors with a lever. The machine

Figure 5.
General view of the military model Enigma.[3]

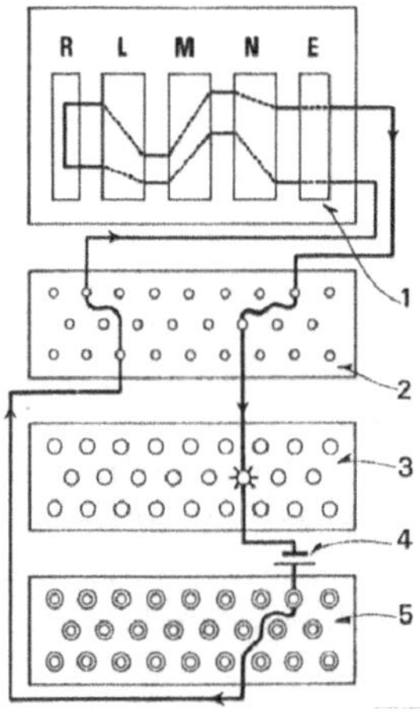

Figure 6.
The path of electric current through Enigma's components. 1: rotors (E: entry ring, L, M, N: main rotors, R: reversing drum), 2: commutator, 3: lamps, 4: battery, 5: keyboard.[4]

was also equipped with a stationary entry ring, which constituted the link between the commutator and the right rotor [Rejewski 1980-1, 1980-2, Deavours 1981].

The three rotors had the letters of the alphabet placed about their rims. The letters were visible in the little windows in the lid. Each rotor had twenty-six fixed contacts on one face and twenty-six spring-loaded contacts on the other. The reversing drum had only spring-loaded contacts connected in pairs on one face. The connections in four rotors constituted the most important part of the ciphering system and the secret of Enigma. Even the organization of connections in the entry ring was a great mystery, and the discovery of that secret by Rejewski at the beginning of 1933 constituted one of the greatest contributions in the work on the machine. The entry ring had the letters connected up in alphabetical order. These connections did not cause any relevant action of the entry ring.

The commutator was in front of the keyboard. Six pairs of plugs connected with wires made possible the interchange of twelve among the twenty-six letters of the alphabet.

The depression of an Enigma key caused the right-hand rotor to rotate through one twenty-sixth of the whole circumference. At the same time, the circuit was closed and current ran from the depressed key through the commutator, the entry ring, all three rotors that moved by rotating a bit, the reversing drum, and back through the rotors, and once again through the commutator. A glow lamp lit under one of the letters, which was always different from the depressed key (Fig. 6). Conversely, if someone struck the key, which was lightened before, a previously depressed letter would appear in the windows with light. The Enigma was constructed both for writing plain texts in order to obtain cipher and, conversely, to transform ciphers into clear messages. When one depressed the successive letters of a plain text, the letters of the bulbs that lit formed the cipher.

[3,4] From [Garliński 1999]

Encipherment procedure
Sending a message, the German clerk first set the rotors in the basic position established for that day and changed the letters in the commutator by placing the plugs in the appropriate sockets. Then he selected the individual key for that message, three letters he enciphered twice. In this way he obtained six letters, which were placed at the opening of the message. Next, he set the rotors to the selected individual key and began to encipher the message. The individual keys for a given day thus had two characteristics: the unknown basic position, and an unknown key for the message, which was enciphered twice. We realize that the first sign meant the same thing as the fourth one, the second was identified with the fifth and the third was compared to the sixth. Let us denote these pairs by *AD*, *BE* and *CF* [Rejewski 1980-1, 1980-2, Bauer 2002]. If we have about eighty messages per day, then all the letters of the alphabet can occur in the keys on the six places. We know from the machine's description that, when we strike a given key, for example "*x*", the lamp "*y*" is to be lit. Then, conversely, striking the "*y*" key will cause the "*x*" lamp to light. It is thus concluded that the permutations *A* through *F* consist of transpositions. Every pair of letters included in the same transposition has one letter coming from the plain text, while the other represents the cipher being associated with the first letter. For instance, the unknown permutation A is expected to be a set of pairs of the type:

$$A = (as)(br)(cw)(di)(ev) \ldots (zu).$$

If the encipherer strikes in the first place the unknown key "*x*" and obtains the letter "*a*", and by striking in the fourth place the same key "*x*" obtaining the letter "*b*", then, by striking in the first place the "*a*" key, he would obtain the letter "*x*", and by striking in the fourth place the "*x*" key, he would obtain the letter "*b*". Thus, there occurs a successive action, first of "*a*" on "*x*", and then of "*x*" on "*b*". The execution of such operations is called the composition of permutations. If we write the letters "*ab*" next to each other we will produce a fragment of the permutation *AD*, which is a product of unknown permutations *A* and *D*.

The cryptologists wrote out separately the six first letters of all the messages from a given day, more precisely, their twice enciphered keys. They chose the arbitrary key and wrote down its first letter, and next to it the fourth. Then they looked for the key, which had as its first letter the fourth letter of the previous key and they wrote the first letter of the second key beside the fourth letter of the first key. They continued seeking for such a key (the third one) that began with the fourth letter of the second key and so on. After a number of steps they returned to the first letter in the first word.

Let us consider the following example. Let

$\underline{d}mq \quad \underline{v}bn$
$\underline{v}on \quad \underline{p}uy$
$\underline{p}uc \quad \underline{f}mq$

designate the chosen openings, that is, the doubly enciphered keys of three of some eighty messages available for a given day. From the first and the fourth letters we can see that "*d*" becomes "*v*", "*v*" becomes "*p*", "*p*" becomes "*f*". In this way we obtain a fragment of a permutation *AD* "*dvpf*". Similarly, from the second and the fifth letters we notice that "*o*" becomes "*u*", "*u*" becomes "*m*" and "*m*" becomes "*b*". We obtain a fragment of the permutation *BE* as "*oumb*". And lastly, we get "*c*" which becomes "*q*", "*q*" which becomes "*n*" and "*n*" which becomes "*y*". Hence, the permutation *CF* begins with "*cqny*". The openings of other messages would permit the complete assembly of the set of permutations *AD*, *BE* and *CF* to appear. For example, *AB*, deciphered from the daily openings was a permutation:

$$AD = \begin{matrix} d & v & p & & o & e & i & j & & t & b & c & r & w & a & s \\ \downarrow & \downarrow & \downarrow & \cdots & \downarrow & \downarrow & \downarrow & \downarrow & \cdots & \downarrow & \downarrow & \downarrow & \downarrow & \downarrow & \downarrow & \downarrow \\ v & p & f & & d & i & j & m & & e & c & b & w & r & a & s \end{matrix} = (dvpfkxgzyo)(eijmunqlht)(bc)(rw)(a)(s)$$

while

$$BE = (blfqveoum)(hjpswizrn)(axt)(cgy)(d)(k)$$

and

$$CF = (abviktjgfcqny)(duzrehlxwpsmo).$$

The set of permutations for *AD*, *BE* and *CF* was called "the characteristic set for a given day". We remember that the permutations contain the enciphered letters without any close connections with the plain text. We wish to separate the permutations *A* through *F*, which, instead, present the associations between the clear text and the cipher.

In the solutions, some theorems were involved. Let us quote the most important of them.

Theorem 1
If two permutations X and Y of the same degree comprise disjunctive transpositions, then their product XY will include disjunctive cycles of the same lengths in even numbers.

We may also prove the converse theorem.

Theorem 2
If a permutation includes disjunctive cycles of the same lengths in even numbers, then the permutation may be regarded as a product XY of two permutations X and Y, composed of disjunctive transpositions.

Analysing the permutation *AD* we can evaluate its contents as two cycles of the length 10, two cycles of the length 2 and two cycles of the length 1.

It may also be shown that:

Theorem 3
Letters entering into one and the same transposition of permutation X or Y, enter always into two different cycles of the same length, which belong to the permutation XY.

The theorems quoted above helped to determine the connections between the plain text and the corresponding cipher. Let us assume that the German clerks had some habits when they arbitrarily created the openings of messages. Suppose, e.g., that the encipherer liked to select three identical letters, such as "*aaa*", "*bbb*" and the like. Since in product *AD* the letters "*a*" and "*s*" form single-letter cycles, then "*a*" and "*s*" should belong to the same transposition (*as*). It means that the plain letter "*a*" has the representing cipher letter "*s*".

Suppose that the enciphered message keys from a given day begin with the letter "*s*":

sug smf
sjm spo
syx scw.

Only the last key starting with "*syx*" could arise from the plain text "*aaa*", for the transposition (*as*) has one representative in the cycle (*a*) and the other in (*s*) – both belonging to *AD*. By analysing the transposition (*ay*), representing the second place of the key, we realize that "*a*" belongs to the cycle (*axt*) while *y* is placed in (*cgy*). Both cycles come from the permutation *BE*. At last, the pair (*ax*) tracing the third position in the key has "*a*" in (*abviktjgfcqny*) and "*x*" in (*duzrehlxwpsmo*). The cycles are parts of the permutation *CF*. Even the strict analysis of the second part in the third message key proves that "*scw*" are ciphers of "*aaa*" according to the rule of double ciphering.

By using the sets *AD*, *BE* and *EF*, when collecting them during a few days, the mathematicians managed to construct the internal connections of the machine.

The set of equations

The unknown permutations *A* through *F* were also found as solutions of the equation set. After a key has been depressed, the current first passes through a series of the machine's components to light a lamp with the letter at last. Each of these components causes a permutation of the alphabet. We denote the permutation caused by the commutator by the letter *S*, the permutations created by the rotors (from right to left) have initials *N*, *M*, *L*, and the permutation caused by the reversing drum is called *R*. Since the letters of the entry ring were linked in alphabetical order, then the permutation *H*, associated with the ring, would be the identity without any relevance. The path of the current will be represented by the product of permutations $SNMLRL^{-1}M^{-1}N^{-1}S^{-1}$, where the sign "–1" denotes an inverse permutation. We also keep in mind that the depression of the key causes a movement of the first right rotor that rotates a bit equal to one twenty-sixth of the circumference. This movement creates the next permutation in which each letter is assigned to the next one. We denote the last permutation by P and write it down as

$$P = \begin{matrix} a & b & c & & z \\ \downarrow & \downarrow & \downarrow & \ldots & \downarrow \\ b & c & d & & a \end{matrix} = (abcdef \ldots xyz)$$

The unknown permutations A–F are represented in the form

$$A = SPNP^{-1}MLRL^{-1}M^{-1}PN^{-1}P^{-1}S^{-1}$$
$$B = SP^2NP^{-2}MLRL^{-1}M^{-1}P^2N^{-1}P^{-2}S^{-1}$$
$$\ldots$$
$$F = SP^6NP^{-6}MLRL^{-1}M^{-1}P^6N^{-1}P^{-6}S^{-1}$$

and the AD, BE and CF products have the presentations

$$AD = SPNP^{-1}MLRL^{-1}M^{-1}PN^{-1}P^3NP^{-4}MLRL^{-1}M^{-1}P^4N^{-1}P^{-4}S^{-1}$$
$$BE = SP^2NP^{-2}MLRL^{-1}M^{-1}P^2N^{-1}P^3NP^{-5}MLRL^{-1}M^{-1}P^5N^{-1}P^{-5}S^{-1}$$
$$CF = SP^3NP^{-3}MLRL^{-1}M^{-1}P^3N^{-1}P^3NP^{-6}MLRL^{-1}M^{-1}P^6N^{-1}P^{-6}S^{-1}.$$

In the set of equations derived above, only the permutation P and its powers are known. By discovering the connections in the drums it was possible to reconstruct the associated pairs "plain text-cipher" in the permutations A–F, which gave a possibility to read the messages during only one day.

The material above, shown as an excerpt from the large documentation of the Enigma description, allows us to understand better the machine construction as well as to realize that enormous efforts were undertaken to break the Enigma's code.

Epilogue

As the title suggests, the first part of the Enigma story, based only on the collected facts, ended in July 1939. After some weeks the war broke up. The British code-breakers at Bletchley Park received an Enigma machine and rotors I to V from the Polish Cipher Bureau on 24 July, 1939. The British recovered rotors VI and VII from the crew of U-33 on 12 February 1940, while rotor VIII was captured in August 1940. In May and June 1940, using clear text and cipher text captured from Schiff 26, Hut 8, a section at Bletchley Park had solved some April Enigma traffic with the aid of the first British bombe. Both the bombe, which was Alan Turing's great device, and further investigations during the war leading to the construction of the computer prototype, constitute a great amount of information, which ought to be discussed separately in another paper.

At the beginning of the war, the three Polish cryptologists were forced to leave the country and to escape, at first to Romania and France and then to England. In 1946, Rejewski returned to Poland from England, where he spent the last years

Figure 7. Leading Polish army officers were present at a ceremony in 2001 when a memorial plaque was unveiled at the tomb of the cryptologist Marian Rejewski (1905–1980). The photo shows some generals, together with Rejewski's daughter and the President of the Polish Mathematical Society. Not many mathematicians have experienced similar honours in life or posthumously.
As a mathematics student, Rejewski had been recruited in 1929 by the Cipher Bureau of the General Staff of the Polish Army. Rejewski then created a mathematical method for breaking the German Enigma code of that time. Long before their competitors, the Polish Cipher Bureau officers realized the potential of mathematics in cryptological research.

of the war. He had worked as a clerk until he retired. In 1980 he died in Warsaw. Zygalski decided to stay in England. He had taught mathematics at the Battersee Technical College before he died in 1978. Różycki was killed during the catastrophe of a French ship in 1942.

The early Polish contributions in the Enigma solution enabled the British codebreakers to make great progress in the further development of methods breaking the Enigma code already in 1940. There is no doubt that the successful effort to break Enigma shortened the war and spared many human lives.

References

Bauer F. L. (2002): *Decrypted Secrets. Methods and Maxims of Cryptology.* Springer Verlag, Berlin-Heidelberg-New York 2002.

Bertrand G. (1973): *Enigma ou la plus grande énigme de la guerre 1939–1945*, Libraire Plon, Paris 1973.

Calvocoressi P. (1977): The secrets of enigma. *The Listener* **20**, (1977), p. 27.

Deavours C. A. (1981): Comments to "How Polish Mathematicians Deciphered the Enigma" by Marian Rejewski. *Annals of the History of Computing* **3** (1981), pp. 229–234.

Freedman, M. (2000): *Unravelling Enigma. Winning the Code War at Station X.* Published by Led Cooper, Barnsley 2000.

Gaj K. (1989): Szyfr Enigmy – metody złamania. Wydawnictwo Komunikacji i Łączności, Warsaw 1989.

Garlinski J. (1999): Enigma – tajemnice drugiej wojny światowej. Wydawnictwo Uniwersytetu Marii Curie-Skłodowskiej, Lublin 1999.

Hodges A. (1983): *Alan Turing: The Enigma.* Simon and Schuster, New York 1983.

Kahn, D. (1967): *The Codebreakers*, Macmillan, New York 1967.

Kozaczuk W. (1967): Bitwa o tajemnice. Książka i Wiedza, Warsaw 1967.

Kozaczuk W. (1976): Złamany szyfr. Wydawnictwo Ministerstwa Obrony Narodowej, Warsaw 1976.

Kozaczuk W. (1977): Wojna w eterze. Wydawnictwa Radia i Telewizji, Warsaw 1977.

Kozaczuk W. (1979): W kręgu Enigmy. Książka i Wiedza, Warsaw 1979.

Kozaczuk W. (1984): *Enigma: how the German machine cipher was broken, and how it was read by the Allies in World War Two.* University Publications of America 1984 (Translation of W kręgu Enigmy).

Kuratowski K. (1980): *A Half-Century of Polish Mathematics: Remembrances and Reflections.* Pergamon Press, Oxford 1980.

Rejewski M. (1980-1): An application of the theory of permutations in breaking the Enigma cipher. *Zastos Mat. (Applicationes Mathematicae)* **16** (1980), no. 4, pp. 543–559.

Rejewski M. (1980-2): Jak matematycy polscy rozszyfrowali Enigmę. *Wiadomości Matematyczne* **23** (1980), pp. 1–28.

Sebag-Montefiore H. (2000): *Enigma - The Battle for the Code*, Weidenfelt & Nicolson, London 2000.

Woytak R. (1979): *On the Border of War and Peace: Polish Intelligence and Diplomacy in 1937–1939 and the Origins of the Ultra Secret*, Columbia University Press, New York 1979.

On the Defense Work of A. N. Kolmogorov during World War II

Albert N. Shiryaev*

In the paper we give a short description of the defense work of Andrei N. Kolmogorov on the Theory of Firing during World War II.

Similarly to his great predecessors P. L. Tschebyshev and A. M. Lyapunov, the academician A. N. Kolmogorov (25.04.1903–20.10.1987) was not only a universal mathematician, proposing and developing fruitful ideas, he also readily answered challenges presented to him in his capacity as an applied mathematician. He had an amazing ability to get to the root of a problem, to determine fundamental issues, and to clarify debatable situations.

A typical example of such work by Kolmogorov are his investigations on **firing dispersion theory** that he performed during World War II.

His paper *Estimation of the center and spread of dispersion for a bounded sample*, [Kolmogorov 1942], was dated September 15th, 1941, i.e., three months after the war began on the Soviet territory. It was published in response to a request from the appropriate military agencies to "give his conclusion on the subject of discrepancies among the existing methods to estimate precision from experimental data." Kolmogorov remarked that this paper presents primarily methodological interest due to critical comparison of distinct approaches. Nevertheless, Kolmogorov and his colleagues from the Mathematical Institute of the Academy of Sciences, Department of Mechanics and Mathematics at Moscow State University, as well as the immediate practitioners from the Institute of Marine Artillery Research, developed largescale *theoretical and computational efforts* toward measuring the efficiency of firing strategies. Simultaneously, he taught a course at the university on firing dispersion theory. He made it *compulsory* for students majoring in probability.

This work concluded in the publication of a separate volume of "Proceedings of the Steklov Mathematical Institute" (1945), nicknamed by Kolmogorov himself *Collection of Articles on the Theory of Firing*, [Kolmogorov 1945]. It included two of his articles, *On the number of hits among several shots and general principles of characterizing efficiency of firing strategies and Artificial dispersion in case of single shot kill and onedimensional dispersion*, as well as two articles by his

* Steklov Mathematical Institute, Gubkina St. 8, 119991 Moscow, Russia.
Email: shiryaev@mi.ras.ru

Translated from the Russian by Galina Shubina.

ИЗВЕСТИЯ АКАДЕМИИ НАУК СССР
BULLETIN DE L'ACADÉMIE DES SCIENCES DE L'URSS
Серия математическая 6 (1942), 3–32 Série mathématique

А. Н. КОЛМОГОРОВ

ОПРЕДЕЛЕНИЕ ЦЕНТРА РАССЕИВАНИЯ И МЕРЫ ТОЧНОСТИ ПО ОГРАНИЧЕННОМУ ЧИСЛУ НАБЛЮДЕНИЙ

Настоящая статья обязана своим появлением двум обстоятельствам:

1° Предполагая в нескольких дальнейших статьях опубликовать изложение своих точек зрения и исследований по вопросу теоретико-вероятностного обоснования математической статистики, автор считает целесообразным предпослать этим статьям детальный критический разбор существующих методов, проведенный на материале какой-либо достаточно простой классической задачи математической статистики. Для этого было вполне естественно остановиться на задаче оценки параметров Гауссовского закона распределения по данным n независимых наблюдений.

2° К автору обратились с просьбой дать свое заключение по поводу разногласий, имеющихся среди артиллеристов, относительно приемов оценки меры точности по опытным данным (см., например, [10], [11], [12]). В связи с рассмотрением этих разногласий автору стала ясной желательность познакомить артиллеристов с результатами Student'а и Фишера, относящимися к малым выборкам. Этими запросами и определился окончательно конкретный материал настоящей статьи.

Из сказанного ясно, что статья претендует по преимуществу лишь на методологический интерес. Новыми с фактической стороны автору представляются в ней определение исчерпывающей статистики и исчерпывающей системы статистик в § 2 и уточнения остаточных членов в предельных теоремах в § 4.

Необходимость критического сопоставления различных подходов к разбираемым задачам привела неизбежно к тому, что статья получилась довольно объемистой по сравнению с элементарностью разбираемых в ней вопросов.

Введение

Допустим, что случайные величины

$$x_1, x_2, \ldots, x_n$$

независимы и подчинены каждая Гауссовскому закону распределения с общим им всем центром рассеивания a и мерой точности h. Как известно, в этом случае n-мерный закон распределения величин x_i определяется следующей плотностью вероятности:

$$f(x_1, x_2, \ldots, x_n \mid a, h) = \frac{h^n}{\pi^{\frac{n}{2}}} \exp(-h^2 S^2), \tag{1}$$

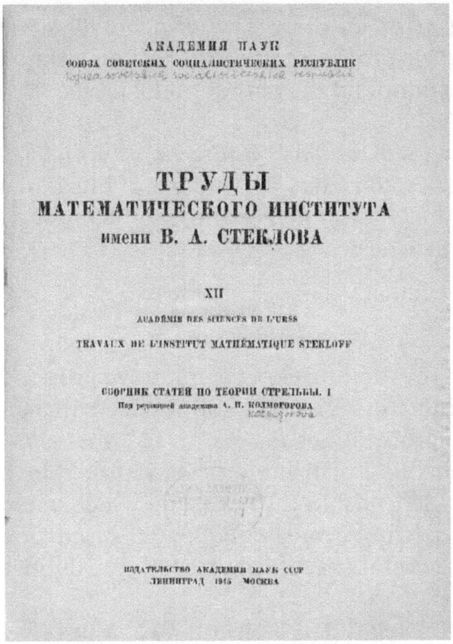

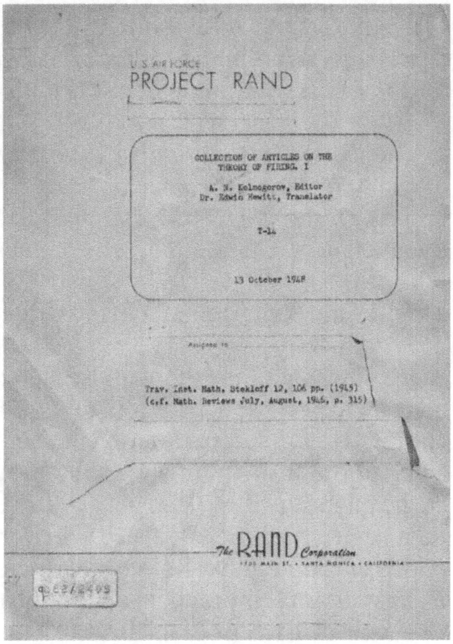

Figure 1.
Kolmogorov's *Theory of Firing*, four front pages 1942 (*opposite page*), 1945 (*above left*), 1948 (*above right*), 2002 (*left*).

The work by Kolmogorov and collaborators on Firing Theory was interesting enough to be translated by the RAND Corporation in 1948 and is still regarded as fundamental in quite recent US military education. It is one of the apparent paradoxes of the relation between mathematics and warfare that an earlier paper by Kolmogorov on the same topic was published in 1942 for everybody to read (including the enemy). Could it be that this early paper (in contrast to the 1945 book) was too mathematical and too general to inform military practice directly?
Kolmogorov himself modestly remarked that the article does not claim anything more than a certain methodological value. On the other hand: was the 1945 work after all too closely linked to a specific problem to inspire further mathematical development? [Editor's note]

collaborators, A. A. Sveshnikov (*Optimal artificial dispersion of firing for some cases*) and I. A. Gubler (*Solution of the firing problem with artificial dis persion for some cases*). Kolmogorov was both its editor and the author of the preface.

The paper *On the number of hits among several shots and general principles of characterizing efficiency of firing strategies* considers the number of hits μ in a group of n shots ($\mu = 0, 1, \dots, n$). Let $P_k = P(\mu = k)$ and E_μ be the expectation of the number of hits. Kolmogorov defines a new term

"efficiency criterion of a firing strategy."

He notes that the "thoughts accepted in the literature on comparative pros and cons of 'expectation criterion' and 'probability criterion' do not provide suffcient clarity." He formulates a problem: can the probabilities $P_0, P_1, \dots, P_n$ that characterize the firing strategy by the hit probability distribution, "be replaced by a single quantity $W = f(P_0, P_1, \dots, P_n)$ that depends on them all, and which could then be declared the efficiency criterion of a firing strategy." Following discussion of this question, A. N. Kolmogorov derives several exact formulas for probabilities P_k as well as their practical approximations.

The next batch of problems this paper touched upon concerns two issues. The first one is the classification of factors that affect firing in order to determine a practical firing strategy. The second is the problem of artificial dispersion (the term itself is discussed in detail also), and is defined as follows.

Let $p_i = p_i(\alpha_i, \beta_i)$ be the probability that the i-th shot is a hit, depending on *azimuth* α_i and *elevation* β_i. Let $(\bar{\alpha}_i, \bar{\beta}_i)$ be a pair (usually unique) of values of α_i and β_i that maximizes the hit probability:

$$\max p_i(\alpha_i, \beta_i) = p_i(\bar{\alpha}_i, \bar{\beta}_i)$$

and

$$\bar{\alpha} = (\bar{\alpha}_1, \dots, \bar{\alpha}_n),\ \bar{\beta} = (\bar{\beta}_1, \dots, \bar{\beta}_n).$$

The question under consideration is whether $\max W = W(\bar{\alpha}, \bar{\beta})$, i.e., is it suffcient to maximize the hit probability of each individual shot in order to maximize our overall firing efficiency?

It is noted in the paper that the property $\max W = W(\bar{\alpha}, \bar{\beta})$ holds in two special cases. The first is when $W = E_\mu$ (where μ is the number of hits out of n shots). In the second case, $W = c_1 P_1 + \dots + c_n P_n$ ($c_i \ge 0$) and the events of hitting the target are independent. Thus in those particular cases, the most practical firing strategy is to *maximize the hit probability for each individual shot.*

However this is not true for some other firing efficiency criteria. In other words, in order to maximize a given criterion, one sometimes has to deliberately choose settings (azimuth and elevation) on a per-shot basis that deviate from the values maximizing the individual hit probability. This is called firing with artificial dispersion. It is practical, for example, in the case when it is "imperative to achieve at least a few hits, the number of hits being possibly much smaller than the total number of shots n."

In the second paper *Artificial dispersion in the case of a single shot kill and onedimensional dispersion* from the *Collection of Articles on the Theory of Firing*, Kolmogorov considers the question of *artificial dispersion* when firing at "small" targets. In an earlier draft of this paper, he sets out the following visual justification for dispersion: "Occasionally, when firing repeatedly at the same target, it is beneficial to artificially increase the dispersion of projectiles in order to maximize the hit probability; similarly a hunter prefers to use pellets with sufficiently wide dispersion cone when aiming at a flying bird. The most dramatic example of this is when it is sufficient to hit once in order to destroy a target."

In considering this very case, Kolmogorov simplifies the problem further by allowing dispersion only in one dimension. A typical example of such a situation would be: bombardment of a narrow stretch (say, a bridge), located perpendicularly to the plane of bombardment (in this case, errors in azimuth are immaterial, and the hit probability depends only on the elevation) or mine shelling at sea (where only the choice of azimuth is essential). Kolmogorov remarks that the formulas derived in this case, "belonging to such a 'onedimensional' case allow for direct *practical* applications."

References

Kolmogorov, A. N. (1942): Sur l'estimation statistique des paramètres de la loi de Gauss. (Russian. French Summary), *Bull. Acad. Sci. URSS. Sér. Math.* [*Izvestia Akad. Nauk SSSR*] **6**, 1/2 (1942), pp. 3–32; MR **4** (1943), 221e (Reviewer: J. Neyman).

Kolmogorov, A. N. (1945) (ed.): *Collection of papers on the theory of artillery fire*. I. (Russian). Trav. Inst. Math. Stekloff 12, Inst. Math. Stekloff, Moscow, (1945); MR **7** (1946), 315h (Reviewer: J. L. Doob). American translation by Edwin Hewitt, RAND Corporation, Santa Monica, California, 1948.

Improbable Warriors: Mathematicians Grace Hopper and Mina Rees in World War II

KATHLEEN WILLIAMS*

The vast expansion of the US Navy in World War II and its increasing reliance on quickly emerging new technologies owed its success, in part, to the previously overlooked work of women. Much of this work was at the level of technician and mechanic, but there were also numbers of women with advanced educational and professional qualifications who held responsible positions during the war and who made high-level contributions in technical fields.

This paper focuses on the war work of two of these women: mathematicians Grace Murray Hopper (1906–1992) and Mina Spiegel Rees (1902–1997). Both were college professors until war service intervened. One joined the navy and the other remained a civilian, but between 1943 and 1945 each held positions from which they influenced the US Navy's ability to wage a modern, math-dependent war.

The vast expansion of the US Navy in World War II and its increasing reliance on rapidly emerging new technologies owed its success, in part, to the work of women. Much of this work was at the level of technician and mechanic, but there were also women with advanced professional qualifications who held responsible positions during the war and who made high-level contributions in technical fields.[1]

My talk focuses on the war work of two of these women: mathematicians Grace Murray Hopper (1906–1992) and Mina Spiegel Rees (1902–1997). Both women grew up in New York City and both had fathers in insurance, perhaps predisposing their daughters to mathematics. Rees's father, however, was a salesman while Hopper's father worked in the family brokerage firm. Both Hopper and Rees taught mathematics in women's colleges – one of the few occupations open to women with advanced degrees in mathematics in the 1920s and 30s. When war service interrupted their careers Hopper joined the navy while Rees remained a civilian, but between 1943 and 1945 each held positions from which they influenced the US Navy's ability to wage a modern, math-dependent war. Both women had been trained in pure mathematics but like many other mathematicians during the national emergency they adapted to the wartime demand for applied math-

* CUNY, New York. Email: kbbwilliam@aol.com

ematics. Hopper and Rees knew of each other during the war, and afterwards their lives intersected frequently as they continued to move in the same professional circles.[2]

The war-service of these two women was different from that of most male scientists and mathematicians because they were both volunteers; neither of them was liable for the draft. Indeed, as teachers of mathematics they were in an occupation essential to the war effort and were strongly discouraged from leaving it. Their case was also different from that of women in Britain and the Soviet Union - among others - as there was no requirement for American women to participate in the war effort in any way, either in the military or as civilians. Indeed, Hopper and Rees chose to serve even though their homeland was neither invaded nor in danger of being overrun.[3] Why, then, did they decide to use their mathematical skills directly in the service of war? What did each contribute to the war effort, and how did they, as mathematicians, perceive the part they played in the conflict?

Grace Hopper

In 1983 Grace Hopper, then seventy-six years old, was made an admiral by special appointment of the President of the United States. In 1987, the US Navy named its new computer center in San Diego for Hopper, and in 1996, four years after her death, it launched the newest Arleigh Burke-class guided missile destroyer *Hopper.* The recipient of numerous medals, awards, and honorary degrees, Grace Hopper was esteemed as a pioneer in the field of computing. As befits a leader instru-

[1] For a comprehensive study of women scientists see Margaret W. Rossiter, *Women Scientists in America: Before Affirmative Action, 1940–1972* (Baltimore, MD: The Johns Hopkins University Press, 1978). For changing women's roles in wartime see among others: Karen Anderson, *Wartime Women: Sex Roles, Family Relations, and the Status of Women During World War II* (Westport, CT: Greenwood Press, 1981); Martin Binkin and Shirley J. Bach, *Women and the Military* (Washington, D.C.: Brookings Institute, 1977); D'Ann Campbell, *Women at War with America: Private Lives in a Public Era* (Cambridge, MA: Harvard University Press, 1984); William Henry Chafe, *The Paradox of Change*, rev. ed. of *The American Woman* (1972), (New York: Oxford University Press, 1991); Sherna Berger Gluck, *Rosie the Riveter Revisited: Women, War, and Social Change* (New York: Meridian, 1987); Maj. Gen. Jeanne Holm, USAF (ret.), *Women in the Military: An Unfinished Revolution* (Novato, CA: Presidio Press, rev. ed., 1992); Doris Weatherfield, *American Women and World War II* (New York: Facts on File, 1990); Nancy Baker Wise and Christy Wise, *A Mouthful of Rivets: Women at Work in World War II* (San Francisco: Jossey-Bass Publishers, 1994).

[2] Much of the following on Hopper and Rees appeared previously in Kathleen Broome Williams, *Improbable Warriors: Women Scientists and the U.S. Navy in World War II* (Annapolis, MD: Naval Institute Press, 2001). Hopper's doctoral thesis was on "The Irreducibility of Algebraic Equations." Yale University, 1934. Rees's doctoral thesis was on "Division Algebras Associated with an Equation Whose Group Has Four Generators," University of Chicago, 1932.

[3] Capt. Grace Hopper interview by Linda Calvert, 3 Sept. 1982–28 Feb. 1983 (hereafter Hopper interview, 1982–83), Women in Federal Government Oral History Project, OH46 (hereafter WFGOH46), Schlesinger Library, Radcliffe Institute, Cambridge, MA (hereafter SL), 10. For Britain see Gail Braybon and Penny Summerfield, *Out of the Cage: Women's Experiences in Two World Wars* (London: Pandora Press, 1987), pp. 157–160. I am indebted to Dr. Reina Pennington for information about Soviet women.

Figure 1. Destroyer USS *Hopper*, 1998. [Courtesy: DoD, Defense Visual Information Center, March ARB, CA]

mental in creating a whole new discipline, her message to everyone was above all to innovate and never to be tied to the old or customary way of doing things.[4] Admiral Hopper never went to sea, but her computer expertise and managerial skills made her a pivotal figure in the navy's path to the computer age. "She's challenged at every turn the dictates of mindless bureaucracy," noted Navy Secretary John Lehman, under whom she served her final years.[5] Even when she reluctantly retired in 1986 – the oldest serving officer in the navy – Grace Hopper continued working as a consultant for Digital Equipment Corporation. She died in 1992, and was buried with full military honors in Arlington National Cemetery.

At the heart of all these accomplishments was Hopper's brilliance in mathematics. Had it not been for World War II, however, she might never have left the idyllic and genteel campus in New York where she was teaching. It really all began that day in December 1943 when Grace Murray Hopper was sworn in to the US Navy.

Hopper had been born in New York City in 1906, and had gone on to major in mathematics at Vassar, one of a group of private women's colleges – the Seven Sisters – catering to the bright daughters of the affluent. In 1930 she earned a masters

[4] Carmen Lois Mitchell, *The Contribution of Grace Murray Hopper to Computer Science and Computer Education* (University of North Texas: University Microfilms, 1994), pp. 1–11, pp. 24–37, pp. 50–51, pp. 63–64; Charlene W. Billings, *Grace Hopper: Navy Admiral and Computer Pioneer* (Hillside, NJ: Enslow Publishers, Inc., 1989), p. 30, pp. 36–38, pp. 47–53, p. 111, p. 115.

[5] Quoted in Billings, *Hopper*, p. 10.

Figure 2.
Lt. Grace M. Hopper, USNR (W), during World War II. [Courtesy: Mary Murram Westcote]

degree and in 1934 a Ph.D. in mathematics at Yale – a rare accomplishment for women in those days – and then she returned to Vassar to teach. By the time of Pearl Harbor, Hopper had worked her way up to associate professor of mathematics at Vassar, but a year later she joined the WAVES – navy women – explaining that she wanted to serve the war effort more directly than by teaching. After graduating from midshipmen's school, Lt. (j.g.) Hopper was sent to the navy's Bureau of Ships Computation Project at Harvard University.[6]

The rapidly expanding data management needs of World War II accelerated the development of modern digital computers, especially in Britain and the United States, although a true American electronic computer – ENIAC – was not completed until after the conflict ended. One electro-mechanical device that was ready

[6] Hopper's sister Mary graduated from Vassar in 1930 and her brother Roger from Yale in 1932. Mary Murray Westcote interview by author, 11 Dec. 1999. Hopper's brother later also earned a doctorate. For accounts of navy women see: *BuPers: The Story of Navy Manpower, U.S. Bureau of Naval Personnel* (Washington, DC: U.S. Government Printing Office, 1949); Winifred Quick Collins, *More Than a Uniform: A Navy Woman in a Navy Man's World* (Denton, TX: University of North Texas Press, 1997); Susan H. Godson, *Serving Proudly: A History of Women in the Navy* (Annapolis, MD: Naval Institute Press, 2001); Virginia C. Gildersleeve, *The "Waves" of the Navy: How They Began* (New York: The Macmillan Company, 1956); Joy Bright Hancock, *Lady in the Navy* (Annapolis, MD: Naval Institute Press, 1972).

in time to be used in the war was the brainchild of Howard Aiken, professor of physics and applied mathematics at Harvard. Frustrated by the tedious and time-consuming mathematical calculations required for his doctoral dissertation, Aiken had designed a mechanism to perform such calculations automatically. Engineers at IBM built Aiken's machine under his guidance. What emerged from this collaboration, the Automatic Sequence Controlled Calculator (ASCC) or Harvard Mark I, was the first functional, large-scale, automatically sequenced, general-purpose, digital computer to be produced in America.[7]

When the Mark I was completed in 1944, IBM gave it to Harvard as a gift. That spring it was installed at the university but was immediately leased for the duration of the war by the US Navy, desperate for gunnery and ballistics calculations. Aiken, a naval reserve officer, was put in charge of the Mark I for the Bureau of Ships.[8]

Aiken's computing machine was in many ways unique. It was fifty feet long, eight feet tall, and eight deep, filling an entire room. It had more than 750,000 parts, used 530 miles of wire, and weighed about five tons.[9] Aiken described the Mark I as a "general arithmetic machine capable of addition, subtraction, multiplication, division, and the transfer of numbers."[10] Most impressive were its speed of computation and its automatic functioning enabling it to proceed through a series of arithmetic operations without human intervention. Automatic sequence control was accomplished according to programmed instructions fed into the machine on punched paper tape while the output was handled either by punched cards or by two electric typewriters.[11]

The multiple-purpose capabilities of the Mark I – the fact that it could be set to accomplish a wide range of different types of numerical calculations – was one of its great strengths and set it apart from other contemporary computing devices. Aiken also estimated that the Mark I was nearly a hundred times more productive

[7] H. H. Aiken, "Proposed Automatic Calculating Machine," edited and prefaced by A. C. Oettinger and T. C. Bartee, *IEEE Spectrum*, (Aug. 1964): pp. 62–69; I. Bernard Cohen, "Babbage and Aiken: With Notes on Henry Babbage's Gift to Harvard, and to Other Institutions, of a Portion of His Father's Difference Engine," *Annals of the History of Computing* **10**, no.3, (1988): pp. 175–177.

[8] Howard H. Aiken and Grace M. Hopper, "The Automatic Sequence Controlled Calculator – I," reprint from *Electrical Engineering* (Aug. – Sept. 1946): 1, box 1, folder 2, Grace Murray Hopper Collection 1944–1965, (hereafter Hopper Papers), Smithsonian Institution, Washington, DC (hereafter Smithsonian).

[9] Emerson W. Pugh, *Memories that Shaped an Industry: Decisions Leading to IBM System 1360* (Cambridge, MA: The MIT Press, 1984), 6; Gregory W. Welch, "Howard Hathaway Aiken: The Life of a Computer Pioneer," *The Computer Museum Report 12* (Boston: Spring 1985):7; "General Purpose Digital Computers,"(n.d.), folder "Bell Telephone Labs," box A-C, UAV 289.2005, Records of the Computation Laboratory, 1944– (hereafter Aiken Correspondence), Harvard University Archives (hereafter HUA).

[10] H. H. Aiken to Dr. Arnold Lowan, 1 Nov. 1944, folder "(dead) BuOrd," box A–C, UAV 289.2005, Aiken Correspondence, HUA.

[11] For the most comprehensive account of Aiken and the Computation Lab see: I. Bernard Cohen, *Howard Aiken: Portrait of a Computer Pioneer* (Cambridge, MA: The MIT Press, 1999), and I. Bernard Cohen and Gregory W. Welch, eds., *Makin' Numbers: Howard Aiken and the Computer* (Cambridge, MA: The MIT Press, 1999).

than a manually operated calculator. Throughout the conflict trained specialists – usually women called "computers" – working on desk calculators were still a major source of the mathematical calculations required for the prosecution of a modern war.[12]

On 2 July 1944 Grace Hopper arrived at the Computation Lab. "That's a computing engine," Aiken snapped at her, pointing to the Mark I. "I would be delighted to have the coefficients for the interpolation of the arc tangent by next Thursday."[13] Aiken knew, of course, that Hopper was there because of her mathematics background and not because she knew anything about computers. Thirty years later, when she was asked how she became interested in computing, Hopper replied that she had had no choice in the matter. "I was ordered to the first computer in the United States by the United States Navy," she said, "and I reported to the Mark I."[14] Twenty-one-year-old Ens. Ruth A. Brendel arrived at the laboratory soon after. She had been an instructor in mathematics at the University of Buffalo and had joined the navy because she thought her technical skills could make a contribution to the war effort. The navy sent both women to Harvard in answer to a direct request from Aiken for WAVES for scientific billets.[15]

The original staff of the Computation Lab was small – fewer than a dozen. All were in the navy, although officers and enlisted alike were all reservists, serving only for the duration. Most of the enlisted men were IBM technicians in peacetime, and the officers were mathematicians; one of them a recent Harvard graduate and another a graduate student of Aiken's. Initially, Hopper was one of only three officers writing programs for the Mark I, and thus was one of the first-ever computer programmers – at that time called coders. Throughout the war this team, which had grown to almost forty by January 1946, ran the Mark I around the clock, churning out essential data for all sorts of military projects.[16]

The Computation Lab completed twenty-three reports for the Bureau of Ships in less than two years. The projects were so secret that even the coders usually identified them only by letters of the alphabet. The machine was much more pow-

[12] "IBM Automatic Sequence Controlled Calculator," (IBM Corporation, 1945), 5, IBM Archives; Douglas R. Hartree, *Calculating Instruments and Machines*, (Urbana, IL: University of Illinois Press, 1949), pp. 74–79; Charles J. Bashe, Lyle R. Johnson, John H. Palmer, and Emerson W. Pugh, *IBM's Early Computers* (Cambridge, MA: The MIT Press, 1986), 29; James W. Cortada, *Before the Computer: IBM, NCR, Burroughs, and Remington Rand and the Industry They Created, 1865–1956* (Princeton, NJ: Princeton University Press, 1993), pp. 190–191; Grace Hopper interview by Christopher Evans, 1976 (hereafter Hopper interview 1976), 12, OH 81, Charles Babbage Institute, University of Minnesota, Minneapolis (hereafter CBI).

[13] Rosario Rausa, "In Profile, Grace Murray Hopper," *Naval History* (Fall 1992):58.

[14] Hopper interview 1976, 1, OH81, CBI.

[15] Ruth Brendel Noller interview by author, 23 Dec.1996; "An Administrative History of the Bureau of Ships during World War II," (hereafter BuShips History), 4 vols., II, pp. 140–142, *U. S. Naval Administrative Histories of World War II No. 89*, 1952, Navy Department Library, Naval Historical Center, Washington, DC, (hereafter NDL/NHC).

[16] Gregory W. Welch, "Computer Scientist Howard Hathaway Aiken: Reactionary or Visionary?" (A.B. thesis, Harvard University, 1986), p. 47; Contract NObs-14966, folder "BuShips Computing Project," box 6, UAV.885.95.2, WW II Government Contract Records, HUA; Noller interview by author, 23 Dec. 1996; Aiken and Hopper, "The Automatic Sequence Controlled Calculator – I," 1, box 1, folder 2, Hopper Papers, Smithsonian.

erful than almost everything else then available to the navy and it ran calculations regarded as essential to the war effort.[17]

The first problem Hopper remembered working on at the lab was the one Aiken thrust at her on the very first day: finding the interpolation coefficients for applications of the arc tangent series. The purpose of the task was for computing rocket trajectories. Instead of shooting only inert objects (projectiles) the navy was now firing self-propelled missiles (rockets) for which there were no firing tables. Another similar early problem, typical of the work the lab did for the navy, was to figure fire-control calculations for 5"/38 caliber anti-aircraft guns. The completed solution accommodated input variables including target bearing, elevation and range, the ship's angle of pitch and roll, drift angle, time of flight, and residual projectile velocity. High capacity projectiles had been developed so much faster than their corresponding range tables that in 1942 the navy was already some five hundred tables behind. The advent of proximity fuses also meant extensive recomputation of existing tables. Until the Mark I became operational, Hopper recalled that the navy had had to rely on "acres of girls down at Aberdeen [Proving Ground] using hand-driven calculators" to create range tables.[18] Another problem Hopper worked on concerned the location of magnetic mines by conducting sweeps with a dipole dragged behind a minesweeper. The sweeper had to know the range of the dipole for an effective sweep, and Hopper made those calculations, in three dimensions, on the Mark I.[19]

Years later, when she was asked about her first days at work at Harvard and how she learned to program the Mark I, Hopper's reply was as straightforward as her approach to the problem must have been. "Well they gave me a code book and told me to do it," she said. "We were all in the navy," she would often say later, and "we didn't have time to react or think or anything. We just had to go ahead and do things."[20] Throughout the course of the war everyone at the lab was involved in an on-going experimental process requiring continual adaptation and development. "Mark I was new, a pioneer;" Hopper wrote some years later, "programs were improved or invalidated as changes were made in the computer's internal circuits to increase its efficiency."[21] After figuring out how to write the machine instructions, Hopper then punched them on tape, put the tape in the computer and hoped it would run. "But it was wartime, and there was never enough time to organize and systematize the information," she remembered.[22]

[17] Aiken, Preface to the *Manual of Operations* (n. p.), BuShip History IV 185, NDL/NHC.

[18] Report No. 7, July 1944, BuShips Computation Project Reports 1944–1945, Naval Research Laboratory, Washington, DC (hereafter NRL); *Annals of the Computation Laboratory of Harvard University*, vol. 26, (Cambridge, MA: Harvard University Press, 1951), p. 5; Hopper interview 1976, 2, OH81, CBI, for quote.

[19] Hopper interview 1982–83, p. 29, WFGOH46, SL.

[20] Hopper interview 1976, p. 11, OH81, CBI.

[21] Grace Murray Hopper, "The Education of Computer," p. 139, in: *Proceedings, Symposium on Industrial Applications of Automatic Computing Equipment*, pp. 139–144, Midwest Research Institute, Kansas City, MO, 8–9 Jan. 1953, from reprint in Hopper, Grace Murray Biographical Files, CBI.

[22] Hopper interview 1976, p. 6, p. 11, OH81, CBI.

Hopper was not particularly interested in hardware and had relatively little concern with it. Instead, she quickly focused on methods to speed the process of writing coding instructions to run individual problems. Hopper claimed to be "...lazy as all get out.[23] I never want to do anything over again." She was always looking for shortcuts, which in the early days meant assembling collections of subroutines. Hopper and her colleagues thus created what were essentially the earliest digital computer programs - the first software. These were stored programs, only independent of the machine itself, stored on punched paper tapes and in programmers' notebooks. Similar work putting together a library of mathematical subroutines was also being undertaken in England at the same time - as Hopper learned later. During the war, however, she was not at all aware of it. There was relatively little information around about different systems (even within the United States) because of security restrictions. Indeed, compartmentalization was the norm at the Harvard lab, each programmer working individually. "You didn't go after anything else unless you had an actual need to know," Hopper pointed out. Besides which, "...you were too busy trying to get the problem you had at the moment on and get the results out to think of anything else or go after anything else."[24]

Aiken, moreover, kept tight control of the work at the lab and it was he who handed out job assignments. Hopper was busy computing the extent of influence of the anti-mine dipole when Aiken put her to work (against her wishes) writing a manual. On Aiken's order she put together a 561-page *Manual* that gave a full and detailed description of the Mark I: all its parts with which she had become intimately familiar, and all its circuits, as well as how to program it.[25]

The Mark I not only ran numbers for a great variety of navy labs, but for the army as well, and for other academic labs.[26] By October, 1944, for example, Aiken's lab was undertaking computations for the Bureau of Ordnance in conjunction with the computing section at Columbia University for spherically and cylindrically symmetric underwater blast problems. At first it had been thought possible to carry out all the blast computations on Columbia's IBM punched-card equipment, but the volume of material involved was too extensive, so the Mark I was used as well.[27]

[23] Grace M. Hopper interview by Philip F. Holmer, 20 Jul. 1979, typed transcript, Oral History Interviews, (hereafter Hopper interview 1979), 8, series II, Sperry Univac Records Accession 1825 (hereafter sperry Acc. 1825), The Hagley Museum and Library, Greenville, DE (hereafter Hagley).

[24] Hopper, "Automatic Programming," 10, Marvin L. Stein Papers, (CBI 10), CBI; Hopper interview 1976, p. 10, OH81, CBI, for quote.

[25] Robert V. D. Campbell and Richard Bloch, interview by author, 12 Sept. 1997. Harvard students Campbell and Bloch, were Mark I's first programmers, soon joined by Hopper. Hopper interview 1982-83, 29, WFGOH46, SL.

[26] Reports No.10, Oct.1944, No.8, Aug. 1944, No.5, Jun.1944, BuShips Computation Project, NRL; Cohen and Welch, Makin' Numbers, 3, 57; Interim Progress Reports to AMP, 21 Aug. 1944, 23 Oct. 1944, 24 Jan. 1945, box 13, Applied Mathematics Panel General Records (hereafter AMP Gen. Recs.), Record Group 227 (hereafter RG227), National Archives 2, College Park, MD (hereafter NA2).

[27] Interim Report to AMP, 24 Jun. 1944, box 13, AMP Gen. Recs., RG227, NA2.

The largest problem run on the Mark I - problem J - was the compilation of tables of Bessel functions requested by the Naval Research Laboratory (NRL). Bessel functions were important in applications as diverse as radio wave propagation, heat flow, and frequency modulation.[28] The Computation Lab also computed tables of Hankel functions. These had been requested by NRL but were also used by other naval research activities, such as the Naval Ordnance Lab, the navy's Radio and Sound Laboratory, and the David Taylor Model Basin, for solution of problems of radio wave propagation and radiation, underwater sound propagation, and the like.[29] The Computation Lab also ran numbers for the celebrated mathematician, John von Neumann, for work regarding the plutonium bomb dropped on Nagasaki on 9 August 1945.[30]

The Computation Lab was far from alone in providing mathematical services to the navy during World War II. Howard Engstrom, for example, who had taught Hopper at Yale (and who was a longtime member of the Naval Reserve), was made a commander and headed a special research group of top mathematicians, engineers, and physicists from all over the country charged with applying mathematics to cryptanalysis. This effort, using versions of the Turing bombe, was part of the innocuously named Communications Supplementary Activities, Washington (CSAW). When she joined the navy, Hopper had initially expected to join Engstrom in whatever exciting and secret work it was that he was engaged in.[31]

In addition to working on the Mark I and writing its first manual of operations, Hopper was instrumental in the development of its successors, the Mark II - also electromechanical - and the Mark III - electronic - which were used by the navy after the war. Reflecting on the significance of military sponsorship of science, Hopper often maintained that there would not have been a computer industry without that early navy support.[32] Surely there would have been, but it would have been slower to develop.

[28] Robert V. D. Campbell interview by William Aspray, 22 Feb. 1984 (hereafter Campbell interview 1984), 23, OH67, CBI; I Bernard Cohen, "Howard Aiken and the Beginnings of Computer Science," CWI Quarterly 3 No. 4 (1990):318.

[29] Chief BuShips from NRL, 3 Jan 1945, folder "(Dead) BuOrd," box A-C, UAV 289.2005, Aiken Correspondence, HUA.

[30] Report No.11, Dec. 1944, BuShips Computation Project, NRL; Letter to Dr. Arnold Lowan from H. H. Aiken, 1 Nov. 1944, folder "(dead) BuOrd," box A-C, UAV 289.2005, Aiken Correspondence, HUA; Letter from Bob Campbell to author, 16 Aug. 1998. Von Neumann also worked on calculations for the hydrogen bomb, Andrew Hodges, *Alan Turing: The Enigma*, (New York: Simon and Schuster, 1983), p. 302.

[31] Colin Burke, *Information and Secrecy: Vannevar Bush, Ultra, and the Other Memex* (Metuchen, NJ: The Scarecrow Press Inc., 1994), pp. 224, 276, 291, 297; David L. Boslaugh, When Computers Went to Sea: The Digitization of the United States Navy (The Institute of Electrical and Electronic Engineers, Inc., 1999), pp. 72, 81-83; Grace Hopper interview by Uta C. Merzbach, Jul. 1968 - Feb. 1969 (hereafter Hopper interview 1968-69), pp. 5-6, folder 7, box 11, Series I, Computer Oral History Collection, National Museum of American History Archives Center, accession #196 (hereafter NMAHAC #196), Smithsonian Institution, Washington, DC.

[32] Grace Hopper interview by H. S. Tropp, 5 July 1972 (hereafter Hopper interview 1972), 68-69, folder 9, box 11, Series I, NMAHAC #196.

After the war, Hopper had wanted to transfer to the regular navy but was over the cut-off age and had to satisfy herself with staying in the Naval Reserve. Instead of returning to teach mathematics at Vassar, where she was offered a full professorship, she remained at the Computation Lab as a Research Fellow. After three years, when she failed to be promoted or to gain tenure - which was the usual case for women - she was forced to leave.[33]

In 1949, after careful consideration of the many job offers she received - including from Engstrom now at Engineering Research Associates - Hopper joined the Eckert-Mauchly Computer Corporation in Philadelphia, because she figured they were the closest to having a commercial computer - the UNIVAC - up and running. She had had an interview at IBM, "but that was back in the days when they still had an IBM flag and sang songs about it, and that was too much for me" she recalled.[34] Hopper remained with the UNIVAC division through Eckert-Mauchly's acquisition by Remington Rand in 1950, and then through the merger which created Sperry Rand (later to become UNISYS) in 1955. She finally retired from the company in 1971.[35]

At UNIVAC in 1952, Hopper developed the first compiler, the A-0, a library of subroutines that the computer itself could assemble into a program. In 1957 she completed FLOWMATIC, the first English language compiler. By 1960, FLOWMATIC became one of the main ingredients in the collaborative creation of COBOL, soon to be widely adopted as a universal computer language.[37]

During all this time Hopper enthusiastically maintained her navy reserve status, serving weekend and annual active duty stints. Finally, however, the years caught up with her and in 1966 she was forced to retire.[38] "It was the saddest day of my life," recalled Hopper.[39]

Only seven months later, however, the navy repented its bureaucratic efficiency and reversed the decision to let Hopper go. With the naval expansion in response

[33] Richard Bloch interview by William Aspray, 22 Feb. 1984 (hereafter Bloch interview 1984), pp. 19, 58, OH66, CBI; Hopper interview 1979, 1, Sperry Acc. 1825, Hagley.

[34] Hopper interview 1968, p. 5, folder 7, box 11, Series I, NMAHAC #196.

[35] Hopper interview 1982–83, pp. 30–31, 47, WFGOH46, SL; Anthony Ralston, ed., *Encyclopedia of Computer Science and Engineering*, 2nd ed. (New York: Van Nostrand Reinhold Co., 1983), pp. 685–686.

[36] "Hopper and Mauchly on Computer Programming," in: *Proceedings of the IEEE*, **72**, no.9 (September, 1984): p. 1216, Reprinted from Grace M. Hopper and John W. Mauchly, "Influence of Programming Techniques on the Design of Computers," *Proceedings of the I. R. E.*, **41**, no.10, (October 1953): pp. 1250–1254.

[37] Hopper interview 1982–83, pp. 32, 34–35, 47–49, WFGOH46, SL; Ralston, *Encyclopedia of Computer Science and Engineering*, pp. 685–686. Hopper credited A.E. (Gene) Smith with being "the first to move to use COBOL all out." Smith had been BuShips liaison to the Computation Lab during the war, and afterwards went to the Bureau of Standards, Hopper interview 1968–69, p. 15, folder 7, box 11, Series I, NMAHAC #196.

[38] Hopper interview 1976, p. 19, CBI.

[39] Quoted in Billings, *Hopper*, p. 87.

to the Vietnam War, and the consequent increasing demand for computerized systems, Hopper's skills were once again invaluable.[40]

On 1 August 1967, Grace Hopper was recalled to active duty with a temporary appointment for six months. She stayed nineteen years. Her most important work was in the standardization of navy computer languages. She implemented a comprehensive program of standardization of COBOL in the navy, replacing the numerous and incompatible versions of the language then in use.[41]

From 1977 until her final, still reluctant, retirement in 1986, Hopper was at the Naval Data Automation Command Headquarters (NAVDAC, now NAVCOMTEL) in Washington, DC. In those final years "Amazing Grace," as her colleagues called her, became the navy's foremost propagandist for its computer program as NAVDAC's representative to learned societies, industry associations, and technical symposia.[42] She had surely repaid the navy in full measure for the opportunity it gave her to pioneer a new discipline. She was eternally grateful, always maintaining that of all the honors she received, the chance to serve in the US Navy was the most precious.[43]

Mina Rees

As Grace Hopper's career illustrates, the technological demands of World War II had an immense impact on the American scientific community, often channeling scientists in new directions. After the war many continued to work in their new fields, profiting from their wartime experiences.[44] Like Grace Hopper, Mina Rees's wartime service greatly affected her postwar career, in her case moving her out of teaching and into administration. In 1923 Mina Rees had earned a bachelor's degree, summa cum laude, from Hunter College in New York, a tuition-free public college for exceptional young women. In 1925 she received a Masters Degree from Columbia University, and in 1931, she was awarded a doctorate in mathematics from the University of Chicago.[45]

[40] Hopper was an exception to the trend among navy women whose numbers actually decreased during the war. They were restricted, as before, to less than a third of the enlisted ratings. Moreover, most WAVES continued to fill clerical, health service, and administrative positions; Godson, *Serving Proudly*, p. 207.

[41] "Captain Grace Murray Hopper, US Naval Reserve," July 1981, Grace Hopper Officer Biography, Box 313, Operational Archives, Naval Historical Center, Washington Navy Yard (hereafter OA/NHC).

[42] Williams, *Improbable Warriors*, p. 212.

[43] Godson, *Serving Proudly*, p. iv.

[44] Among other works see, for example: Harvey Brooks, *The Government of Science* (Cambridge: M.I.T. Press, 1968), Daniel S. Greenberg, *The Politics of Pure Science* (The University of Chicago Press, 1999), Daniel M. Hart, Forged Consensus: Science, Technology and the Economic Policy of the United States, 1921–1953 (Princeton University Press, 1998), Penick, Pursell, Sherwood, and Swain, eds., *The Politics of American Science, 1939 to the Present* (Cambridge: M.I.T Press, 1972), Michael D. Reagan, *Science and the Federal Patron* (New York: Oxford University Press, 1969), Daniel J. Kevles, *The Physicists: The History of a Scientific Community in Modern America* (New York: Knopf, 1978).

Figure 3.
Dr. Mina Rees at the office of naval research, march 1946. [Courtesy: US Navy]

Like Hopper, with doctorate in hand Rees returned to teach mathematics at her alma mater, Hunter College. For the next twelve years she worked her way from assistant to associate professor while assiduously keeping up with new developments in her field. She became active in the American Mathematical Society, sitting on its major committees, and getting to know mathematicians from all over the country. This gave her an overview of what was going on in the discipline, and meant that in a community of mathematicians which probably numbered only 3,000 or so, she was one of the best known of the women.[46]

Before World War II, beyond encouraging states to establish institutions of higher education, the federal government had done little to develop science. With the fall of France in 1940, however, it seemed to some key American scientists that U.S. participation in the war was inevitable. Recognizing the need to organize the scientific war effort, a group of them began to mobilize. They were led by

[45] Martha J. Bailey, *American Women in Science: A Biographical Dictionary* (Denver, CO: ABC-CLIO, 1994), 322; Louise S. Grinstein and Paul J. Campbell, *Women of Mathematics: A Bibliographic Sourcebook* (New York: Greenwood Press, 1987), 175; Mina Rees interview by Nina Cobb, 16 Nov. 1983 - 20 Jan. 1984 (hereafter Rees interview 1983-84), 1-15, 19-24, Women in Federal Government Oral History Project, OH40 (hereafter WFGOH40), Schlesinger Library, Radcliffe Institute, Cambridge, MA (hereafter SL).

[46] Rees interview 1983–84, pp. 52–59, 78–81, WFGOH40, SL.

Vannevar Bush, an electrical engineer and president of the Carnegie Institution. At the urging of Bush and a number of prominent colleagues, President Roosevelt established the National Defense Research Committee (NDRC) in June 1940. Bush believed that science had advanced so far since the last conflict that military leaders did not have a good grasp of what might be scientifically possible. Therefore, instead of reverting to the former pattern of having the military request from science the weapons it wanted, science should acquaint itself with the needs of the military so that it could advise on what was possible. Much of the initiative for scientific military research thus swung towards civilian scientists.[47]

Soon, academic and research institutions all over the country were drawn into war work, the NDRC engineering a massive exchange of personnel. Much was accomplished in one year to create a civilian-led mechanism for harnessing the nation's scientists to the war effort and to consult with the military without falling under their control. By June 1941, Bush had recruited 6,000 physicists, chemists, mathematicians, and engineers, a number that grew to 30,000 by the end of the war.[48]

By mid-1941, the NDRC had already grown to such an extent that its direction was given to a new, supervisory and administrative body, the Office of Scientific Research and Development (OSRD).[49] In December 1942, the work of the NDRC was divided among eighteen new divisions and two panels which continued to operate until June 1946. The divisions were concerned with such subjects as ballistics research, ordnance accessories, missiles, subsurface warfare, fire control, explosives, absorbents and aerosols, chemical engineering, radar, radio, optics, and metallurgy. The two panels were the Applied Psychology Panel and the Applied Mathematics Panel.[50]

With NDRC's reorganization, Warren Weaver (1894–1978), mathematician and director for the natural sciences at the Rockefeller Foundation, was appointed chief of the Applied Mathematics Panel (AMP). It was already clear by then that the demand for sophisticated mathematical studies had increased rapidly as had the need for mathematical assistance throughout the NDRC. Divisions requiring extensive mathematical investigations, computing, or statistical work had to designate at least one member to sit on the AMP. The AMP, for its part, was to provide expert consulting services to the divisions, and to conduct authorized studies. Projects could also come to the AMP from the army, the navy, allied governments, or from the OSRD itself.

[47] Irvin Stewart, *Organizing Scientific Research for War: The Administrative History of the Office of Scientific Research and Development* (Boston: Brown, Little and Company, 1948), p. 6; James Phinney Baxter, *Scientists Against Time* (Boston: Brown, Little and Company, 1946), p. 14.

[48] Stewart, *Organizing Scientific Research*, p. 34; Baxter, *Scientists Against Time*, p. 24. For a less sanguine view of the cooperation see Harvey M. Sapolsky, *Science and the Navy: The History of the Office of Naval Research* (Princeton University Press, 1990), pp. 18–19, 29–30.

[49] Stewart, *Organizing Scientific Research*, pp. 35–37.

[50] Stewart, *Organizing Scientific Research*, pp. 85–97.

The NDRC selected, farmed out to research establishments, supervised, and funded all projects through a contract system. In most cases, once a contract was approved technical aides for each division undertook the oversight of the work. Weaver defined the duties of AMP technical aides to include: "reading a large amount of report material, both British and American, visiting contracts, consulting on studies, suggesting material and contacts, and establishing inter-relationships between Panel personnel and the services." With her considerable charm, her aptitude for solving, not creating problems, and her skill at making people feel like valuable members of a team, Mina Rees was very well qualified for the job of technical aide.[52]

After Pearl Harbor, Rees had felt a great sense of frustration, believing she could be useful in the war, but not knowing how to go about making a contribution. Looking back she wrote that she had really wanted, "as most young people did, I think, at that time, to have some part in trying to save the world."[52] Fortuitously for her, Warren Weaver found he could not run the AMP alone, especially as he still maintained his position at the Rockefeller Foundation and was trying to juggle the two jobs. The AMP agreed that he needed somebody full-time to take care of central operations. Weaver immediately called Rees, offering her the dual job of technical aide to the panel, and executive assistant to himself. Rees knew that being a woman "it would be harder to get a job that would be really significant in the war effort," so she had no hesitation in accepting Weaver's offer. Assuming that Weaver had turned to her because of the extensive connections she had cultivated among mathematicians around the country, Rees began work on 1 August 1943, as soon as she was approved for a civil service appointment.[53] Her role at the AMP was succinctly described in a citation years later recognizing her deep involvement "in deploying and sustaining the efforts of this country's mathematicians in the war effort."[54]

Weaver's mathematicians, who eventually numbered several hundred, worked on AMP problems in groups set up at eleven universities across the country, including Princeton, Columbia, NYU, the University of California at Berkeley, Brown, Harvard, and Northwestern. Weaver was advised by an executive committee composed of some of the nation's leading mathematicians, including Richard Courant, Thornton C. Fry, Oswald Veblen, and Samuel Stanley Wilkes. Rees served as secretary to this group, and like everyone on the AMP had an identifica-

[51] Mina Rees Diary (hereafter Rees Diary), AMP Executive Committee Meeting (hereafter AMP ExCo), 31 Jan. 1944, box 1, AMP Gen. Recs., RG227, NA2; author's interview with Norma Kenigsberg, 15 June 2000, for insights into Rees's management style. Kenigsberg worked with Rees at City University of New York (CUNY).

[52] Rees interview, 1983–84, pp. 107–111, WFGOH40, SL.

[53] Rees interview, 1983–84, pp. 107, 111, WFGOH40, SL.

[54] "Citation for Mina Rees," Hunter College Commencement, June 5, 1973, typescript, "Awards and Honors," box 1, Rees Collection, CUNY Graduate School and University Center Archives, New York City.

tion badge reading "Mathematical Studies relating to Warfare."[55] By the end of the war the AMP had undertaken 194 studies, almost half of them the result of direct requests from the armed services.[56]

In the early days of the AMP, Rees sometimes had difficulty persuading scientists to accept panel contracts. She knew they were jealous of their freedom to pursue research independently of direction, yet in most cases during the war mathematicians had to solve other people's problems which was very trying for them. This forced technical aides to develop formidable skills as intermediaries; they were the government source of funding standing between the contractor (the scientists) and the client (usually the military), interpreting the project back and forth between them and making its completion possible.[57]

When problems came in from the military Rees transmitted them to the executive committee of the AMP after doing some preliminary work on them. She provided the background information with which the committee could make a decision. Generally this involved reformulating the problem in mathematical terms, and then seeing if there were people available who had the right kind of mathematical specialty to reach the answer to it. Rees was largely responsible for identifying a suitable person or group of people to carry out the work, so she needed to know what field people were working in and how busy they were.[58] She also had to keep current with all projects undertaken that might in any way involve mathematics - in the navy especially, but also in the army and in other OSRD divisions - in order to avoid duplication of effort.[59]

According to Rees, one of the strengths of the system thus developed was that many of the most able mathematicians in the country were identified with the AMP, either in the contract places or as members of the panel itself.[60] To be sure, both the army and the navy also recruited civilian mathematicians, while others were associated with British and Canadian research efforts. Still others held posts in industry. However, the NDRC, and especially its Applied Mathematics Panel, greatly influenced wartime mathematics thereby shaping postwar mathematics as well.[61]

[55] Mina Rees, "The Mathematical Sciences and World War II," in: *American Mathematical Monthly* **87**, No.8 (October 1980): p. 609; for the ID badge, Rees Diary, AMP ExCo, 11 Sept. 1944, box 1, AMP Gen. Recs., RG227, NA2. The Executive Committee had its first meeting on 27 Jan. 1943, before Rees was brought on board.

[56] Warren Weaver, *Scene of Change: A Lifetime in American Science* (New York: Charles Scribner's Sons, 1970), pp. 88–89; U.S., OSRD, NDRC, *Summary Technical Report of the Applied Mathematics Panel*, 3 vols. (confidential) (Washington, D.C., 1946) (hereafter OSRD, *Summary Technical Report of AMP*), vol.2, p. vii; "Meeting of the ExCo of AMP," 3 May 1943, box 1, AMP Gen. Recs., RG227, NA2.

[57] Weaver to Dr. Irvin Stewart, 9 Sept. 1943, box 8, AMP Gen. Recs., RG227, NA2; Rees interview, 1983–84, p. 131, WFGOH40, SL.

[58] Rees interview, 1983–84, pp. 108, 112–114, WFGOH40, SL.

[59] Rees interview, 1983–84, pp. 108, 112–113, WFGOH40, SL.

[60] Rees interview, 1983–84, pp. 110, 113, 118, WFGOH40, SL.

[61] Rees, "The Mathematical Sciences and World War II", pp. 608–609; Rees, "Mathematics and the Government,", p. 4, "Publications," box 1, Rees Collection, CUNY.

One of the first problems the navy brought to the AMP was how to determine what kind of torpedo barrage to lay down against Japanese vessels to maximize the chance of hitting them. The navy had no information on how fast Japanese vessels could go in a straight line, how rapidly they could turn, etc. They, did, however, have very good photographs of many Japanese vessels. The people at NYU knew that in 1887 British physicist Lord Kelvin had established that the speed of a ship moving in a straight line is indicated by the spacing of the cusps along the bow waves. Most of the navy photographs, though, showed Japanese ships in the process of turning, so Lord Kelvin's analysis had to be extended to ships in turns. This was rather simply done and the data on the speed of Japanese ships was figured from pictures of their bow waves or wakes. A test of the mathematical results in an experimental run with a new destroyer off the coast of Maine found an extremely good agreement of theory and observation.[62] As Rees pointed out, "This and similar experiences won over the armed services to the notion that mathematics could be of great help to them."[63]

Rees noted that some of the tasks taken on by the AMP involved what was called "classical applied mathematics." At NYU, for example, work in gas kinetics concerned the theory of explosions in the air and underwater, which required extensive study of the kinds of shock fronts created by explosions. A Bureau of Aeronautics request for assistance in the design of nozzles for jet motors initiated Study No.137, an analysis of gas flow in nozzles and supersonic gas jets involving the development of jet and rocket theory. It was hoped to use rockets to assist in carrier plane take-offs. Rees remembers, in particular, one trip she made to Pasadena in 1944, accompanying Richard Courant and Kurt Friedrichs to check the problem Cal Tech was having launching their rockets. Courant and Friedrichs were able to make some suggestions and the experiments begun after their departure were successful.[64] By work of this sort AMP mathematicians gained expertise in new and important fields putting them at the forefront of postwar developments.

The Naval Research Lab frequently sought help for computations involving fire control experiments as did MIT physicist Philip Morse of the Anti-Submarine Warfare Operational Research Group for a number of probability problems concerning the lethal range of depth charges and airborne submarine searches.[65]

As technical aide to the AMP one of Rees's areas of responsibility was the effort to expand computing power, making use of all the various types of computing equipment including differential analyzers, relay computers, and other devices developed by National Cash Register and IBM. Rees also assigned projects to

[62] Rees, "The Mathematical Sciences and World War II,", p. 617; Rees Diary, AMP ExCo, 1 May 1944, box 1, AMP Gen. Recs., RG227, NA2.

[63] Rees, "The Mathematical Sciences and World War II,", p. 617.

[64] Rees Diary, AMP ExCo, 21 Aug. 1944, box 1, AMP Gen. Recs., RG227, NA2, for Study No. 137; for Courant in Pasadena, Rees Diary, ExCo, box 1, AMP Gen. Recs., RG227, NA2.

[65] Rees Diary, AMP ExCo, 1 May 1944, box 1, AMP Gen. Recs., RG227, NA2; "Diary of ExCo Meeting," 8 Mar. 1943, box 1, AMP. Gen. Recs., RG227, NA2; "Meeting of the ExCo of AMP," 3 May 1943, box 1, AMP Gen. Recs., RG227, NA2; "Notes on ExCo Meeting, 7 Jun. 1943," box 1, AMP Gen. Recs., RG227, NA2; Rees Diary, AMP ExCo, 27 Sept. 1943, box 1, AMP Gen. Recs., RG227, NA2.

Howard Aiken's Mark I at Harvard, where she heard of Grace Hopper for the first time. Rees thus became very familiar with the most up-to-date work in computers, helping to qualify her for her first postwar position.[66]

To be sure, Rees was not the only woman of science at the OSRD/NDRC, but there were not many others, and even fewer who were in positions as influential as hers.[67] Before going to the AMP, however, Rees had had little administrative experience and she watched Weaver closely, learning from his great skill in dealing alike with military officers, the Washington bureaucracy, and fellow scientists. In many ways her talents were very similar to his. Weaver wrote perceptively of himself that he "lacked that strange and wonderful creative spark that makes a good researcher." Rees, too, acknowledged that she did not have the kind of mathematical creativity that would have qualified her to do research for the AMP. Her job had been basically an administrative one, although it was one that also required a deep knowledge of mathematics. For their war work both Weaver and Rees received the King's Medal for Service in the Cause of Freedom from the British government, and Weaver received the Medal for Merit of the United States, while Rees was awarded the U.S. President's Certificate of Merit.[68]

Rees had joined the AMP in August 1943 and remained until its last meeting in April 1946. In a glowing letter Weaver wrote of her accomplishments that she had "made a large and distinguished contribution to the war effort."[69] She never returned to teaching, however. Her wartime experience had set her on the administrative path that defined the rest of her life. In 1946 Rees accepted an invitation

[66] Weaver to BuOrd, BTL, MIT, 23 Dec. 1943, box 25, AMP Gen. Recs., RG227, NA2; Mina Rees, "The Computing Program of the Office of Naval Research, 1949–1953," *Annals of the History of Computing* **4**, no. 2 (April 1982): p. 104; "Final Report of Committee on Computing Aids for Naval Ballistics Laboratory," 28 Apr. 1944, box 25, AMP Gen. Recs., RG227, NA2; Rees to Dr. A. N. Lowan, 4 Dec. 1945, box 26, AMP Gen. Recs., RG227, NA2; Rees Diary, AMP ExCo, 1 Nov. 1943, AMP Gen. Recs., RG227, NA2.

[67] There was, for example, Dr. Dorothy W. Weeks, of the OSRD Liaison Office, Dr. Gladys Anslow, head of the physics department at Smith College who was appointed to the Office of Field Service in 1944, and Dr. Gladys Anderson Emerson, a biochemist working for OSRD on the effects of vitamin deficiencies. There were also a few other very able women on the administrative side, like Margaret Moses and Dr. Louise Kelley in the office of the Chairman, NDRC. A number of prominent women scientists worked on the Manhattan Project, and among mathematicians there two very important statisticians working with Jerzy Neyman at Berkeley. One of them, Elizabeth "Betty" Scott, replaced him as chair of the department of mathematical statistics when Neyman retired. Margaret Moses to Rees, 26 Jan. 1944, and Rees to Moses, 28 Jan. 1944, box 23, AMP Gen. Recs., RG227, NA2; Minutes of Conference on Summary Technical Report, 30 Aug. 1945, box 3, AMP Gen. Recs., RG227, NA2; for Gladys Emerson see Edna Yost, *Women of Modern Science* (New York: Dodd, Mead & Company, 1959), pp. 140–155; Rees interview, 1983–84, pp. 139, 150–152, WFGOH40, SL; Rees to Dr. Dorothy W. Weeks, 8 Mar. 1944, box 23, AMP Gen. Recs., RG227, NA2; MR to WW, 29 July 1944, AMP Gen. Recs., RG227, NA2, for Weeks and Anslow.

[68] Weaver quoted in Mina Rees, "Warren Weaver, 1898–1978," National Academy of Sciences, *Biographical Memoirs* **57** (1987): p. 501; for Rees see Donald J. Albers and G. L. Alexanderson, eds., *Mathematical People: Profiles and Interviews* (Basle, Boston, Berlin: Birkhäuser, 1985), p. 258; Minutes of Conference on Summary Technical Report, 30 Aug. 1945, box 3, AMP Gen. Recs., RG227, NA2; Mina Rees c.v., box 1, Rees Collection, CUNY.

[69] U.S., OSRD, *Summary Technical Report of AMP*, vol 2, Foreword, v; Weaver to Pres. George N. Shuster of Hunter College, 29 Jan. 1946, box 8, AMP Gen. Recs., RG227, NA2.

to head the mathematics branch of the newly established Office of Naval Research (ONR). Building on the broad contacts she had made through the AMP, she took up the challenge to design and develop a mathematical program melding civilian and navy science in peacetime. With the end of the OSRD, the ONR became the chief government office to continue subsidizing scientific research after the war. A powerful incentive, of course, in the developing Cold War environment, was the awareness that applied mathematics was being vigorously pursued abroad, especially in Russia, Germany, England, and the Netherlands.[70]

The easy wartime availability of federal research dollars had irrevocably changed the scientific climate and Rees's war service had placed her on the ground floor of the new developments. With that head start she soon created a flourishing program at ONR.[71] Later, as chief of the Mathematical Sciences Division at ONR, Rees accomplished the distinction of being the only woman at division-head level in the navy's whole research program. In her final year at ONR, Rees became the deputy research director, responsible for the entire research activity.[72] For her work at ONR, Rees has been called "the architect of the first large-scale, comprehensively planned program of support for mathematical research; she pioneered its style, scale and scope." Both the American Mathematical Society and the Institute of Mathematical Statistics adopted resolutions of appreciation for her work at ONR.[73]

In 1953 Rees returned to New York as Dean of Faculty at Hunter College. She went on to a brilliant career as an academic administrator, eventually serving as the first president of the City University of New York Graduate Center. Questioning the timing of her move an interviewer asked Rees if she left Washington because Eisenhower came in. While not replying directly, she did admit "when Mr. Eisenhower was elected president, the budget was cut...the Republicans cut budgets."[74]

[70] Rees interview, 1983–84, p. 160, WFGOH40, SL; Rees, "The Mathematics Program of the Office of Naval Research: *Bulletin of the American Mathematical Society* **54**, no.1 (January 1948): p. 4; Reagan, *Science and the Federal Patron*, pp. 5–6, 264–266; Samuel S. Snyder, "Influence of U.S. Cryptologic Organizations on the Digital Computer Industry," 3, Special Research History - 003, box 4, RG457, NA2.

[71] Rees, "Establishment of CUNY," p. 6–7, box 1, Rees Collection CUNY; Rees, "The Mathematics Program of the ONR," pp. 2–3; Sapolsky, *Science and the Navy*, p. 8.

[72] "Who's Who in Naval Research," Office of Naval Research, *Research Reviews* (Feb. 1952): pp. 13–14, box 1, Rees Collection, CUNY.

[73] F. J. Weyl, who served with Rees at ONR is quoted in Grinstein and Campbell, *Women of Mathematics*, p. 176; Albers and Alexanderson, *Mathematical People*, p. 263; Rees, "The Computing Program of the Office of Naval Research 1949–1953," *Annals of the History of Computing* **4**, no.2 (April 1982): p. 119.

[74] Rees interview, 1983–84, pp. 230–231, 236, WFGOH40, SL.

New Mathematical Disciplines and Research in the Wake of World War II

Tinne Hoff Kjeldsen*

This paper focuses on the significance of the Second World War for the rise and establishment of new disciplines in applied mathematics as well as for the renewed interest and growth in some related subjects in pure mathematics. The mathematical topics involved are mathematical programming, operations research, game theory, the theory of convexity, and the theory of systems of linear inequalities. Connections and interactions between different branches of mathematics on the one hand and between different kinds of driving forces in the development of mathematics on the other hand are discussed. Special emphasis is devoted to the significance of the interplay between practical problem solving and basic research in mathematics proper as a consequence of World War II and the post-war organization of science support in the USA.

1 Introduction

The significance of the Second World War for the directions of – and growth in – scientific research in the USA was immense. In the case of mathematics the war gave rise to the emergence and establishment of new disciplines in applied mathematics as well as a renewed interest and growth of research in related subjects in pure mathematics.

The purpose of this paper is to discuss and analyze how mathematical programming, game theory and operations research in applied mathematics,[1] and the theories of convexity and systems of linear inequalities in pure mathematics either came into being or benefited from the scientific mobilization and the subsequent post-war organization of science support in the USA.

It is a story that shows how these particular fields of mathematics evolved through interactions between practical problem solving and basic research brought about by the American post-war military-science symbiosis. It is also a story that illustrates the influence of prominent scientists on the kind of research done and thereby also the kind of knowledge gained.[2]

* IMFUFA, Department of Mathematics, Roskilde University, P. O. Box 260, DK 4000 Roskilde, Denmark. Email: thk@ruc.dk

[1] Whether operations research is a field of applied mathematics or not will be dicussed in Sect. 5.

[2] I would like to thank Barbara Green and David Volckmann for correcting my English.

2 The Struggle to Involve Mathematics in the War Effort

The scientific mobilization

The American mobilization of science in World War II was spearheaded by Vannevar Bush, who served as President of the Carnegie Institution from 1939, James Conant and Karl T. Compton, President of Harvard and MIT respectively, and Frank Jewett from AT&T's Bell Laboratories. In June 1940 Bush discussed his plan for the organization of scientists for the war effort with President Roosevelt. The result was the establishment of the National Defense Research Committee (NDRC) whose scientific personnel besides Bush, Conant, Jewett, and Compton also included the physicist Richard Tolman from Caltech. The Army and the Navy were represented by Brigadier Admiral George V. Strong and Rear Admiral Harold G. Bowen respectively. The last person in the committee was Coe who was the commissioner of patents.[3]

NDCR operated independently of the military, which meant that it could initiate research projects that had not been requested by the armed forces. In 1941 the Offce of Scientific Research and Development (OSRD) was established with Bush as the leader. OSRD was financed by the Congress and unlike NDRC it had the power not only to initiate research projects but also to actually construct new types of weapons and develop new types of defense systems. Bush wanted to overcome the shortcomings he found in the research as it was conducted within the military which he found ineffective and second rate, ruled by military people with very little appreciation for science and restricted by internal rivalries between the different military departments. His vision was to build a cooperation between the military, industry, and universities that would render scientific research in defense and weapon development more effcient. The way it worked was that the scientists employed by OSRD remained civilians. They were bound to OSRD by contracts and they did their work not in military laboratories but in the universities and in industry.[4] This organizational structure reflects Bush's wish that the science and the scientists should not be controlled by the military. OSRD functioned as a mediating link between the military and the scientists.

The Applied Mathematics Panel

The people who took the initiative and were responsible for the organization and mobilization of civilian scientists had academic backgrounds in electrical engineering, chemistry, and physics – mathematicians were conspicuous by their absence. There were no mathematicians represented, and there was originally no section for mathematics in either NDRC or OSRD. This does not mean that no

[3] [Zachary, 1997].

[4] [Zachary, 1997].

mathematician contributed to the war effort; there just was no joint coordination of the services of mathematicians. There were mathematicians working in different military establishments, but it was not until the end of 1942 – well into the war – that an Applied Mathematics Panel (AMP) was created within NDRC. Warren Weaver, the director of the Division of Natural Sciences of the Rockefeller Foundation, was appointed the leader of the panel. Weaver had experience with applied mathematics from his time as professor in the mathematics department at the University of Wisconsin. Before he became the leader of the AMP he had been head of the fire control analysis section of NDRC.[5]

The importance and the success of the AMP has been reported several places, but as pointed out by Larry Owens in his essay *Mathematicians at War: Warren Weaver and the Applied Mathematics Panel, 1942–1945* the panel also had its failures. The story Owens tells shows "the empirical ambitions of those who, like Vannevar Bush, James Conant, and Warren Weaver, took the lead in the mobilization of wartime science."[6] The relatively late involvement of mathematicians should not be taken as a sign that mathematicians weren't interested in offering their services. On the contrary the leaders of the two mathematical societies, AMS and MAA, were very much aware that the expected benefits that would eventually accrue to the different sciences depended on their perceived relative contributions to the war effort. In February 1941 Marston Morse, who was the president of the AMS, presented a paper *Mathematics in the Defense Program* before the National Council of Teachers of Mathematics where he talked about the importance of mathematics in a war situation and how mathematics could be applied. He directed attention towards the first step made by mathematicians to bring mathematics to the service of the country: "About a year ago these societies [AMS and MAA] appointed a committee known as the War Preparedness Committee, to prepare the two societies to be useful to our nation in time of war."[7] The War Preparedness Committee included such able mathematicians as John von Neumann and Norbert Wiener.

Despite attempts to bring the effort of the AMS to the notice of OSRD, mathematics was still not implemented in OSRD, and in March 1942 Morse and Marshall Stone from Harvard University presented Bush, Conant, and Jewett with the memo *Mathematics at War.* As a result a Joint Committee of Mathematics with Marston Morse as chairman was established under the National Academy of Science and the National Research Council. But mathematicians still suffered from lack of recognition, and it got even worse when *Time Magazine* in 1943 – that is after the establishment of the AMP – quoted the science writer George Gray as saying that "the U.S. has been severely handicapped by its shortage of topflight mathematicians."[8] Morse got very angry and wrote back that "The actual fact is that the deficiency lies [...] in the failure of the civilian authorities to use mathe-

[5] [Owens, 1989], [Rees, 1980].

[6] [Owens, 1989, p. 289].

[7] [Morse, 1941, pp. 293–294].

[8] *Time Magazine*, November 29, 1943, quoted in [Owens, 1989, p. 293].

matics at an early time, in adequate numbers and in the proper way."[9] When mathematics through AMP finally was implemented in NDRC the leaders of NDRC asked Weaver to be the chief of the panel without first discussing it with either the AMS or the Joint Committee. Karl T. Compton wrote to Weaver in October 1942 asking him to take on the appointment as chairman of AMP and gave the following assesment of the Joint Committee:

> [...] thus far the mathematicians as a group have not been brought into the war picture very effectively. For example; the committee headed by Marston Morse has not been effective; I think this is because it is too "pure."[10]

The fact is, according to Owens, that "the leaders of OSRD [...] seem to have expected little from the committee [the Joint Committee chaired by Morse] and doubted that a comfortable working relationship with the AMS was generally possible."[11] The reason why the leaders of OSRD apparently did not expect very much of the Joint Committee headed by Morse seems to be buried in the dichotomy between pure and applied mathematics. Ward Davidson from NDRC reported after a first meeting with the Joint Committee that he could not "bridge the wide gap between the view of an engineer and those of a 'pure' mathematician. My imagination just didn't go far enough to understand how problems that seem to me to be rather practical could be handled effectively in the quite rare atmosphere of abstract mathematical thinking."[12]

Whether these problems of overcoming the breach between applied and pure mathematics were self-fulfilling or not, it seems that Weaver had some problems handling the more eccentric ones of the mathematicians. In Owens' essay a discussion between Weaver and Stone about the priorities between suitable personality and good mathematics illustrates some of the administrative problems Weaver faced as leader of the AMP. There is the story about how he "bent over backwards" to apply Wiener's expertise, the problems he had with John von Neumann, and with Jerzy Neyman, whose contract in the end was terminated by the AMP. There were more than hurt feelings at stake here on the part of the AMS. The expected flow of money to mathematics after the war was dependent on how well mathematics was represented in the war effort. There was also a concern that applied mathematics would benefit at the expense of pure mathematics.[13] Before the war only a very few mathematicians in the USA working in academia were interested in applied mathematics and the few who were, were not considered to be doing first class mathematics. The state of affairs before the war can be summarized by the words of Professor Prager, who gave the following description in 1972:

[9] *Time Magazine*, December 20, 1943, quoted in [Owens, 1989, p. 293]

[10] Quoted in [Williams, 2001, p. 168].

[11] [Owens, 1989, pp. 294–295].

[12] Letter from Davidson to Conant, quoted from [Owens, 1989, p. 295]

[13] See [Owens, 1989, pp. 291–292].

> [...] their number [professional mathematicians interested in applications] was extremely small. Moreover, with a few notable exceptions, they were not held in high esteem by their colleagues in pure mathematics, because of a widespread belief that you turned to applied mathematics if you found the going too hard in pure mathematics.[14]

Despite the problems and disagreements the AMP did succeed in making mathematics play an important role in the war. One of its more celebrated accomplishments was its program for educating mathematicians to serve in operations research groups. This will be discussed below in Sect. 3.

A programming planning problem

Linear programming was a new branch of applied mathematics that – in the USA – came into being as a direct consequence of mathematicians' war work. It was not done under contract with the AMP but by some of the mathematicians employed directly by the armed forces.[15] The source was a concrete practical problem within the US Air Forces,[16] a logistic problem that eventually led to the mathematical theory of linear programming, and from there to mathematical programming.[17] The person normally associated with the origin of linear programming in the USA is George B. Dantzig. Dantzig was one of the mathematicians hired directly by the armed forces for the war effort. In 1941 he egan working at the Combat Analysis Branch of the United States Air Force Headquarters Statistical Control under the leadership of Tex Thornstons. During the war Dantzig worked on what was called "programming planning" methods to calculate Air Force programs. An Air Force program was a kind of activity plan. In 1951 Dantzig gave the following explanation of the nature and purpose of such a program:

> The levels of various activities such as training, maintenance, supply, and combat had to be adjusted in such a manner as not to exceed the availability of various equipment items. Indeed, activities should be so carefully phased that the necessary amounts of these various equipment items were available when they were supposed to be available, so that the activity could take place.[18]

Dantzig's job during the war was to teach Air Force staff how to compute these programs.

Programming planning was a practical problem in the Air Forces. After the war it entered – with the help of the military – into the universities where it became

[14] Quoted in [Rees, 1980, p. 607].

[15] For a history of the Russian contribution see [Brentjes, 1976b], [Charnes and Copper, 1961].

[16] The U.S. Air Force was not established until after the war; Air Forces refers to the aviation branches of the Army, Navy, and Marine Corps. See [Shell, forthcomming].

[17] The history of the development of linear programming has been reported several places. See e.g. [Dantzig, 1982, 1991], [Lenstra et al., 1991], [Grattan-Guiness, 1970, 1994]

[18] [Dantzig, 1951, p. 18].

subject to mathematical theorizing. It turned into the subject of applied mathematics called linear programming and expanded from there to the broader field of mathematical programming. This will be discussed in Sect. 4

3 The Struggle to Include Operations Research in OSRD

Operations research (OR) and its significance during the war are often mentioned as one of the great successes of the AMP. OR was "invented" in Britain in connection with responses to the growing German air force. In 1936 the British Army and Air Force established a joint research center – the Bawdsey Research Station – for the development of radar as a tool for air defense.[19] In July 1938 the British completed a test which revealed that while radar was technically an effcient tool in air defense, there was nevertheless an important operational problem:

> the Superintendent of Bawdsey Research Station, A. P. Rowe, announced that although the exercise had again demonstrated the technical feasibility of the radar system for detecting aircraft, its operational achievements fell far short of requirements. He, therefore, proposed that research into the operational – as opposed to the technical – aspects of the system should begin immediately.[20]

– and this launched one of the first operations research groups in England.

One of the characteristics of the operations research groups was their mixed composition of expertise. They consisted of representatives from various scientific fields like mathematics, chemistry, biology, and physics. Also different kinds of engineers joined these groups. The job of OR groups during the war was not to invent new kind of weapons but to analyze what went on in the field and suggest ways to optimize the use of existing military equipment. The first groups in England focused primarily on the use of radar equipment in air defense and antisubmarine warfare. By the summer of 1941 their accomplishments were generally valued and accepted by the Royal Air Force (RAF) who by then set up operations research groups to be spread out over the various units of the RAF. Eventually these groups were also implemented in the British army and navy and their work expanded to include strategic and logistic planning as well.[21]

In the course of the war, operations research was imported into the USA and – eventually – incorporated in OSRD in October 1943 through the formation of Offce of Field Service. Even today members of American operations research groups tend to emphasize Vannevar Bush and OSRD as central agents in bringing about operations research in the USA. Contrary to this flattering picture painted by some of these members themselves, Eric P. Rau suggests in his paper *The Adoption*

[19] See [Larnder, 1979, pp. 4–5] and [Rosenhead, 1989, p. 89].

[20] [Lardner, 1979, p. 8].

[21] [Fortun and Schweber, 1993, p. 602], [Rau, 2000, p. 59].

of Operations Research in the United States During World War II that the implementation of operations research groups in the USA military actually happened not because of Bush and OSRD but rather in spite of him and his organization.[22] From the beginning of 1941 Bush already knew for certain about the significance of the English operations research groups. NDRC had by then established an offce in London with people who had direct contact to the English operations research groups from which they reported back to Bush.[23] Two of NDRC's own people, the Princeton physicist H. P. Robertson and J. E. Burchard, an architect from MIT, brought back to the USA information about the data on blast damage collected by an English OR group. Robertson strongly urged that similar kinds of work should be started in the USA.[24] Bush however did nothing to introduce OR groups into the American military services. Rau convincingly demonstrates the resistance put forward by Bush against the introduction of OR groups in the OSRD. According to Rau the nature of OR did not fit into Bush's very carefully constructed system for scientific mobilization.

The conflict was due to "incompatible strategies for organizing research and development for the war effort."[25] Bush had very carefully structured the OSRD in such a way that scientists were shielded from governmental influence. Scientists were not under military command; they were under the leadership of civilians independent of the government. The reason for this was of course to protect the civilian research institutions against future governmental interference. The consequence, in effect, was the creation of a gap between the scientists who developed the new types of weapons and the military people who were going to use these inventions. In Britain OR originated as a conscious attempt to bridge this gap, but the incorporation of OR in OSRD would necessarily break down the boundary between users and developers, which would shake the very foundation of Bush's organization.

Within the first year after the USA entered the war, the military services had already begun to establish their own OR groups and OSRD began to receive requests for scientific personnel to participate in OR groups. Bush held the opinion that it was up to the armed forces themselves to administer and recruit people for the OR groups they wanted.[26] As it turned out the recruitment was a problem, and Bush's own leaders within the various sections under OSRD began to oppose Bush on the OR-issue. For example the physicist John T. Tate who chaired the NDRC section for sub-surface warfare supported Captain Baker's formation of the Antisubmarine Warfare Operations Research Group (ASWORG) through a contract with Columbia University in New York. ASWORG became one of the best known OR groups. It was headed by Philip Morse, a physicist from MIT, who became very instrumental in the establishing of OR as an academic discipline

[22] See [Rau, 2000].

[23] [Rau, 2000, p. 62], [McCloskey, 1987, p. 911], [Fortun and Schweber, 1993, p. 603].

[24] [Rau, 2000, p. 62], [Fortun and Schweber, 1993, p. 603].

[25] [Rau, 2000, p. 57].

[26] [Rau, 2000, p. 70].

in the universities in the USA in the immediate post-war period.[27] Also Warren Weaver disagreed with Bush on the question of OR, and along with other NDRC sections the AMP began to develop their own programs for educating scientists to become future OR personnel. In the end Bush relented and in October 1943 the Offce of Field Service (OFS) was created as a section under OSRD. In the USA operations research seems to have been closely connected to mathematics during the war. Half of the people in Morse's ASWORG group were mathematicians and the AMP's program for educating OR personnel is counted as one of the Applied Mathematics Panel's great successes.[28] According to the mathematician Barkley Rosser OR was regarded as mathematics in the military:

> The Air Force Generals and Navy Admirals thought it [operations research] was wonderful stuff. You could not have convinced one of them that it was not mathematics.[29]

But it was not only military people who linked OR with mathematics. As we shall see in Sect. 5, Philip Morse also thought of it as a new branch of applied mathematics.

4 The Post-War Organization of Science

The end of the war also meant the end of OSRD. There was a common concern that the vitality and flourishing of wartime research would dissolve in the post-war period. The scientists would go back to the kind of work they did before the war with the consequence that the research cooperation within the military-university-industry complex, which had proved itself so productive during the war, would simply disappear. Not surprisingly there was a shared belief that the USA had to be strong scientifically in order to be strong militarily.[30] The secretary of the Navy, James V. Forrestal, brought up the question of how to continue the cooperation between the universities and the military in peacetime in his annual report to President Truman in 1945:

> In peace, even more than in war, scientists owe to their nation an obligation to contribute to its security by carrying on research in military fields. The problem which began to emerge during the 1944 fiscal year is how to establish channels through which scientists can discharge this obligation in peace as successfully as they have during the war [...] The Navy believes that solution to this problem is the establishment by law of an independent agency devoted to longterm, basic military research, securing its own founds from Congress and responsive to, but not dominated by, the Army and Navy [...] The Navy so firmly believes in the importance of this solution to the future welfare of the

[27] On the formation of ASWORG see [McCloskey, 1987], [Miser, 1986], [Morse, 1986].

[28] [Rees, 1980, p. 617].

[29] [Rosser, 1982, p. 510].

[30] [Rees, 1977a, p. 104].

> country that advocacy of it will become settled Navy policy [...] The Navy feels so deeply about the importance of the solution of this problem, that it requests your intervention, guidance and support on this problem, which transcends the responsibility and authority of any single department.[31]

Also Bush prepared a strategy for the organization of post-war research and education. His proposal *Science: The Endless Frontier* was delivered to President Truman in 1945. Basically it contains a plan for government support of science organized in the spirit of OSRD – through a National Science Foundation – based on contracts to secure the independence of the universities and the industry.[32] A main point in Bush's report was to stress the significance of basic – not necessarily military – research:

> basic research leads to new knowledge. It provides scientific capital. It creates the fund from which the practical applications of knowledge must be drawn. [...] today it is truer than ever that basic research is the pacemaker of technological progress. [...] A nation which depends upon others for its new basic scientific knowledge will be slow in its industrial progress and weak in its competition in world trade, regardless of its mechanical skill.[33]

The National Science Foundation was not established until 1950 and in the meantime the military services initiated different channels for supporting scientific research. There were two primary places where the new mathematical techniques that emerged during the war became the subject of military funded basic research, Project RAND and the Offce of Naval Research (ONR).

RAND and game theory

Game theory was not "invented" during the war, and it doesn't seem to have played any significant role in the mathematics used and developed for the war effort. However, it benefitted tremendously from the importance attached to mathematics and operations research after the war.[34]

In the 1920s von Neumann developed a theory for two-person zero-sum games. His theory was published in 1928 in the paper *Zur Theorie der Gesellschaftsspiele* and – as the title indicates – the object under consideration at that point was parlor games. On the first page of the paper von Neumann pointed towards the possibility of a much wider scope for game theory because – as he phrased it – under the right circumstances "any event [...] may be regarded as a game of strategy if one looks at the effect it has on the participants."[35] In a footnote he characterized the game

[31] Quoted in [Rees, 1977a, p. 104].

[32] [Dupree, 1986, p. 213].

[33] Quoted in [Schweber, 1988, p. 14].

[34] According to the historian Robert J. Leonard ASWORG should have used game-theoretic analysis in two applications; see [Leonard, 1992, p. 65].

problem as "the principal problem of classical economics: how is the absolutely selfish "homo economicus" going to act under given external circumstances?"[36] The theory was built up on the concept of strategy. Von Neumann discussed at length what should be understood by a "best" or an "optimal" strategy. The main result in the paper is the so-called minimax theorem, which proves the existence of optimal strategies to every finite zero-sum two-person game. For such games a value V can be assigned, which is the average gain that one player can expect to win from the other player, regardless of what strategy the opponent chooses.[37]

The really significant development in game theory took place during the war when von Neumann and another refugee, the Austrian economist Oskar Morgenstern, co-authored the book *Theory of Games and Economic Behavior.*[38] The book was written with economists in mind. Morgenstern, and von Neumann explicitly stated so in the opening phrase of the book and in Morgenstern's diary one can follow his attempts to bring the theory to the attention of economists. Comments in Morgenstern's diary show that the economists were at best just not interested in game theory even though they expressed a hostile attitude in the first decade after the book was printed.[39] Even though the book was written and parts of the theory were developed in the mist of the war, it seems that they did not think of war applications in the process. There are no references to possible war applications in Morgenstern's diaries. Yet game theory became the main subject of mathematical research at the RAND Corporation, the nest of military support of science. Why was that?

According to the historian Philip Mirowski, the disregard shown by economists brought von Neumann to search for another "home" for game theory. Given the time, the place, and the concept of optimal strategies for winning a game, which fitted perfectly into the war context, and given von Neumann's multiple connections, reputation, and influence within the military-science complex during the war, the military context was an obvious choice.[40] Project RAND in Santa Monica, California became the most important home for game theory. This project originated in March 1946 by the initiative of Army Air Force Chief of Staff Henry H. "Hap" Arnold and Donald Douglas, the president of Douglas Aircraft. In the beginning the project functioned as a subsidiary of Douglas Aircraft but in 1948 Project RAND became a free-standing nonprofit corporation, a so-called "think-tank".[41]

In the first decade after the war RAND was the center for mathematical research in game theory. The first mathematicians working there were recruited

[35] [von Neumann, 1928, (1959, p. 13)].

[36] [von Neumann, 1928, (1959, p. 13, footnote no. 2)].

[37] For a historical account of von Neumann's conception of the minimax theorem see [Kjeldsen, 2001]. For the history of game theory see also [Weintraub, 1992].

[38] [von Neumann and Morgenstern, 1944].

[39] [Mirowski, 1991, p. 239].

[40] [Mirowski, 1991].

[41] See [Hourshell, 1997, pp. 241–242], [Smith, 1969].

mainly from the Applied Mathematics Panel. It counted people like John Williams, Morris Girschich, Olaf Helmer, and Ed Paxton among its theorists. John von Neumann himself was a consultant for RAND as well as Weaver and Wilks, who had been the leader of AMP's Statistical Research Group at Princeton. Wilks also referred his student Ted Harris to RAND. This group at RAND was the first established group of game theorists and they all either came from the war work or had connections to mathematicians who had been involved with OSRD. The group at RAND held lengthy summer sessions in game theory and collaborated with another military financed project – the logistic project – in Princeton which included people like Albert Tucker, John Nash, Martin Shubik, and Harold Kuhn.[42] This project will be discussed below.

Besides the promise that game theory could provide optimal strategies for military problems, there were other connections as well. It turned out that a two-person zero-sum game is mathematically equivalent to the mathematical model of the programming planning problem or – as it soon became known as – a linear programming problem. This connection had a lot of impact on further mathematical research both applied and pure, and again – as will be elaborated in the next section – von Neumann was instrumental in the series of events that started that line of development.

The mathematics program of the Offce of Naval Research

In 1946 the Navy established the Offce of Naval Research (ONR) which in the first four years of its existence became the main sponsor of government supported research in the USA. ONR continued the research practice of OSRD by supporting research done at the universities through contracts with individual scientists.[43]

Mina Rees, who had served as technical aid to Warren Weaver at the AMP, was asked by the ONR to take on the position as head of the mathematics department of ONR and as such she became a very influential figure in the post-war mathematical community in the USA.[44] Mina Rees has reported her recollections regarding her time at ONR in several places, and it seems that she at the outset was skeptical about the idea of military funding of mathematical research in peacetime. She simply didn't think mathematicians would let the military finance their peacetime research. Nevertheless she accepted the position at ONR because – according to her recollections – she found it extremely important for the further development of mathematics in the USA to be actively engaged in the ONR program. Judging from her own writings she was very concerned about whether the Navy would support basic research in mathematics proper without any relevance to the Navy. She didn't want the ONR program to fragment the community of mathematicians who

[42] See [Mirrowski, 1991] and [Leonard, 1992].

[43] For historical accounts on ONR see [Sapolsky, 1979], [Schweber, 1988], [Old, 1961].

[44] For a historical account on Rees and her role in the history of the U.S. government's role in funding research in the mathematical sciences see [Shell, forthcoming]. Also [Williams, 2001] has a chapter on Mina Rees.

feared that applied mathematics would benefit from the wartime at the expense of pure mathematics.[45]

In a note in the Bulletin of the AMS from January 1948 Mina Rees announced "the philosophy which has determined the mathematical research projects which ONR is sponsoring":

> The Offce of Naval Research is committed primarily to the support of fundamental research in the sciences, as contrasted with development, or with applications of known scientific results. [...] It is natural, however, that the most obvious types of mathematical research which would seem to warrant Navy support would be research in applied directions.[46]

This "philosophy" is reflected in the budget where 4/5 of the annual expenditure went to research in applied mathematics and computing devices. But in the note Mina Rees emphasized that basic research in mathematics proper was deemed important by the Navy, and actually more than 1/3 of the contracts dealt with projects with theoretical objectives.

Scientific computation of optimum programs

In the meantime the Air Force had re-employed George Dantzig to continue the work on calculating Air Force programs. The emergence of the computer had a profound influence on this work because it made it possible to build some kind of objective into the programs, such as the possibility of choosing the "best" program among feasible ones with respect to some objective, for example, the least expensive program.[47] The Air Force then established project SCOOP, which stood for Scientific Computation of Optimum Programs. The main persons in the group were Dantzig and Marshall Wood – an expert on programming procedures. Later John Norton and Murray Geisler joined the project. In 1949 Marshall Wood gave the following description of the project:

> Early in 1947 the Air Comptroller's Offce undertook a concerted attack on this problem, establishing the Planning Research Division. [...] The work of this Division, now designated as PROJECT SCOOP [...], was directed to four main problem areas:
> a) The systematic and comprehensive identification and quantitative evaluation of interrelationships among Air Force activities, objectives, and limitations, usually expressed in the form of planning factors;
> b) The development of a system of equations, expressing these interrelationships explicitly in mathematical form;
> c) The development of mathematical computing techniques for the solution of these systems of equations, so as to construct a program which will accom-

[45] See [Albers and Alexanderson, 1985], [Rees, 1977a, 1977b].

[46] [Rees, 1948, p. 1].

[47] [Dantzig, 1991].

plish our objectives to the maximum extent possible within the external limitations of funds, industrial capacity, etc.;

d) The development and construction of high speed electronic computing machines adequate to perform in a few days the computations required for the equations for a complete Air Force program.[48]

Through the military-academic complex this problem got introduced into the world of university mathematics and it appears that the person who pulled the strings was – again – John von Neumann who served both as a consultant for the various military establishments and was on the advisory board for ONR.

An ONR project in linear programming and game theory

The Air Force programming group had built a mathematical model for the programming problem which resulted in the mathematical problem of optimizing a linear function subject to linear inequality constraints.[49] Dantzig was encouraged to seek the advice of von Neumann on the problem of finding an algorithm that could solve such a programming problem, and in the fall of 1947 Dantzig introduced von Neumann to the Air Force programming problem.[50] In the book *Theory of Games and Economic Behavior* von Neumann had based the theory of two-person zero-sum games on the theory of convexity and linear inequalities, and – according to Dantzig – von Neumann immediately suggested that a two person zero sum game is equivalent to a programming problem.[51]

A month later von Neumann circulated the note *Discussion of a Maximum Problem* in which he transformed a problem of maximizing a linear function constrained by linear inequalities into the problem of finding a solution to a system of linear inequalities.[52] This shows that von Neumann was interested in the mathematics underneath the programming problem. A couple of months later – that is in the spring of 1948 – Dantzig revisited von Neumann at the Institute for Advanced Study to discuss the possibility of setting up a university-based project to study further the linear programming problem, game theory, and the underlying mathematical structure.[53]

Mina Rees has described the occurrence of this university-based project as follows:

> [...] when, in the late 1940's the staff of our offce became aware that some mathematical results obtained by George Dantzig, who was then working for the Air Force, could be used by the Navy to reduce the burdensome costs

[48] Quoted in [Brentjes, Ph.D. thesis, p. 177].

[49] [Dantzig, 1949]

[50] [Dantzig, 1982, 1988].

[51] [Dantzig, 1982, p. 459].

[52] [von Neumann, 1947]

[53] [Albers and Alexanderson, 1985, pp. 342–343].

> of their logistics operations, the possibilities were pointed out to the Deputy Chief of Naval Operations for Logistics. His enthusiasm for the possibilities presented by these results was so great that he called together all those senior offcers who had anything to do with logistics, as well as their civilian counterparts, to hear what we always referred to as a "presentation". The outcome of this meeting was the establishment in the Offce of Naval Research of a separate Logistics Branch with a separate research program. This has proved to be a most successful activity of the Mathematics Division of ONR, both in its usefulness to the Navy, and in its impact on industry and the universities.[54]

The project began in the summer of 1948 as a trial project with Albert W. Tucker, a mathematician from Princeton, as the principal investigator. Together with two students, Harold W. Kuhn and David Gale, Tucker read von Neumann and Morgenstern's book on game theory and studied von Neumann's note on a maximum problem.[55] Their work resulted in the paper *Linear Programming and the Theory of Games* which they presented at the first conference on linear programming held in Chicago in the summer of 1949.[56]

The project continued with support from ONR until 1972[57] and it gave rise to a substantial amount of research both in what traditionally will be counted as applied as well as pure mathematics. This project is a prime example of the effect of the military-university cooperation formed by the post-war organization of science support. By initiating research on the programming problem and game theory by university mathematicians usually engaged in pure mathematical research, the field expanded, and basic research in mathematics proper – pure and applied – was the result.

In their first work on linear programming and game theory, Kuhn, Tucker, and Gale proved the main theoretical theorem in linear programming, the so-called duality theorem. This result and its connection to the important minimax theorem in two-person zero-sum games were interesting from a mathematical point of view and changed the scientific status of linear programming.[58] Its connection to game theory, to the theory of systems of linear inequalities, and to convex analysis broadened the field and made linear programming an interesting potential mathematical research area. As I have argued elsewhere, I find that this change in scientific status was very important for the further development of the theory of mathematical programming.[59] Until Tucker and his group got involved, the driving force behind the development had been from a practical point of view, the solving of the Air Force programming problem. Tucker, Kuhn, and Gale's work on the other hand was done within a university context of theoretical mathemati-

[54] [Rees, 1977a, p. 111].

[55] Personal interview with H. W. Kuhn, Princeton, April 23, 1998.

[56] [Gale et al., 1951].

[57] Personal interview with H. W. Kuhn, Princeton, April 23, 1998.

[58] For an analysis of how this changed the scientific status of linear programming see [Kjeldsen, 1999, 2000b].

[59] See [Kjeldsen, 1999, 2000b].

cal research. Kuhn and Tucker got stuck on the project and they began to work on an extension of the duality theorem to the more general nonlinear case. They published their joint work in 1950 in the paper *Nonlinear Programming* – a classic in mathematical programming – that launched the theory of nonlinear programming.[60] Kuhn and Tucker did not prove a duality result for nonlinear programming but they did prove the so-called Kuhn-Tucker theorem about necessary conditions for a point x_0 to be an optimal solution to a nonlinear programming problem.[61] The different circumstances under which linear and nonlinear programming originated is a remarkable consequence of the military-university cooperation. In contradiction to what was the case with linear programming, there was no direct external, applicational, motivation for the development of nonlinear programming. The subsequent practical usefulness of nonlinear programming notwithstanding, it originated as an interest in pursuing a purely mathematical issue of generalization and understanding.

The work on game theory within Tucker's group can be followed in a series of papers published in several volumes of the *Annals of Mathematics Studies* under the title *Contributions to the Theory of Games*.[62] The cooperation between RAND and the logistic project under ONR is reflected in these volumes, since approximately half of the papers in each volume were done under contract with the ONR while the other half were written by RAND people.

Research in pure mathematics

Subjects of more pure character were also pursued within the military context. Theodore Motzkin's thesis *Beiträge zur Theorie der linearen Ungleichungen*[63] was translated into English both at RAND by D. R. Fulkerson and in the logistics project by S. Bargmann.[64] In 1956 a series of eighteen papers exploring "various aspects of one mathematical theme, the theory of linear inequalities" was published, and again part of the research was supported by ONR and part of it was done at RAND. As is explicitly stated in the introduction the research was motivated by the developments of game theory and linear programming.[65]

The theory of convexity is another field of theoretical mathematics that gained renewed interest as a consequence of the Logistic Research Project financed by ONR. Von Neumann had based the theory of two-person zero-sum games in

60 [Kuhn and Tucker, 1950a].

61 See [Kjeldsen, 2000a] for the argumentation for the claim that the work Kuhn and Tucker published in their *Nonlinear Programming* paper was motivated by a wish to extend the duality result for linear programming.

62 See [Kuhn and Tucker, 1950b], [Kuhn and Tucker, 1953], [Dresher and Tucker, 1957], and [Luce and Tucker, 1959].

63 [Motzkin, 1936].

64 See the introduction to Motzkin's collected works in [Cantor et al., 1983], and [Billera and Lucas, 1978, p. 5].

65 [Kuhn and Tucker, 1956, p. v].

Theory of Games and Economic Behavior on the geometrical theory of convexity. The theory of convex functions also seemed to be a promising tool in the newly emerged field of nonlinear programming. As mentioned above Kuhn and Tucker did not succeed in expanding their duality theorem for linear programming to the general nonlinear case but they were able to prove that if the involved functions are concave (and differentiable) there will be complete equivalence between the nonlinear programming problem and the saddle value problem for the corresponding Lagrangian function, suggesting the possibility of a duality result for concave functions.[66]

One of the experts on the theory of convexity at that time was Werner Fenchel from the University of Copenhagen in Denmark. He happened to be visiting the USA in the academic year of 1950/51 with extended visits first in California at the University of Southern California and Stanford. He ended his US visit in Princeton, first as a member of the Institute for Advanced Study and then as a visiting professor at Princeton University.[67] Tucker invited Fenchel to give a series of lectures on convex sets and functions within the logistic project.[68] The lecture notes from Fenchel's course became a source of inspiration and influenced widely the further research in the theory of convexity. Significant for the theory of mathematical programming was the fact that Fenchel was able to derive a duality result for "a generalized programming problem" as he phrased it.[69] The background for this result, which is often termed Fenchel duality, was a small paper published in 1949 where Fenchel introduced the concept of conjugate convex functions.[70] Fenchel originally introduced this concept with the purpose of examining the mathematical structure underlying various inequalities – like the Hölder inequality – that appears in analysis. In Princeton Fenchel was introduced to the new field of mathematical programming, and by applying his concept of conjugate convex functions on this type of questions Fenchel derived the first duality result in nonlinear programming.[71] Fenchel did not explore this any further, but his lectures and the notes had quite an influence on the following development of the theory of convexity in the USA and in the developing of convex programming. R. T. Rockafellar in particular was very much inspired by Fenchel's Princeton lectures. He used the Fenchel duality to build a duality theory for convex programming based on Fenchel's concept of conjugate convex functions.[72]

66 [Kuhn and Tucker, 1950a], [Kjeldsen, 2000a].

67 [Fuglede, 1989, p. 167].

68 [Fenchel, 1953, Acknowledgement], [Fenchel, Bidrag til de konvekse funktioners teori, BOX 2, Folder: Manuskripter om konvekse mængder og funktioner. Non-dated, but from the years 1953/54.], [Letter from Tucker to Fenchel, June 11, 1951, BOX 1.].

69 [Fenchel, 1953, p. 105].

70 [Fenchel, 1949].

71 [Fenchel, 1953, pp. 105–106].

72 See [Rockafellar, 1970].

5 The Interactions between Operations Research, Mathematical Programming, and Game Theory in the Post-War Period

The discussion in Sect. 3 focused on the success of and importance ascribed to operations research during the war. After the war both the military and the academics who had been involved with operations research during the war were very anxious to secure the field in the post-war period and transform it into an academic discipline. Both game theory and mathematical programming were almost immediately considered as essential items in the toolkit of operations research and, as such, benefited from the importance attached to operations research as a direct consequence of the war work.

The establishment of operations research as an academic discipline

One of the exercises was to "move" operations research into peaceful applications. In 1947 Charles Kittel published a paper about "The Nature and Development of Operations Research" in *Science* where he gave the following definition of operations research:

> Operations Research is a scientific method for providing executive departments with a *quantitative basis for decisions*. Its object is, by the analysis of past operations, to find means of improving the execution of future operations.[73]

and expressed his hope for its future:

> It is hoped that the publication of this paper will serve to stimulate the establishment of operations research groups in the United States for the advancement of peaceful objectives. This powerful new tool should find a place in government and industry. [74]

As we saw in Sect. 3, operations research was popular within the military and the Office of Naval Research supported its move into the universities through the logistic project. According to Fred Rigby, who was ONR's leader of the logistic project, ONR exerted an immense influence on operations research and its development:

> We did indeed influence the introduction of operations research into business schools. The subdiscipline called management science is our invention, in quite a real sense. That is, we and our contract researchers recognized its potentials, planned its early growth, and, as it turned out, set the dominant pattern in which it has developed. [75]

[73] [Kittel, 1947, p. 150].

[74] [Kittel, 1947, p. 153].

[75] Quoted in [Rees, 1977a, p. 111].

Also the National Research Council contributed to the growth of operations research by forming a committee under its section of applied mathematics to "further its [operations research] development and applications outside the armed forces."[76] This was done in practice by the granting of Ph.D. scholarships, financial support to conferences etc.

Philip Morse from MIT was a key person in the shaping of operations research as an academic discipline. Already from 1948 he had two courses running at MIT and he also introduced MIT's summer sessions in operations research, two-week courses for civil servants and people working in industry. From 1952 Johns Hopkins University had a program in operations research and from 1954 it was possible to earn a Ph.D in the field.[77]

Different opinions on the role of mathematics in operations research

Mathematical programming and game theory benefitted from Morse's strong influence on the early establishment of operations research because Morse continued to emphasize the importance of mathematical research for operations research. Not everyone agreed with him on the importance of mathematics. In the journals of the Operations Research Societies of America (ORSA) and Britain respectively as well as in the proceedings from the international conferences there was a continuing debate about what operations research actually was. Morse tried at some point to stop the arguing by simply defining operations research to be what operations research people do:

> We should no longer have trouble explaining the scope and methods of operations research to the layman. We already can say: operations research is the activity carried on by members of the Operational Research Society; its methods are those reported in our journal.[78]

Morse was a persistent advocate for the inclusion of game theory and linear programming in operations research. In 1953 he wrote about *Trends in Operations Research* in ORSA's journal where he pointed towards game theory as a very important tool that should be further developed. Morse wrote that:

> Linear programming is rapidly becoming an important theoretical tool in economics; it deserves equal or greater exploitation in operations research.[79]

And two years later, in 1955, he maintained that "Just as with any other field of science, we are finding that we need our own kind of mathematics"[80] in operations research. How serious this issue was for Morse can be seen from his appeal to a growing generation of operations researchers:

[76] [Fortun and Schweber, 1993, p. 611].
[77] [Morse, 1956], [Roy, 1956].
[78] [Morse, 1953, p. 159].
[79] [Morse, 1953, p. 169].
[80] [Morse, 1955, p. 383].

> But linear programming is only one part of a larger theory of optimal programming, which covers such subjects as dynamic programming, some aspects of search theory and, probably, of game theory. It is hard to visualize, right now, all the mathematical aspects of this broader subject, because they haven't been investigated as yet in any detail. [...]
>
> It is not hard to foresee the considerable usefulness of this general theory in solving many operations problems, particularly those concerned with planning. But a great deal of basic research will be needed before the theory will be able to answer our practical needs. Some of the fundamental mathematics has not yet been developed and a great number of the algorithms for solving specific problems have not yet been worked out. Much of this basic work can probably best be done as a long-term study, not subject to the short-term deadlines and crises which occur in the study of immediate, practical problems. It needs a good many man-years of concentrated work. Operations research needs this sort of research. No branch of science can continue to grow unless its underlying theory is continuously being expanded.[81]

Morse regarded mathematics as not only the theoretical basis of operations research, but actually more than that. In 1948 he characterized operations research as a "new field of applied mathematics ..."[82]

A look at the papers published in *Operations Research* during the first two decades of its existence reveals that mathematical programming, especially linear programming, played an important part in operations research. This is confirmed in the proceedings from the first international meetings and from the content of the early textbooks where mathematical programming is presented.[83]

Not everyone shared the opinion of Morse regarding the role of mathematics in operations research. Many expressed their discontent with the importance attached to mathematics. Already in 1953 Norman Hitchman warned against this tendency in ORSA's journal:

> One main caution seems to stand out glaringly. It concerns the emphasis which we place on certain specialized fields as to their value in the operations research team. It is not an uncommon observation of today to note the very great emphasis given to mathematical and physical sciences. Our new society, ORSA, for example, is playing a preeminent part in creating the impression that mathematics and physics are almost synonymous with operations research itself.[84]

In 1956 W. N. Jessop warned against "the placing of emphasis on mathematical methods and on highly abstract treatments of general situations" also in ORSA's journal. When it comes to linear programming Jessop also held the opinion that

[81] [Morse, 1955, p. 383].

[82] [Morse, 1948, p. 621].

[83] See [Banbury and Mailand, 1961], [Dunod, 1963], [Hertz and Melese, 1966], [Churchman et al., 1957], [Sasiene et al., 1959], and [Hillier and Lieberman, 1974].

[84] [Hitchman, 1953, p. 242].

there was too much focus on the developing of theory "a subject [linear programming] so delightful to the pure mathematician that many papers appear to have had their origin in sheer exuberance unsullied by any thought of a factual situation."[85]

Despite the quarrels linear programming became - and still is - a widely used tool in operations research. The simplex algorithm developed by Dantzig for solving linear programs was and is used in many sorts of branches of industry.

The usefulness of game theory in operations research

Game theory was also imbedded in operations research, and again John von Neumann seems to have played a crucial role. During the war von Neumann functioned - and was often used - as a consultant for Morse's operations research group (ASWORG) and according to Fortun and Schweber, he introduced game theory into operations research:

> By the end of the war the new game theoretic methods that had been developed by von Neumann and Morgenstern were added to the toolkit and mathematical techniques that operations research scientists deployed. These proved very valuable, and game theoretic approaches took on great importance after the war.[86]

Morse considered game theory an important tool in operations research, just as he advocated mathematical programming. The first courses in operations research at MIT also included game theory,[87] and that was also the case in operations research education at Johns Hopkins.[88] An analysis of the content of the early textbooks as well as a comparison of bibliographies of game theory and operations research reveals - as phrased by R. Rider - that operations research and game theory "shared much territory".[89] There were, however, criticisms of the value of game theory. In her paper *Operations Research and Game Theory: Early Connections* Rider questions the usefulness of game theory in practical problem solving in operations research. Churchman, Ackoff, and Arnoff wrote in their textbook from 1957 in a conclusion to their chapter on game theory that:

> Very little has been accomplished by way of applying the theory.[90]

The same conclusion was reached more than twenty years later by George R. Lindsey in his talk *Looking Back over the Development and Progress of Operational Research* held at the Eighth International Conference in Operations Research

[85] Quoted from [Rider, 1992, p. 231 and 234].

[86] [Fortun and Schweber, 1993, p. 604].

[87] [Rider, 1992].

[88] [Roy, 1956].

[89] [Rider, 1992, p. 227].

[90] [Churchman et al, 1957, p. 557].

where he gave the following evaluation of the significance of game theory in operations research:

> Direct application of the theory of games to the solution of real problems has been rare, and its chief uses have been to offer some insight and understanding into the problems of competition (without actually solving them), and to provide mathematicians with new fields to conquer. Many important real problems involve more than two opponents, are not zero-sum, and exceed the bounds of the most developed versions of game theory.[91]

6 Conclusion

Mathematical programming and operations research have their origin directly in scientific work for the war effort. Their continued growth and final establishment as scientific disciplines in American universities in the post-war period developed as a consequence of the military financing of science after the war. Much of this work was done in projects directly financed and established by military means like the Project RAND and the ONR's logistic project. One of the major journals – the *Naval Research Logistics Quarterly* – was also financed by the military.

The construction of the post-war organization of science with the military-university cooperation inherited from the war allowed mathematics of a more pure character to benefit from the military science support and to experience a growth in research. The theory of convexity and the theory of linear inequalities are two examples of that.

Mina Rees, the head of the mathematics department of the ONR, originally did not think that mathematicians would let the military pay for their peacetime research, but she was indeed very successful in the way she constructed and managed the mathematical department of ONR. When asked, Kuhn expressed the opinion that the military had no direct influence on the mathematical work he did within the logistic project. The military did not demand him to work on the solution of concrete problems. The work he did was not classified but published in journals, proceedings, and books. The financial support from ONR was spent on summer salaries, conferences, guests, and traveling expenses. Kuhn characterizes the support from ONR as just an improvement of the working conditions at the universities. In the day to day work it made no difference where the money was coming from. He was free to investigate the aspects of mathematical programming and game theory that he found theoretically interesting, regardless of potential applications.[92] Actually in the first four years of its existence the ONR supported promising scientific research projects without regard to direct applicational value or possible military applications. Instead the support from ONR was justi-

[91] [Lindsey, 1979].

[92] Personal interview with Kuhn, Princeton, April 23, 1998.

fied within the scientific discipline itself and often there were no explicit strings attached to the support.[93] That changed gradually after 1950.

But the logistic project at Princeton was not – at the outset – initiated by Tucker but by the Armed Forces. Even though the mathematicians in the group did not work on explicitly formulated problems given by the military, the theoretical mathematical research questions they pursued within the project had to belong to or be justified by issues in the field of mathematical programming and game theory. By giving money to promote research in certain areas, the military had a huge influence on the growth and direction of mathematical research after the war.

One may also wonder whether the selection of theorems the mathematicians chose to prove was influenced by the military context. According to Philip Mirrowski that was the case with game theory. In his paper *When Games Grow Deadly Serious* he argues that the military connections went all the way into the core of game theory and shaped its content and he claims that "a social-constructivist history of game theory is not only possible; it is the only plausible candidate."[94] Reading Mirrowski one gets the impression that since the economists were not interested in game theory, von Neumann turned to the military context and actively placed game theory at the doorstep of RAND.[95] The military connection created an artificial focus on – among other things – the minimax solution and made it the ultimate solution concept in game theory,[96] which makes sense in a military context of "capitalists versus communists, USA versus USSR."[97] Mirrowski concludes that the military influence hindered the theory in its "proper" development, that it postponed, for example, the realization that the minimax solution was not *the* solution concept but only one of many possible solution concepts. John Nash, who worked as a consultant at RAND in the early 1950's, published his Equilibrium Point Theorem (which is the central solution concept in game theory) in 1950, that is quite early in the development of game theory, indicating that the restraint was not too damaging. The military context might have influenced the content of game theory, but there is another level, I think, namely the question of what can be proven – given the axioms and the logical rules – within such a mathematical system, and that is indifferent to the context.[98] Eventually game theory moved into economics where it is very much alive today.

From the discussion in Sect. 5 it follows that game theory was not able to meet its promises as a useful tool in operations research. Also the mathematical programming people felt a need to have their own home. In 1970 they began publishing their own journal *Mathematical Programming* and two years later they organized the Mathematical Programming Society. During the 1970s mathematical programming became established with a professional society, journal, prizes, and a growing number of monographs and textbooks.

[93] [Sapolsky, 1979, p. 386], [Schweber, 1988, p. 8].

[94] [Mirrowski, 1991, p. 233].

[95] [Mirrowski, 1991, p. 240–242].

[96] [Dresher, 1961, p. 79], [Mirrowski, 1991, p. 244].

[97] [Mirrowski, 1991, p. 246].

[98] The context might influence the choice of the axioms and maybe also the logical rules.

This story also demonstrates the influence of individual scientists like John von Neumann and Philip Morse. von Neumann was involved with military work from 1937 when he became a consultant to the Ballistics Research Laboratory of the Army Ordinance Department.[99] From then on his connections with the military expanded rapidly, and he died as one of the most powerful scientists in the USA.[100]

References

Albers, D. J. & Alexanderson, G. L. (1985) (eds.): *Mathematical People, Profiles and Interviews*. Boston: Birkhäuser 1985.

Banbury, J. & Maitland, J. (1961) (eds.): *Proceedings of the 2th International Conference of Operational Research*. Paris: Dunod 1961.

Billera, L. J. & Lucas, W. F. (1978): Delbert Ray Fulkerson: August 14, 1924 – January 10, 1976. In: M. L. Balinski & A. J. Hoffman (eds.): *Polyhedral Combinatorics: Dedicated to the memory of D. R. Fulkerson*. Mathematical Programming Study **8**, 1978, pp. 1–16.

Brentjes, S. (1976): Der Beitrag der sowjetischen Wissenschaftler zur Entwicklungen der Theorie der linearen Optimierung. *NTM-Schriftenreihe zur Geschichte, der Naturwissenschaften, Technik und Medizin*, **13**, 1976, pp. 105–110.

Brentjes, S. (Ph.D.): *Untersuchungen zur Geschichte der linearen Optimierung (LO) von ihren Anfängen bis zur Konstituierung als selbständige mathematische Theorie – eine Studie zum Problem der Entstehung mathematischer Disziplinen im 20. Jahrhundert*. Ph.D dissertation, Leipzig, DDR.

Cantor, D., Gordon, B., & Rothschild, B. (1983) (eds.): *Theodore S. Motzkin: Selected Papers*. Boston: Birkhäuser 1983.

Charnes, A. & Cooper, W. W. (1961): On some Works of Kantorovich, Koopmans and Others. *Management Science* **8**, 1961, pp. 246–263.

Churchman, C. W., Ackoff, R. L. & Arnoff, E. L. (1957): *Introduction to Operations Research*. New York: John Wiley and Sons, Inc. 1957.

Dantzig, G. B. (1949): Programming in a Linear Structure. *Econometrica* **17**, 1949, pp. 73–74.

Dantzig, G. B. (1951): Linear Programming. In: *National Bureau of Standards, NBS. Applied mathematics series* **15**, Washington, D.C. 1951, pp. 18–21.

Dantzig, G. B. (1982): Reminiscences about the Origins of Linear Programming. *Operations Research Letters* **1**, 1982, pp. 43–48.

Dantzig, G. B. (1991): Linear Programming. In: Jan Karel Lenstra, Alexander H. G. Rinnooy Kan & Alexander Schrijver (eds.): *History of Mathematical Programming, A Collection of Personal Reminiscences*. Amsterdam: North-Holland 1991, pp. 19–31.

Dresher, M. (1961): *Games of Strategy*. Englewood Cliffs, N. J.: Prentice-Hall 1961.

[99] [von Neumann, 1963, p. 499]. See Collected Works. volume 7. Oxford: Pergamon Press.

[100] [Heims, 1980, p. 275]. Besides being involved with mathematics during the war he was also involved with the Manhattan Project.

Dresher, M. & Tucker, A. W. (1957) (eds.): *Contributions to the Theory of Games, Vol. III.* Princeton, New Jersey: Princeton University Press 1957.

Dunod (1963) (ed.): *Proceedings of the 3rd International Conference of Operational Research.* London: English University Press Ldt. 1963.

Dupree, A. H. (1986): National Security and the Post-War Science Establishment in the United States. *Nature* **323**, 1986, pp. 213–216.

Fenchel, W. (1949): On Conjugate Convex Functions. *Canadian Journal of Mathematics* **1**, 1949, pp. 73–77.

Fenchel, W. (1953): *Convex Cones, Sets, and Functions.* Lecture Notes, Department of Mathematics, Princeton University 1953.

Fortun, M. & Schweber, S. S. (1993): Scientists and the Legacy of World War II: The Case of Operations Research (OR). *Social Studies of Science* **23**, 1993, pp. 595–642.

Fuglede, B. (1989): Werner Fenchel. *Det Danske Videnskabernes Selskab, Oversigt over Selskabets Virksomhed, 1988–1989.* København: Munksgaard 1989, pp. 163–171.

Gale, D., Kuhn, H. W. & Tucker, A. W. (1951): Linear Programming and the Theory of Games. In: T. C. Koopmans (ed.): *Activity Analysis of Production and Allocation.* Cowles Commission Monograph **13**, New York: Wiley 1951, pp. 317–329.

Grattan-Guinness I. (1970): Joseph Fourier's Anticipation of Linear Programming. *Operational Research Quaterly* **21**, 1970, pp. 361–364.

Grattan-Guinness, I. (1994): "A New Type of Question": On the Prehistory of Linear and Non-Linear Programming, 1770–1940. In: E. Knobloch & D. Rowe (eds.): *The History of Modern Mathematics. III.* Boston: Academic Press 1994, pp. 43–89.

Heims, S. J. (1980): *John von Neumann and Norbert Wiener.* Cambridge, Massachusetts: The MIT Press 1980.

Hertz, D. B. & Melese, J. (1966) (eds.): *Proceedings of the 4th International Conference of Operational Research.* New York: Wiley-Interscience 1966.

Hillier, F. S. & Lieberman, G. J. (1974): *Operations Research.* San Francisco: Holden-Day, Inc. 1974.

Hitchman, N. (1953): What is the Mission of Operations Research? *Journal of the Operations Research Society of America* **1**, 1953, pp. 242–243.

Hourshell, D. (1997): The Cold War, RAND, and the generation of knowledge, 1946–1962. *Historical Studies in the Physical and Biological Sciences* **27**, 1997, pp. 237–267.

Kittel, C. (1947): The Nature and Development of Operations Research. *Science* **105**, 1947, pp. 150–153.

Kjeldsen, T. H. (1999): A Contextualized Mathematico-historical Analysis of Nonlinear Programming: Development and Multiple Discovery. (in Danish), *IMFUFA*, Text 372, Roskilde University 1999.

Kjeldsen, T. H. (2000a): A Contextualized Historical Analysis of the Kuhn-Tucker Theorem in Nonlinear Programming: The Impact of World War II. *Historia Mathematica* **27**, 2000, pp. 331–361.

Kjeldsen, T. H. (2000b): The Emergence of Nonlinear Programming: Interactions between Practical Mathematics and Mathematics Proper. *The Mathematical Intelligencer* **22**, Summer 2000, 3, pp. 50–54.

Kjeldsen, T. H. (2001): John von Neumann's Conception of the Minimax Theorem: A Journey Through Different Mathematical Contexts. *Archive for History of Exact Sciences* **56**, 2001, pp. 39–68.

Kuhn, H. W. & Tucker, A. W. (1950a): Nonlinear Programming. In: J. Neyman (ed.): *Proceedings of the Second Berkeley Symposium on Mathematical Statistics and Probability.* Berkeley 1950, pp. 481–492.

Kuhn, H. W. & Tucker, A. W. (1950b) (eds.): *Contributions to the Theory of Games, Vol. I.* Princeton, New Jersey: Princeton University Press 1950.

Kuhn, H. W. & Tucker, A. W. (1953) (eds.): *Contributions to the Theory of Games, Vol. II.* Princeton, New Jersey: Princeton University Press 1953.

Kuhn, H. W. & Tucker, A. W. (1956) (eds.): *Linear Inequalities and Related Systems.* Annals of Mathematics Studies **38**, Princeton, New Jersey: Princeton University Press 1956.

Lardner, H. (1979): The Origin of Operational Research. In: K. B. Haley (ed.): *Operations Research '78.* Proceedings of the Eighth IFORS International Conference on Operational Research, Toronto, Canada. Amsterdam: North-Holland Publishing Company 1979, pp. 3–12.

Lenstra, J. K., Rinnooy Kan, A., & Schrijver, A. (1991) (eds.): *History of Mathematical Programming, A Collection of Personal Reminiscences.* Amsterdam: North-Holland 1991.

Leonard, R. J. (1992): Creating a Context for Game Theory. In: E. Roy Weintraub (ed.): *Towards a History of Game Theory.* Durham and London: Duke University Press 1992.

Lindsey, G. R. (1979): Looking back over the Development and Progress of Operational Research. In: K. B. Haley (ed.): *Operational Research '78.* Amsterdam: North-Holland Publishing Company 1979, pp. 13–31.

Luce, R. D. & Tucker, A. W. (1959) (eds.): *Contributions to the Theory of Games, Vol. IV.* Princeton, New Jersey: Princeton University Press 1959.

McCloskey, J. F. (1987): U.S. Operations Research in World War II. *Operations Research* **35**, 1987, pp. 910–925.

Mirowski, P. (1991): When Games Grow Deadly Serious: The Military Influence on the Evolution of Game Theory. In: D. G. Goodwin (ed.): *Economics and National Security,* Annual Supplement to Volume 23, History of Political Economy. Durham and London: Duke University Press 1991, pp. 227–255.

Mirowski, P. (1992): What Were von Neumann and Morgenstern Trying to Accomplish? In: E. Roy Weintraub (ed.): *Toward a History of Game Theory.* Durham and London: Duke University Press 1992, pp. 113–147.

Miser, H. J. (1986): Introductory Note. *Operations Research* **34**, 1986, p. 10.

Morse, M. (1941): Mathematics in the Defense Program. *American Mathematical Monthly,* **48**, 1941, pp. 293–302.

Morse, P. M. (1948): Mathematical Problems in Operations Research. *Bulletin of the American Mathemtical Society* **54**, 1948, pp. 602–621.

Morse, P. M. (1953): Trends in Operations Research. *Operations Research* **1**, 1953, pp. 159–165.

Morse, P. M. (1955): Where is the New Blood? *Journal of the Operations Research Society of America* **3**, 1955, pp. 383–387.

Morse, P. M. (1956): Training in Operations Research at the Massachusetts Institute of Technology. *Operations Research* **4**, 1956, pp. 733–735.

Morse, P. M. (1986): The Beginnings of Operations Research in the United States. *Operations Research* **34**, 1986, pp. 10–17.

Motzkin, T. (1936): *Theorie der linearen Ungleichungen.* Jerusalem: Azriel 1936.

Old, B. S. (1961): The Evolution of the Offce of Naval Research. *Physics Today* **14**, 1961, pp. 30–35.

Owens, L. (1989): Mathematicians at War: Warren Weaver and the Applied Mathematics Panel, 1942–1945. In: D. E. Rowe & J. McCleary (eds.): *The History of Modern Mathematics, vol.II: Institutions and Applications.* San Diego: Academic Press, Inc. 1989, pp. 287–305.

Rau, E. (2000): The Adoption of Operations Research in the United States during World War II. In: A. C. Hughes & T. P. Hughes (eds.): *Systems, Experts, and Computers,* Dibner Series, Jed Z. Buchwald (ed.), Cambridge, MA: MIT Press 2000.

Rees, M. S. (1948): The Mathematics Program of the Offce of Naval Research. *Bulletin of the American Mathematical Society,* 1948, pp. 1–5.

Rees, M. S. (1977a): Mathematics and the Government: The Post-War Years as Augury of the Future. In: D. Tarwater (eds.): *The Bicentennial Tribute to American Mathematics, 1776–1976.* The American Association of America, Buffalo, NY 1977, pp. 101–116.

Rees, M. S. (1977b): Early Years of the Mathematics Program at ONR. *Naval Research Reviews* **30**, 1977, pp. 22–29.

Rees, M. S. (1980): The Mathematical Sciences and World War II. *Mathematical Monthly* **87**, 1987, 6, pp. 607–621.

Rider, R. (1992): Operations Research and Game Theory: Early Connections. In: E. Roy Weintraub (ed.): *Towards a History of Game Theory.* Durham and London: Duke University Press 1992, pp. 225–240.

Rockafellar, R. T. (1970): *Convex Analysis.* Princeton, New Jersey: Princeton University Press 1970.

Rosenhead, J. (1989): Operational Research on the Crossroads: Cecil Gordon and the Development of Post-war OR. *Journal of the Operational Research Society* **40**, 1989, pp. 3–28.

Rosser, J. B. (1982): Mathematics and Mathematicians in World War II. *Notices of the American Mathematical Society* **29**, 1982, pp. 509–515.

Roy, R. H. (1956): Operations Research at the Johns Hopkins University. *Operations Research* **4**, 1956, pp. 735–738.

Sapolsky, H. M. (1979): Academic Science and the Military: The Years Since the Second World War. In: N. Reingold (ed.): *The Sciences in the American Context: New Perspectives.* Washington, D. C.: Smithsonian Institution Press 1979, pp. 379–399.

Sasieni, M., Yaspan, A. & Friedman, L. (1959): *Operations Research, methods and problems.* New York: John Wiley and Sons, Inc. 1959.

Schweber, S. S. (1988): The Mutual Embrace of Science and the Military: ONR and the Growth of Physics in the United States after World War II. In: E. Mendelsohn, M. R. Smith & P. Weingart (eds.): *Science, Technology and the Military.* Dordrecht, The Netherlands: Kluwer Academic Publishers 1988, pp. 3–45.

Shell, A. E. (Forthcoming): Mina Rees and the Funding of the Mathematical Sciences. *The Mathematical Intelligencer*, forthcoming.

Smith, B. (1969): *The RAND Corporation, Case Study of a Nonprofit Advisory Corporation.* Cambridge, Massachusetts: Havard University Press 1969.

von Neumann, J. (1928): Zur Theorie der Gesellschaftsspiele. *Mathematische Annalen* **100**, 1928, pp. 295–320.

von Neumann, J. (1947): Discussion of a Maximum Problem. In: A. H. Taub (ed): *von Neumann Collected Works* **6**, Oxford: Pergamon Press 1963, pp. 89–95.

von Neumann, J. (1959): On the Theory of Games of Strategy. (English translation of [von Neumann, 1928]) in: A. W. Tucker & R. D. Luce (eds.:) *Contributions to the Theory of Games, Vol. IV.* Princeton, New Jersey: Princeton University Press 1959, pp. 13–42.

von Neumann, J. (1963): *von Neumann Collected Works 7*, ed. by A. H. Taub. Oxford: Pergamon Press 1963.

von Neumann, J. & Morgenstern, O. (1944): *Theory of Games and Economic Behavior.* Princeton: Princeton University Press, Princeton 1944.

Weintraub, E. R. (1992) (ed.): *Towards a History of Game Theory.* Durham and London: Duke University Press, 1992.

Williams, K. B. (2001): *Improbable Warriors.* Annapolis, Maryland: Naval Institute Press 2001.

Zachary, P. G. (1997): *Endless Frontier: Vannevar Bush, Engineer of the American Century.* New York: The Free Press 1997.

Mathematics and War in Japan

SETSUO FUKUTOMI*

The first, shorter part of the article describes the anti-war movement among Japanese mathematicians during the Vietnam War. In the second, longer part, the author decribes the decoding program of the Japanese Army during World War II, in which he worked himself as a drafted soldier, at the defeat, all written evidence regarding this program and its collaborators was destroyed. Today, the only surviving participant in the project apart from the author is Commander KAMAGA, its initiator and leader (alias KATO Masataka, his pen name as a cryptographic author).

1 The Movements of Japanese Mathematicians against the Vietnam War

In April 1965, Professor Laurent SCHWARTZ was in Japan. He informed the Japanese mathematicians that Steven SMALE, mathematics professor at the University of California at Berkeley, had created a protest movement against the Vietnam War, and invited the Japanese mathematicians to support Professor SMALE's initiative. He received 183 signatures from Japanese mathematicians in solidarity with Professor SMALE's anti-war campaign, which was announced publicly at the annual meeting of the Mathematical Society of Japan in May 1965.

At that moment, a civic movement against the American war in Vietnam was already in existence, arranged by the Japan "Peace for Vietnam" Committee (BEHEIREN in Japanese abbreviation, henceforth JPVC).

While the U.S. "Vietnam Day Committee" (of which Professor SMALE was one of the principal organizers) was founded in Berkeley, Makoto ODA, president of the JPVC and writer, agreed with Professor SMALE to organize a coordinated manifestation against the Vietnam War on the same day in Japan and in the United States, namely on May 22, 1965. They decided to establish a lasting collaboration between the two countries against this war in Vietnam, thus making the JPVC one of the foci of the Japanese movements against the Vietnam War. The anti-war movement of Japanese mathematicians ("Committee of Japanese Mathematicians Against the War in Vietnam", henceforth CJMAWV) often collaborated with the JPVC.

The mathematicians of the University of Kyushu (South Japan) demonstrated against the Vietnam War three times each month, the 10th, the 20th and the 30th ("demonstrations of the ten"). The mathematicians were leading in the movements

* Professor of mathematics (retired) at the Tokyo University of Agriculture and Technology.

of Japanese scientists against the Vietnam War and the American invasion of Vietnam.

In February 1966, CJMAWV made an appeal and organized a demonstration against the American B52 bombings of North Vietnam, asking at the same time the Japanese population to express solidarity with its activities. An appeal was also made to the mathematicians of the whole world to organize a meeting on the occasion of the International Congress of Mathematicians (ICM) in Moscow in August 1966. Indeed, this meeting was held in Moscow thanks to the efforts and the collaboration of the mathematicians L. SCHWARTZ, E. B. DYNKIN, S. IYANAGA, Chandler DAVIS and others, who were joined by Professor SMALE. Professor SCHWARTZ and Professor DYNKIN also invited the mathematicians of the whole world to sign an appeal against the Vietnam War. The appeal voiced the strong opposition of the mathematicians gathered in Moscow to the war that was waged by the American armed forces and was becoming increasingly cruel every day, and stated that President JOHNSON's strategy of "escalation" could lead to no equitable solution; the appeal affirmed the sovereignty of the Vietnamese people and its right to self-determination, and finally expressed its solidarity with those American mathematicians who fought against this Vietnam War which dishonoured their country.

The appeal was published in Tokyo and Paris with 628 signatures of mathematicians from 26 countries (Japan excluded) and 1333 signatures of Japanese mathematicians.

Afterwards, several dozens of Japanese mathematicians went on organizing manifestations against the invasion of Vietnam by the American armed forces on the occasions of the Annual Meetings and the Autumn Meetings of the Mathematical Society of Japan in the cities where these meetings took place.

The (rather vague) notions of a professional community of mathematicians, shaped in research and teaching activities, allowed the mathematicians to continue these movements against the Vietnam War lastingly. These were thus civic movements carried by a community that was defined professionally and not geographically. The participants did not think of themselves primarily as mathematicians, nor was it much discussed whether and how mathematics (and mathematicians) were involved in the war...

In spite of this ambiguity, the mathematicians were the only Japanese scientists that organized meetings and manifestations against the French nuclear tests in the Pacific (Mururoa), against the terrorist attacks in New York, and against G. W. BUSH's wars of revenge in Afghanistan.

2 Japanese Mathematicians' Military Research and Collaboration with the Military during World War II

The following treats of the author's own experience in cryptographic research, in particular in decoding (cryptanalysis), in the Japanese armed forces during the 1940s.

Toward June 1943, Commander Kazuo KAMAGA from the General Staff of the Army requested the collaboration of mathematicians with the armed forces for studies of the mathematical foundations of military coding. At first his idea was to create an undecipherable (unbreakable) cryptography, but later he concentrated on mathematical research in decoding (cryptanalysis). He asked for the collaboration of the Mathematics Department of the Imperial University of Tokyo.

Commander KAMAGA started by giving preparatory cryptography courses to mathematicians, mainly coming from the Imperial University of Tokyo. This resulted in the official establishment of the Mathematical Seminary of the Army on April 3, 1944, formed by mathematicians from the Imperial University of Tokyo, of Tohoku, and of Osaka.

Studies of Euler squares were considered important for the creation of a new code and for the error-free transmission of coded messages; so were studies of the theory of Galois fields. Professor Y. KAWADA undertook profound studies of the distribution of consecutive letters in English phrases applying the theory of Markov chains. Professor Kunihiko KODAIRA (winner of the Fields Medal 1954) studied some types of permutations for the analysis of the Enigma. Commander KAMAGA himself concentrated on studies of error-correcting encoding. His studies in this domain are precursors of the modern theory of such codes, for instance the encoding theory of R. W. HAMMING.

With KAMAGA, a team headed by the mathematician Kōichi YAMAMOTO, engineer of the General Staff of the Army, worked much on constructing Euler squares of order n, aiming first of all at the creation of a new cryptography. After the war, this engineer-mathematician became Professor of mathematics at several universities, making fruitful teaching and research in the theory of combinatorics.

Before the war, the Japanese Army had "stolen" the American strip-cipher code (i.e., code book). One had secretly photographed the diplomatic code-books for communication between Washington, Tokyo, and Tch'ang-Tch'ouen. This code was based on several tens (between 50 and 100) of bars, each carrying the same permutation of the alphabet twice. Each day, a number of bars (from 20 to 30) were chosen in order to create a clear text, picking one letter from one vertically positioned bar, another one from another bar, etc., ordering them afterwards from left to right so as to produce a text horizontally. In a certain distance upwards or downwards, if reading a line of letters from left to right, one obtains an apparently illegible text as encoded text (see Figure 2).

This encoding was very efficacious and resistant to decoding. We were no longer able to decode it after the Americans changed the code-books that had been "stolen". We worked much on breaking its keys. The studies of the distribution of consecutive letters of the alphabet in English phrases served the same purpose. The author (S. FUKUTOMI) used indeterminant equations for breaking strip cipher.

Concomitantly, the Japanese Army had decided in 1943 to recruit graduates in mathematics for work in cryptanalysis. Its interests concentrated on the "strip-cipher" and the cipher machine "M209". The machine M209 was furnished with six coaxial disks arranged on a horizontal bar. They carried, respectively, 26, 25, 23, 21, 19 and 17 letters of the alphabet (pairwise relatively prime integers). If one

Figure 1.
US Army Hagelin M-209 cipher machine. [Photo: Masataka Kato [Kazuo Kamaga]]

writes a clear text turning the handle carrying the alphabet, the first disk begins to rotate, dragging the neighbouring disk by means of a pin which it carries, and this second disk drags again by a pin which it carries, and so on. These motions of the six disks thus produced random numbers with a period of c. 100 millions (26×25×23×21×19×17), finite but enormous, for producing an encoded text on a paper tape, transcribed according to a pre-arranged position of the pins of the six disks in agreement with a cryptographic key for M209 (see Figure 1).

These M209 encoding-machines were in general use at the front divisions of the American armed forces. The General Staff of the Japanese Army had bought a prototype of this machine in Sweden before the war. The military engineer Yamamoto mentioned above guessed that "M209" was an adaptation of this Swedish prototype. He could even determine the details of the modifications of the prototype that had produced M209.

At that time, I was serving as a soldier working at the General Staff. I noticed that the keys of the codes of the M209 were so-called "double keys", and I succeeded in breaking these double-keys. After that, a team including a number of drafted officers including myself were sent to the Philippines. We managed to obtain some good decoding results, but the way the Americans made the daily change of coding keys was such that we were unable to break their codes every day, which strongly restricted the benefits we earned from our work.

By the way, the American "Strip-cipher" codes we speak of were also given to us by the Germans, who at the moment were our allies, and our work was based on these documents. Now, the material which was brought from Germany by the last U-Boot arriving in Japan in 1945 I found inconsistent with the preceding documents in our possession, in spite of the commentaries that attested to the authenticity of these documents. I understood that beyond purely tactical relations the confidence even among allies was not necessarily unlimited.

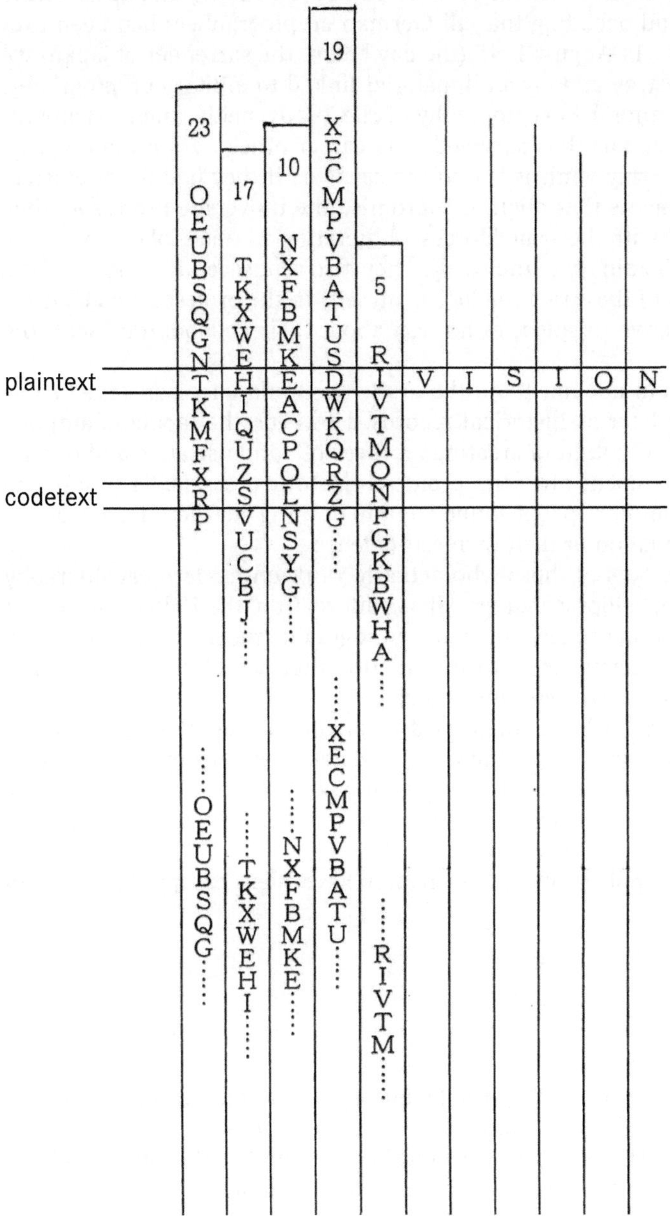

Figure 2. Strip cipher principle. (1) Each of the bars that carries twice the same permutation of the alphabet is numbered. (2) Every day, some 20 to 30 bars are chosen for the codes of this day. The keys of the cryptography are these numbers and their order for each day of the year providing two-sided equivalents in the form of a code book. (3) On the image, the plaintext THE DIVISION ... is indicated on top and the codetext RSLZN ... below. (4) With two permutations of the alphabet on each bar, it was always possible to find the necessary letters.

After the German capitulation there were rumours running among us Army workers on coding and decoding that all German cryptographers had been executed. Therefore, after 14 August 1945 (the day before the surrender of Japan) we received orders to destroy and burn all material linked to military cryptography. No document concerning the cryptography of the 1940s, neither new cryptography nor code breaking, has thus survived. The career officers from the cryptography section of the Army vanished from the scene as if they had never existed. They burnt all documents that might compromise the university professors they had made collaborate with the armed forces, obliterating all traces of the existence of the Mathematical Seminary of the Army. They also sent several members of the cryptography section of the Army (including myself) to the professors concerned, giving them the directive to speak in no way about their collaboration with the armed forces.

I now believe certain documents in the archives should have conserved so as to make them public later as historical records. However, the Japanese authorities of the times had no notion of creating archives of administrative and official documents at some moment after the events as historical assets. Therefore the "memoirs" of former military people often contain nothing but accretions of deformations and disinformation or their own self-defence.

In any case, in the eyes of those who actually work on modern cryptography such as "public key" or "elliptic coding", all our efforts from the 1940s almost look like fairy tales from former times. The new cryptographic theories are much more deeply linked with mathematics, with information theory and with informatics, and speaking of them exceeds my competence.

Commander KAMAGA, who worked hard to create and develop the cryptography of the Japanese Army before the war (and, not least, during the war) is 85 years old today and is in good health. His gigantic work *Treatise of Cryptography* (published under the name of Masataka KATO) has a unique status within classical cryptography.

All those who are spoken of in this chapter have died, except Kamaga and myself.

3 Conclusions

In American bombers shot down by the Japanese, tables of numbers were found for the "space-probing pilotage" destinated to overseas bombing flight by astronomical observations. The Japanese armed forces recruited mathematics students to prepare numerical tables of the same kind. I have heard about mathematicians recruited by the armed forces for ballistic calculations, but I am not informed about the details of this.

It was impossible to refuse orders or demands of the armed forces during the regime of the imperial powers. Orders from the armed forces were treated as direct orders of the emperor himself. So were the orders to mathematicians to collaborate with the armed forces. One may easily imagine that the pressure was even greater on scientists belonging to so-called applied disciplines.

The idea to make use of mathematicians to improve the efficiency of warfare did not come early to the armed forces – much later indeed than the other scientific domains. Even for cryptographic research, linguists were first recruited. Commander KAMAGA himself loved mathematics. He presented himself to Teiji TAKAGI, emeritus professor of mathematics at the Imperial University of Tokyo, the patriarch of the Japanese mathematical community and known throughout the world for his monograph *Algebraic Number Theory – Generalities and Class Field Theory* (MR 14, 953a). Commander KAMAGA sought the collaboration of mathematicians for research in military cryptography within the Army. Initially, Professor TAKAGI was rather sceptical about the utility of mathematics within the domain of military coding. He finished by promising, however, to examine together with the mathematicians of the Imperial University of Tokyo the possibility of cooperating with the armed forces. This was the origin of the Mathematical Seminary of the Army which was mentioned in Section 2.

At the Imperial University of Tokyo, Professor Shokiti IYANAGA was the key figure of this seminary. Being a soldier in the Army and a member of the cryptography study group of the General Staff of the Army, I myself belonged on the side of those who invited the mathematicians.

In 1985, this Professor Shokiti IYANAGA referred publicly in a meeting of the "Association of Japanese Mathematicians Aiming to Eliminate All Nuclear Weapons" to his remorse for having supported the war in his capacity as a mathematician. He ended his talk with the words "I collaborated with the armed forces during the war as a mathematician; which I should not have done. Since then, my conscience as a scientific worker has never been released from these bitter memories. With NAKASONE's ascent to power, I feel something threatening to peace in Japan. One must make an effort to make sure Japan will not again find itself on the slippery road leading to war".

I was deeply touched by his sincere statement. Many university professors collaborated with the armed forces during the war. And yet I know of no university professor in Japan except him who publicly regretted his collaboration as a scientist with the armed forces for the war.

Reference

KATO Masataka, *Fundamental Theory of Cryptography.* I. *For Security in Informatics* (in Japanese). (Information and Computing). Tokyo: Saiensusha, 1989 [LC Control No 90115183].

Discovery of the Maximum Principle in Optimal Control

REVAZ V. GAMKRELIDZE*

Discovery of the Maximum Principle for the needs of optimal control and its subsequent development give a classical example of a theory, which initially emerged as an effective device for solving a purely engineering problem not amenable by existing methods, and eventually developed into a mathematical theory of major significance.

I present a short history of the discovery of the Maximum Principle in Optimal Control by L. S. Pontryagin and his associates.

1 Formulation of the Time-Optimal Problem

In 1970, at the World Congress in Nice, Prof. Pontryagin gave a plenary talk on differential games, which was motivated by pursuit-evasion strategies of aircrafts for a very simplified model of behavior. During the after-talk discussions, A. Grothendieck put a rhetorical question to Pontryagin. He said that though the listeners witnessed a beautiful piece of mathematics, still he would like to know whether the speaker feels himself morally responsible for supporting military trends in the society. Pontryagin's answer was quite definite and blunt. He was convinced, he said, that, on an intellectual level, any intellectual problems could be discussed openly in a developed society, and if we would follow to the logical end Prof. Grothendieck's recommendation, we should be prohibited from speaking openly about some topics of abstract Algebra, since Cryptography, which has much deeper correlations with military problems than the differential game considerations he spoke about, is completely based on the theory of finite fields.

Lev Semenovich Pontryagin was one of the leading figures in 20th century algebraic topology and topological algebra, but in mid-1950s he abandoned topology, never to return to it, and completely devoted himself to purely engineering problems of mathematics. He organized at the Steklov Mathematical Institute a seminar on applied problems of mathematics, often inviting theoretical engineers as speakers, since he considered a professional command over the engineering part of the problem under investigation to be mandatory for an adequate mathematical development.

* Member of the Russian Academy of Sciences and Member of the Steklov Mathematical Institute, Moscow. Email: gam@ipsun.ras.ru

The activity in the seminar culminated very soon in the formulation of two major mathematical problems. One of them developed into the general theory of singularly perturbed systems of ordinary differential equations. The second problem brought the discovery of the Maximum Principle and the emergence of optimal control theory.

Pontryagin was led to the formulation of the general time-optimal problem by an attempt to solve a concrete fifth-order system of ordinary differential equations with three control parameters related to optimal maneuvers of an aircraft, which was proposed to him by two Air Force colonels during their visit to the Steklov Institute in the early spring of 1955. Two of the control parameters entered the equations linearly and were bounded, hence from the beginning it was clear that they could not be found by classical methods, as solutions of the Euler equations. The problem was highly specific, and very soon Pontryagin realized that some general guidelines were needed in order to tackle the problem. I remember he even said half-jokingly, "we must invent a new calculus of variations." As a result, the following general time-optimal problem was formulated.

Consider a controlled object represented in the n-dimensional state space $\mathbb{R}^n$ of points

$$x = \begin{pmatrix} x^1 \\ \vdots \\ x^n \end{pmatrix}$$

by a system of n autonomous differential equations with r control parameters

$$u = \begin{pmatrix} u^1 \\ \vdots \\ u^r \end{pmatrix} \in U \subset \mathbb{R}^r ,$$

$$\frac{dx}{dt} = f(x, u), \quad f = \begin{pmatrix} f^1 \\ \vdots \\ f^n \end{pmatrix} . \tag{1}$$

Initially it was supposed that the control vector u attains its values from an open set $U \subset \mathbb{R}^r$. Necessity of a closed U, the most important case for applied problems, was evident from the very beginning, though could be handled only later. To denote control parameters, the letter "u" was chosen, as the first letter of the Russian word "control" – "управление".

Formulation of the problem

Given initial and terminal states $x_0, x_1 \in \mathbb{R}^n$, *find a control function* $u(t) \in U \; \forall t \in [t_0, t_1]$, *such that it minimizes the transition time of the state point x, moving* *from x_0 to x_1 according to the non autonomous equation*

$$\frac{\mathrm{d}}{\mathrm{d}t}x = f(x(t), u(t)) \cdot$$

Thus, we come to the time--optimal control $u(t)$ and the corresponding time-optimal trajectory $x(t)$, $t_0 \leq t \leq t_1$, which satisfies the boundary value problem

$$\frac{\mathrm{d}}{\mathrm{d}t}x(t) = f(x(t), u(t)), \;\; x(t_0) = x_0, \;\; x(t_1) = x_1$$

and minimizes the transition time,

$$t_1 - t_0 = \min .$$

It should be noticed that the general optimal problem with an arbitrary integral-type functional is easily reduced to the formulated time-optimal problem, so that by solving the time-optimal problem with fixed boundary conditions we actually overcome all essential difficulties inherent in the general case.

The first and the most important step toward the final solution was made by Pontryagin right after the formulation of the problem, during three days, or better to say, during three consecutive sleepless nights. He suffered from severe insomnia and very often used to do math in bed all night long. As a result, he completely disrupted his sleep in his later years and systematically took barbiturates in great quantities.

Thanks to his wonderful geometric insight, he derived from very simple duality considerations about the first order variational equation the initial version of necessary conditions, introducing an auxiliary covector-function $\psi(t) = (\psi_1(t), \ldots, \psi_n(t))$ subject to the adjoint system of differential equations,

$$\left.\begin{aligned} &\frac{\mathrm{d}\psi_i}{\mathrm{d}t} = \sum_{\alpha=1}^{n} \psi_\alpha \frac{\partial f^\alpha}{\partial x^i}(x(t), u(t)), \quad i = 1, \ldots, n, \\ &\Updownarrow \\ &\frac{\mathrm{d}\psi}{\mathrm{d}t} = -\psi \frac{\partial f}{\partial x}(x(t), u(t)). \end{aligned}\right\} \qquad (2)$$

This was the first appearance in optimal control theory of the adjoint system, which turned out to be of crucial importance for the whole subject. Actually, Pontryagin constructed for the first time, for the needs of optimization, what is usually called the *Hamiltonian lift* of the initial family of vector fields on the state space of the problem into its cotangent bundle, the phase space of the problem, cf. n° 5.

2 Initial Formulation of Necessary Conditions

Initial formulation of necessary conditions, reported by Pontryagin at the seminar right after they were derived, is expressed in formulas

$$\left.\begin{aligned}
&\frac{\mathrm{d}x(t)}{\mathrm{d}t} = f(x(t),u(t)),\ \ x(t_0)=x_0,\ x(t_1)=x_1, && (3.1)\\
&\frac{\mathrm{d}\psi(t)}{\mathrm{d}t} = -\psi(t)\frac{\partial f}{\partial x}(x(t),u(t)), && (3.2)\\
&\psi(t)\frac{\partial f(x(t),u(t))}{\partial u^j} = 0, && (3.3)\\
&u(t)\in U,\ \forall t\in[t_0,t_1],\ j=1,\dots,r.
\end{aligned}\right\} \quad (3)$$

They assert that if $x(t)$, $u(t)$, $t_0 \le t \le t_1$, *is an optimal solution, then there exists a nonzero covector-function* $\psi(t)$ *such that* $\psi(t)$, $x(t)$, $u(t)$, $t_0 \le t \le t_1$, *is a solution of the system of differential equations (3.1)–(3.2), and along the solution, for every* t, $j = 1, \dots, r$, *"finite" equations (3.3) are satisfied.*

This formulation supposes that the set U of admissible values of control is open, though, as I already mentioned above, from the very beginning it was clear that the ultimate result should be applicable to closed sets as well.

I shall describe now Pontryagin's very simple and straightforward geometric arguments which directly lead to equations (3).

Consider an arbitrary admissible variation of the optimal control $u(t)$,

$$\delta u(t) = \begin{pmatrix} \delta u^1(t) \\ \vdots \\ \delta u^r(t) \end{pmatrix},\ \ u(t)+\delta u(t)\in U,\ \ t_0\le t\le t_1.$$

If we ignore quadratic and higher order terms with respect to δu in the Taylor expansion of the corresponding trajectory variation, we obtain the *first (linear) variation*

$$\delta x(t) = \begin{pmatrix} \delta x^1(t) \\ \vdots \\ \delta x^n(t) \end{pmatrix},\ \ t_0\le t\le t_1,$$

of the optimal trajectory, which satisfies the standard *linear variational equation*

$$\left.\begin{array}{c}\frac{\mathrm{d}}{\mathrm{d}t}\delta x=\frac{\partial f}{\partial x}(x(t),u(t))\,\delta x+\frac{\partial f}{\partial u}(x(t),u(t))\,\delta u(t),\\ \delta x(t_0)=0,\, i=1\ldots,n.\end{array}\right\} \tag{4}$$

The mapping $\{\delta u(t), t_0 \leq t \leq t_1\} \mapsto \delta x(t_1)$ is a linear operator from the space of variations $\delta u(t), t_0 \leq t \leq t_1$ into the state space $\mathbb{R}^n$. Since the set of admissible values of u is assumed open, the admissible variations $\delta u(t)$ are arbitrary (piecewise continuous) functions. Hence the set

$$L = x(t_1) + \{\delta x(t_1) | \delta x(t_0) = 0\} = x(t_1) + \Gamma \subset \mathbb{R}^n \tag{5}$$

is a plane through $x(t_1)$ in $\mathbb{R}^n$, Γ is the corresponding subspace of $\mathbb{R}^n$. Since $x(t)$, $t_0 \leq t \leq t_1$,is optimal, we obtain the relation

$$\dim L = \dim \Gamma \leq n-1 ,$$

which is easily derived from the implicit function theorem. Hence, there exists a (nonzero) covector $\chi = (\chi_1, \ldots, \chi_n)$ orthogonal to Γ,

$$\chi \delta x(t_1) = \sum_{\alpha=1}^{n} \chi_\alpha \delta x^\alpha (t_1) = 0 \;\forall \delta x(t_1) \in \Gamma.$$

This is the geometric condition, from which the equations (3) follow immediately, if we express $\delta x(t_1)$ through $\delta u(t)$, i.e., integrate the variational equation (4). For this purpose we introduce the fundamental matrix $\Phi(t)$ of the corresponding homogeneous equation,

$$\frac{\mathrm{d}}{\mathrm{d}t}\delta x = \frac{\partial f}{\partial x}((x(t), u(t))\,\delta x,$$

and the inverse $\Psi(t) = \Phi^{-1}(t)$. They satisfy matrix differential equations

$$\frac{\mathrm{d}}{\mathrm{d}t}\Phi = \frac{\partial f}{\partial x}((x(t), u(t))\,\Phi,\quad \frac{\mathrm{d}}{\mathrm{d}t}\Psi = -\Psi \frac{\partial f}{\partial x}((x(t), u(t)). \tag{6}$$

Solution of the nonhomogeneous equation (4) with the initial condition $\delta x(t_0) = 0$ is represented as

$$\delta x(t) = \Phi(t) \int_{t_0}^{t} \Psi(\tau) \frac{\partial f}{\partial u}(x(\tau), u(\tau))\, \delta u(\tau)\, d\tau,\;\; t \in [t_0, t_1],$$

hence

$$\chi \delta x(t_1) = \chi \Phi(t_1) \int_{t_0}^{t_1} \Psi(\tau) \frac{\partial f}{\partial u}(x(\tau), u(\tau))\, \delta u(\tau)\, d\tau = 0 \;\; \forall \delta u(\tau). \tag{7}$$

The n-dimensional covector

$$\psi(t) = (\psi_1(t), \dots, \psi_n(t)) = \chi \Phi(t_1) \Psi(t), \;\; t_0 \le t \le t_1,$$

is nonzero and satisfies, according to (6), the differential equation

$$\frac{\mathrm{d}}{\mathrm{d}t}\psi(t) = -\psi(t)\frac{\partial f}{\partial x}(x(t), u(t)),$$

which coincides with the adjoint system (3.2). Finally, the equation (7) attains the form

$$\int_{t_0}^{t_1} \psi(\tau)\frac{\partial f}{\partial u}(x(\tau), u(\tau))\,\delta u(\tau)\,d\tau = 0.$$

Since the control variation $\delta u(t), t_0 \le t \le t_1$, is an arbitrary vector function, we obtain equations (3.3) and come to the optimality conditions (3) formulated above. They easily imply the Euler-Lagrange equations for the Lagrange problem of the classical calculus of variations.

3 The Second Variation

As soon as the equations (3) were obtained, Pontryagin recognized the decisive role of the covector-function $\psi(t)$ and the adjoint system (2) for the whole problem. He considered, in the generic case, r finite equations (3.3) as conditions, which eliminate r control parameters $u^1, \dots, u^r$ from system (3), thus making it possible to solve uniquely the $2n$-th order system of differential equations (3.1)–(3.2) with a given initial condition $x(t_0) = x_0$ and an arbitrary (nonzero) initial condition for ψ. All such solutions were declared as *extremals of the problem*, from which the optimal solutions were to be derived.

Pontryagin's idea about a universal procedure of elimination of control parameters, which reduces the problem of determining extremals to solving ordinary differential equations with given boundary conditions, found its ultimate realization in the maximum principle, which was formulated by him several months later after his first report at the seminar and was supported by the subsequent advancements obtained meanwhile at the seminar.

After his talk in the seminar, Pontryagin suggested to V. Boltyanski and me, his former students and close collaborators at that time, to join him in his investigations of the problem. V. Boltyanski held a formal position at the Steklov Institute as Pontryagin's assistant, helping him in everyday computations and manuscript editing; I was a young member of the department of the Steklov Institute headed by Pontryagin.

Pontryagin's vision of the problem at this early stage of development could be described as follows.

Instead of considering the boundary value problem with fixed endpoints for the controlled system (3.1), we should only fix the initial point x_0, take an arbitrary initial value $\psi(t_0) = \psi_0 \neq 0$, and solve the system of $2n + r$ equations (3.1)–(3.3) with $2n + r$ unknowns x^i, ψ_j, u^k, proceeding along an arbitrary extremal through x_0. This should be possible, since the r control parameters u^k are, "in general", successfully eliminated by r conditions (3.3), hence only $2n$ unknown parameters are left, x_i, ψ_j, subject to the system of $2n$ differential equations (3.1)–(3.2) and the initial conditions $x(t_0) = x_0, \psi(t_0) = \psi_0$. Since the adjoint system (3.2) is linear in ψ, the function $\psi(t)$ is defined up to a nonzero constant factor, hence we can normalize the initial value $\psi(t_0)$, obtaining thus the $(n - 1)$-dimensional sphere of the initial values of ψ, which should generate an $(n - 1)$-parametric family of extremal trajectories of the problem emanating from the point x_0.

According to this initial picture, the final goal of the program consisted in the study of extremals $\psi(t), x(t), x(t_0) = x_0$, as solutions of the system (3.1)–(3.3), parametrized by the initial value $\psi_0 \neq 0$. Today we recognize in the given formulation the problem of controllability in its most rudimentary setting. Certainly at that stage, before the maximum principle was not even formulated, it was practically impossible to obtain in this direction any nontrivial results.

I was fascinated by Pontryagin's geometric approach and got an idea how to apply this picture to investigate the problem up to the second-order approximation. So, we decided to split our further advancements in two directions. Pontryagin, together with Boltyanski, pursued the problem in the controllability direction, I started to investigate the second variation of the problem. As it turned out, this latter direction led to the final formulation of the maximum principle.

Necessary conditions of optimality, expressed by equations (3), are derived from purely first-order approximation. They are independent of "general position" considerations. My second-order considerations demanded from the very beginning general position assumptions, which were overcome only in the final version of the proof by Boltyansky. The set of admissible values of the control parameters U was still assumed to be open.

Very briefly, the idea of the second-order approximation could be described in the following way.

Take an arbitrary "generic" solution of the optimal problem, $x(t), u(t), t_0 \leq t \leq t_1$, which means that the plane L in (5) is of maximal possible dimension,

$$\dim L = \dim \Gamma = n - 1,$$

and the trajectory $x(t)$ intersects L at $x(t_1)$ transversally (is not tangent to L). Hence L divides $\mathbb{R}^n$ in distinguishable half-spaces, $\mathbb{R}^n_-$ – before $x(t)$ intersects L, $\mathbb{R}^n_+$ – after the intersection. Every variation $\delta u(t)$ displaces the endpoint $x(t_1)$ in the first order into the hyperplane $L, x(t_1) + \delta x(t_1) \in L$. The real displacement $\Delta x(t_1)$ is certainly nonlinear in δu and, generally, stays off the hyperplane,

$$x(t_1) + \Delta x(t_1) \in \mathbb{R}^n_- \quad \text{or} \quad x(t_1) + \Delta x(t_1) \in \mathbb{R}^n_+ .$$

Denote from now on the first variation $\delta x(t)$ by $\delta_1\, x(t)$, and let $\mathbb{K}$ be the kernel of the linear operator from the space of control variations into the space of first variations of $x(t)$ for $t = t_1$; the operator is given by

$$\{\delta u(t), t_0 \le t \le t_1\} \mapsto \delta_1 x(t_1) = \Phi(t_1) \int_{t_0}^{t_1} \Psi(\tau) \left\| \frac{\partial f^i}{\partial u^j}(x(\tau), u(\tau)) \right\| \delta u(\tau)\, d\tau \,.$$

Define the *second variation* $\delta_2 x(t), t_0 \le t \le t_1$, of $x(t)$ as the solution of the linear nonhomogeneous equation

$$\frac{\mathrm{d}}{\mathrm{d}t}\delta_2 x = \frac{\partial f}{\partial x}(x(t), u(t))\, \delta_2 x + Q(\delta u(\tau), \delta u(\tau)),\ \ \delta_2 x(t_0) = 0,$$

where Q is a vector-valued integral quadratic expression in $\delta u(t), t_0 \le t \le t_1$, which is easily computed if we take in the Taylor expansion of $\Delta x(t)$ only quadratic terms with respect to $\delta u(\tau)$. The obtained equation differs from (4) only by the nonhomogeneous part. The displacement of the endpoint $x(t_1)$ up to the second order is given by the vector $\delta_1 x(t_1) + \delta_2 x(t_1)$.

The key geometric fact for a generic optimal trajectory $x(t), t_0 \le t \le t_1$, *consists in the assertion that the second order displacement of its endpoint, considered on the kernel* $\mathbb{K}$, *belongs to the half-space* $\mathbb{R}^n_-$,

$$x(t_1) + \delta_1 x(t_1) + \delta_2 x(t_1) \in \mathbb{R}^n_- \;\Leftrightarrow\; x(t_1) + \delta_2 x(t_1) \in \mathbb{R}^n_- \;\; \forall \delta u \in \mathbb{K}.$$

Hence we come to the conclusion that, additionally to the system (3.1)–(3.3), the following integral quadratic form in δu is nonpositive, provided the covector $\psi(t_1)$, which is transversal to L, is correctly normalized (directed toward the half-space $\mathbb{R}^n_+$),

$$\psi(t_1)\, \delta_2 x(t_1) = \int_{t_0}^{t_1} \psi(\tau) Q(\delta u(\tau)), \delta u(\tau), \delta u(\tau))\, d\tau \le 0 \;\; \forall \delta u \in \mathbb{K}.$$

After some elaborate investigation of this integral quadratic form, I came to the conclusion that from its nonpositivity on $\mathbb{K}$ follows the nonpositivity of its singular part on $\mathbb{K}$, which easily implies the final form of the second order optimality condition, the pointwise (in $t \in [t_0, t_1]$) nonpositivity of the quadratic form in $v \in \mathbb{R}^r$

$$v * \left\| \psi(t) \frac{\partial^2 f}{\partial u^i \partial u^j}(x(t), u(t)) \right\| v \le 0, \;\; \forall v \in \mathbb{R}^r, \forall t \in [t_0, t_1], \tag{8}$$

satisfied, together with the first-order conditions (3), by every generic optimal solution $x(t), u(t), t_0 \le t \le t_1$,.

4 Final Form of the Maximum Principle

Collecting all necessary conditions (3.1)–(3.3), (8) together, we immediately recognize that a certain stable combination of symbols reappears in all of them, the scalar function of three arguments ψ, x, u,

$$H(\psi, x, u) = \sum_{\alpha=1}^{n} \psi_\alpha f^\alpha(x, u) = \psi f(x, u). \tag{9}$$

It enables us to rewrite the system (3.1)–(3.2) as a Hamiltonian system (10.1) with the Hamiltonian function (9), together with additional conditions (3.3), (8), written as (10.2)–(10.3):

$$\left.\begin{aligned} \frac{dx(t)}{dt} &= \frac{\partial H}{\partial \psi}(\psi(t), x(t), u(t)), \\ \frac{d\psi(t)}{dt} &= \frac{\partial H}{\partial x}(\psi(t), x(t), u(t)), \end{aligned}\right\} \tag{10.1}$$

$$\frac{\partial H}{\partial u^j}(\psi(t), x(t), u(t)) = 0, \ \ \forall t \in [t_0, t_1], \ \ j = 1, \ldots, r, \tag{10.2}$$

$$v^* \left\| \frac{\partial^2 H}{\partial u^i \partial u^j}(\psi(t), x(t), u(t)) \right\| v \leq 0, \ \ \forall v \in \mathbb{R}^r. \tag{10.3}$$

They assert that generic extremals are solutions of the Hamiltonian system (10.1), and, according to (10.2), their points are stationary points of the Hamiltonian (9) with respect to the control parameter u; furthermore, according to (10.3), along regular extremals, for which the form (10.3) is definite, the function H attains its local maximum with respect to u.

We can unite two independent conditions (10.2)–(10.3) into one condition and write

$$H(\psi(t), x(t), u(t)) = \max_{v \in O_t} H(\psi(t), x(t), v), \tag{10.4}$$

where O_t is a neighborhood of $u(t)$. Furthermore, the equations (10.1)–(10.2) imply,

$$\frac{dH}{dt}(\psi(t), x(t), u(t)) = \frac{\partial H}{\partial \psi} \cdot \frac{d\psi}{dt} + \frac{\partial H}{\partial x} \cdot \frac{dx}{dt} + \frac{\partial H}{\partial u} \cdot \frac{du}{dt} \equiv 0.$$

It is also easy to show that $H(\psi(t), x(t), u(t))$, as a function of t, is continuous, even if the control function $u(t)$ has jumps. Hence, taking into account the generic character of the solution – the trajectory $x(t)$ is transversal to L at $x(t_1)$ – we obtain

$$H(\psi(t), x(t), u(t)) \equiv const = \psi(t_1) f(x(t_1), u(t_1)) > 0. \tag{11}$$

After the equations (10.1)–(10.3) were written, Pontryagin realized that the universal elimination method of the control parameters he was searching for, was found. He replaced the local maximum condition (10.4) by the global maximum over the whole set U, the "*Pontryagin maximum condition*" (12), which made any restrictive assumptions about the admissible set U superfluous,

$$H(\psi(t), x(t), u(t)) = \max_{u \in U} H(\psi(t), x(t), u) \equiv const \geq 0. \tag{12}$$

Thus, he came to the final formulation of the maximum principle, combining the Hamiltonian system (10.1) with the maximum condition (12) and dropping off any assumptions about genericity of the solutions or the nature of the admissible set U.

The Maximum Principle. *Suppose a controlled equation is given,*

$$\dot{x} = f(x, u), \quad x = \begin{pmatrix} x^1 \\ \vdots \\ x^n \end{pmatrix} \in \mathbb{R}^n, \quad u = \begin{pmatrix} u^1 \\ \vdots \\ u^n \end{pmatrix} \in U \subset \mathbb{R}^r,$$

where the admissible set U is arbitrary. We introduce the Hamiltonian function of the problem

$$H(\psi, x, u) = \psi f(x, u) = \sum_{\alpha=1}^{n} \psi_\alpha f^\alpha(x, u), \tag{13.1}$$

which depends on three arguments – the covector $\psi = (\psi_1, \dots, \psi_n)$ and the vectors x, u. If $u(t), t_0 \leq t \leq t_1$, is a time-optimal control, $x(t), t_0 \leq t \leq t_1$, the corresponding time-optimal trajectory,

$$\frac{\mathrm{d}}{\mathrm{d}t} x(t) = f(x(t), u(t)), \quad t_0 \leq t \leq t_1; \; t_1 - t_0 = \min,$$

then there exists a nonzero covector function $\psi(t)$ such that the triple

$$\psi(t), x(t), u(t), \quad t_0 \leq t \leq t_1,$$

is a solution of the Hamiltonian system (13.2), and the maximum condition (13.3) holds,

$$\left.\begin{aligned} \frac{\mathrm{d}x(t)}{\mathrm{d}t} &= \frac{\partial H}{\partial \psi}(\psi(t), x(t), u(t)), \\ \frac{\mathrm{d}\psi(t)}{\mathrm{d}t} &= -\frac{\partial H}{\partial x}(\psi(t), x(t), u(t)), \end{aligned}\right\} \tag{13.2}$$

$$\left.\begin{aligned} H(\psi(t), x(t), u(t)) = \max_{u \in U} H(\psi(t), x(t), u) \equiv const \geq 0, \\ \forall t \in [t_0, t_1] \end{aligned}\right\} \tag{13.3}$$

In this formulation, the maximum condition (13.3) could be viewed not only as a universal elimination method, but also as a generalization of the Legendre transformation from the state-space variables (x, u)tothe phase-space variables (ψ, x).

5 Symplectic Invariance of the Maximum Principle

It took approximately a year before a full proof of the maximum principle was found. It is not my intention here to go into the details of the history of this proof. I shall only mention that in full generality the proof of the PMP, as formulated above, became possible after V. Boltyansky's invention of the so-called "needle variations" of the control function. These variations are zero everywhere on the time-interval, except on several segments with a small total length, where they can attain arbitrary admissible values, and they have an important property of admitting an operation of convex combination, regardless of the shape of U. This was Boltyansky's major contribution to the subject.

I should like to make now some final remarks about the nature of the maximum principle and its significance for optimal control theory and calculus of variations.

The role of the Pontryagin Maximum Principle, discovered for the needs of optimal control, was twofold in mathematics. First, it created a really working device for solving important applied problems of optimization, not amenable by classical calculus of variations, and second, it gave a much broader perspective on necessary conditions of extremality in mathematics in general. Though formulated in 1955, the maximum principle was never changed, nor even slightly improved, since then. All first-order advancements were directed toward generalizations of the optimal problem itself, especially toward developing nonsmooth optimization, with corresponding first-order necessary conditions shaped after the maximum principle.

This could be explained by the very nature of the maximum principle. Despite its seemingly hard analytic form, it is deeply geometric and symplectic invariant already in its initial formulation. It prescribes a canonical way of deriving differential equations of extremals in the phase space of the problem, the cotangent bundle of the state-space manifold, on which the variational problem is formulated. Since the equations are obtained canonically, they are invariantly connected with the cotangent bundle, hence with the canonical symplectic structure of the bundle, and any information about the interrelations of the equations with the symplectic structure is a nontrivial information about the variational problem itself. We can even say that a real mathematical investigation of the variational problem starts after the equations of extremals are derived, and not after the formulation of the variational problem. The latter could be considered only as a "letter of intent", whereas the equations of extremals give us the "contract" itself, which should bring the mathematical dividends.

To reformulate the maximum principle in an explicitly invariant form, let me rewrite the initial control system in the "state-invariant" form on a smooth manifold M,

$$\left.\begin{aligned}&\frac{\mathrm{d}x}{\mathrm{d}t}=f(x,u),\, x\in M, u\in U,\\ &f_u : x\mapsto f(x,u)\in TM,\, x\in M, u\in U\end{aligned}\right\}$$

and consider the family of vector fields f_u as a family of scalar-valued functions H_u on the cotangent bundle $T^*M\xrightarrow{\pi}M$, which are linear on fibers,

$$f_u\approx H_u\in C^\infty(T^*M),\ H_u \text{ is linear on fibers.}$$

The family of scalar-valued functions H_u generates on T^*M a family of Hamiltonian vector fields $\vec{H}_u$ according to the standard relation

$$i_{\vec{H}_u}\omega=\mathrm{d}H_u,$$

where ω is the canonical symplectic form on T^*M. We thus obtain the Hamiltonian system of the maximum principle (13.2). The field $\vec{H}_u$ is the canonical lift into the cotangent bundle T^*M of the vector field f_u defined on the base manifold M, i.e., it generates on T^*M a fiber-preserving flow, mapping an arbitrary fiber T_x^*M linearly onto the fiber $T^*_{e^{tf_u}(x)}M$,

$$e^{t\vec{H}_u}: T_x^*M\to T^*_{e^{tf_u}(x)}M,\ x\in M.$$

The maximum principle asserts that the extremals of the problem are trajectories $\xi(t)$ of the nonstationary Hamiltonian vector field $\vec{H}_{u(t)}$ such that the maximum condition holds,

$$\frac{\mathrm{d}\xi(t)}{\mathrm{d}t}=\vec{H}_{u(t)}(\xi(t)),\, H_{u(t)}(\xi(t))=\max_{v\in U}H_v(\xi(t)),\ t_0\le t\le t_1.$$

If the maximum condition

$$H_{u(t)}(\xi)=\max_{v\in U}H_v(\xi),\ \xi\in O\subset T^*M,$$

eliminates the parameter u from the family H_u in some region $O\subset T^*M$, and as a result of this elimination, we obtain a *smooth* scalar-valued function (without parameters) H, the *master-Hamiltonian* of the problem, then the optimal problem considered on O is reduced to studying trajectories of a fixed Hamiltonian vector field $\vec{H}$:

$$\frac{\mathrm{d}\xi(t)}{\mathrm{d}t}=\vec{H}(\xi(t)),\ \pi\xi(t_0)=x_0,\ H(\xi(t))\equiv const\ge 0.$$

Regular problems of the calculus of variations are typical examples of this situation. Actually, this picture was envisaged by Pontryagin in his initial attempt to solve the problem.

It is remarkable that the described geometric approach permits us, practically in all interesting cases, including nonregular cases, to construct canonically from

Figure 1. According to information received from Samara State Aerospace University, the launchings of Sputnik 1 and Sputnik 2 in 1957 were made without recourse to the Pontryagin's Maximum Principle. Later tasks - bringing down cosmonauts safely, and guaranteeing that intercontinental missiles surviving a first strike would not miss New York by more than the radius of efficiency of a hydrogen bomb - did use the Maximum Principle, which was in the public domain well before that. For Sputnik 1 and 2, ideas due to Pontryagin were used to find the correct weight of the space craft (same source). [Editors' note; photo: Samara State Aerospace University Museum]

the given optimal problem a nonlinear connection on T^*M, which produces new important infinitesimal invariants of the optimal problem that are nontrivial already in the regular case. In particular, we can obtain the curvature tensor of the optimal problem, cf. [4]. If we try to derive from here global invariants of the state manifold M, for example, try to express its Euler characteristic through the curvature of the optimal problem (a possible generalization of the Gauss-Bonnet-Chern formula), we should inevitably come to generalizations of some classical relations concerning characteristic classes due to Pontryagin and Chern, which were obtained in the special case when the standard Riemannian length on the manifold M is minimized. Thus, two major achievements of L. S. Pontryagin, based on completely different ideas and obtained in different periods of his activity, might be intimately related.

I have reached the end of the story, which started with two humble Air Force engineers who ignited the mathematical genius of L. S. Pontryagin almost fifty years ago.

Summing up, we should consider the discovery of the Maximum Principle and the subsequent development of optimal control as a classical example of a theory that initially emerged as an effective device for solving a concrete engineering problem not amenable by existing methods, and eventually developed into a mathematical theory of major significance.

References

Among the first publications on the maximum principle I can indicate here [1]–[3]. In [4], construction of invariant connections is described.

[1] V. G. Boltyanski, R. V. Gamkrelidze, and L. S. Pontryagin, On the Theory of Optimal Processes, *Doklady Akad. Nauk SSSR* **110** (1956), pp. 7–10. (Russian)

[2] V. G. Boltyanski, R. V. Gamkrelidze, and L. S. Pontryagin, The Theory of Optimal Processes I. The Maximum Principle, *Izvest Akad. Nauk SSSR, Ser. Mat.* **24** (1960), pp. 3–42. (Russian)

[3] L. S. Pontryagin, *Optimal Processes of Regulation*, Proc. of the International Math. Congress, Edinburgh, 14–21. August, 1958, Cambridge UP (1960). (English)

[4] A. A. Agrachev, R. V. Gamkrelidze, Feedback--Invariant Optimal Control Theory and Differential Geometry I. Regular Extremals, *Journal of Dynamical and Control Systems* **3**(1997), No. 3, pp. 343–389. (English)

Mickey Flies the Stealth

Philip J. Davis*

This article describes in a vivid and popular way the intimate relationship between mathematics, computer graphics, entertainment and warfare.

Entertainment, mathematics and war: what a strange trio! The mixture of brilliance and glory, brutality and suffering that inheres in its story would require an Aeschylus to get the proportions right.

I recently attended a large conference of the Society for Industrial and Applied Mathematics, an international organization. The range of topics discussed by the conferees was enormous, going from fluid dynamics to biomathematical problems to econometrics to computer vision to mathematical cryptography. A major talk by Christoph Bregler on computer animation of human movements including facial expressions feeds right into the needs of this strange trio.

In talking to the conference participants, I learned that the *marriage* of war, mathematics and entertainment is now taken for granted just as they take for granted the marriage of entertainment, mathematics and medical imaging. The link in both cases: computer graphics, animation, and the formation of virtual objects.

The threefold combination first entered my consciousness forcibly when I read a 1997 report of the National Research Council entitled *Modeling and Simulation: Linking Entertainment and Defense.* This report was based on a workshop held in October 1996 that was attended by people from the film, video game, and theme-park industries and also by people from the defense department and their contractors. The object of the conference was to discuss common interests and how they might help each other.

The notions of entertainment, mathematics, and war together with their ambiguities and misapprehensions, are sufficiently understood so I shall build up to the threefold combination in stages by beginning with the twofold.

Mathematics and entertainment: This combination is ancient. Simple puzzles or games of chance and of strategy often have a mathematical underlay. More recently, sports of all kinds have experienced increasing mathematization through the accumulation of statistics and the decisions based on them. Mathematics itself

* Emeritus Professor of Applied Mathematics, Brown University, Providence, Rhode Island, U.S.A. Email: Philip_Davis@brown.edu

is often considered a game. In his introduction to Johan Huizinga's famous *Homo Ludens: a Study of the Play Element in Culture*, George Steiner wrote

> Of all human activities – particularly pure mathematics – comes closest to Huizinga's own standards of elevated play.

In education, mathematics is often promoted on the basis that "mathematics can be fun". Computer graphics, which, as we shall see, has a substantial mathematical base, is used for the simulation of humans and environments, both static and animated, and is combined with very elaborate and sometimes interactive story lines.

Mathematics and War: This also is very old, going back at the very least to Archimedes (220 BC). Niccolo Tartaglia (*La Nova Scientia*: 1537) wrote on the mathematical science of ballistics and warned that such knowledge, if widely known, could be dangerous. Since the beginning of the 20th Century, war has become increasingly mathematized in all its aspects and research in both pure and applied mathematics have been supported in a great measure out of governmental offense/defense budgets.

War and Entertainment: Today's computer games are mostly simulated battles. Though virtual war flourishes via the computer, it did not require the computer to link war and entertainment. Homer was one of the first to provide the mix. Think of the gladiators in the Roman arenas or the mock naval battles that amused the court of the French Louies. Comic books are filled with war situations. Live reenactment of old battles currently flourishes. I recently saw on TV a reenactment of a 10th Century battle in Ireland, in full costume, between the Irish and the Vikings. Spectators were watching the engagements from the sidelines while the children ate ice cream.

> Ever since words existed for fighting and playing, men have been wont to call war a game. [Huizinga]

Combining all three elements we have: Entertainment, mathematics and war: Though it would be easy to give simple examples from antiquity, chess for example, I'll limit my discussion to recent instances where the possibility of applications to armed conflict is the principle motivation. The link I want to stress is simulated actions and environments achieved through computer graphics. On the entertainment side, computer graphics appear increasingly in Hollywood productions. Many of us have seen *The Perfect Wave*. Computer graphics is used increasingly for military instruction and strategic matters. And computer graphics is solidly based on mathematical theories and constructions (algorithms) combined with incredible developments in computer hardware.

* * * *

What precisely are the areas of overlap between entertainment and defense? The principal ones, according to the 1997 report of the National Research Council are

1. Virtual reality technologies; the creation of simulated immersive environments against which defensive-offensive action takes place.
2. Rapid communication between many – possibly thousands – of simultaneous players (i.e., combatants).
3. Computer-generated characters "that model human behavior in activities such as flying an aircraft, driving a tank, or commanding a battalion such that the participants cannot tell the difference between a human-controlled force and a computer-controlled force."

What are the differences between the entertainment and the military cultures? The goal of the military is to win battles, to defend one's country. The goals of entertainment are to create products that capture the public's imagination strongly, hook the public on such products, and make money. But, according to Capps, McDowell, and Zyda of the Naval Postgraduate School, the final products of both groups are "ultimately judged on immersiveness, ease of use and realism."

It would appear that with the investment of billions of dollars in computerized filmography and video games, the entertainment industry has been ahead of the military in all three of the above areas of mutual interest despite the computer sophistication of the military. E.g., the Internet was initially a military communication arrangement that rapidly exploded into worldwide usage and misusage.

The military therefore can communicate its simulation needs to the entertainment world and subcontract for the latter's specialized products – a very good thing for sales. Reciprocally, assuming freedom of communication, the computer entertainment world can benefit from the techniques developed independently by the military.

Simulated warfare can be bought or downloaded from the web and played. According to Capps, McDowell and Zyda, "the Navy recently began training prospective student aviators using a commercial off-the-shelf- flight simulator and found that it improves their performance." While the general goals of the trio of war, mathematics and entertainment are in the open literature, the mathematical or programmatic details of products are either company or militarily confidential.

The principal computer games have scripts that set up conflicts between the human player or numerous other players and the computer. The rules of engagement can be made as complex as desired. Thus, a description (you can find it on the Web) of a game called *Mech Commander 2* says, in part

> On the full 3D (three dimensional) battle field, you will control movement, targeting, and engagement tactics of your mechs [i.e., equipment]. Call in support elements such as artillery or air strikes, scout choppers and sensor probes, even a salvage team to salvage downed mechs. Capture resource buildings or resource trucks for more 'support points'.... It's all up to you.

Figure 1. On the first evening of the *International Scientific Meeting on "Mathematics and War"*, organized by Prof. de Gosson of Blekinge Institute of Technology and held in the historic Swedish military port of Karlskrona, August 29–31, 2002, the author gave a public lecture in the local Naval Warfare Museum. [Courtesy: Ch. de Gosson]

The military can outline its own specialized scenarios for training or logistical purposes, e.g., dealing with an ambush, freeing hostages. In order to make a forceful training impact, the virtual environment against which the scenarios are played must be tremendously realistic. Game producers are now talking about "haptic" applications (i.e., pertaining to the sense of touch) that go beyond sight and sound by wiring the players. Even without such advanced technology, cyberartists have been able to produce violent reactions such as nausea or convulsions in viewers.

The achievement of realism has a strong mathematical base. In modeling: The characters in computer games (people, equipment, are built from graphical prototypes, such as spheres and cylinders. Other primitives are built up from meshes and mathematical spline curves, the latter having been developed during World War II and saw early employment in the design of airplane, ship, and automobile surfaces. I understand from talking to mathematical technologists in the auto industry that this industry now does a good fraction of its design work using graphic packages and immersion into virtual reality constructions.

In articulation and animation: Motion through space is guided by mathematically defined spline curves. Differential calculus applied to the spline curves can, for verisimilitude, determine e.g., how to compute the direction of the tires in a moving truck. Martha Gregg and Paul Davis note how a realistic image of the surface of a waving flag is made:

> A sine wave plus noise is imposed on the vectors normal to the triangular mesh elements defining the surface.

In shading and reflections: These use textured micropolygons and surface normals. Capps, McDowell and Zyda point out that producers are claiming 66 million textured polygons per second, whereas 4.8 billion are required to make computer images absolutely indistinguishable from reality. It is predicted that advances in computer hardware will reach this goal shortly.

I would like to add parenthetically, that my field as a professional mathematician is known as "approximation theory". This concerns itself with the representations of arbitrary curves or data through mathematical formulas. By training up students in the techniques of this field, I had a small finger in the early development of CAGD (computer aided geometric design). At the time I could hardly have foreseen the widespread uses of this mathematical technology.

The triangle is now closed: mathematics-entertainment-war-mathematics, and we can move back and forth around its sides. The products of these links can be employed for military training on the ground, in the air, or on the sea, or for the study of strategic and tactical alternatives.

For recreational purposes, the links can provide history buffs or partisans with the ability to sit in front of their terminals and by merely clicking, refight the Peloponnesian War, the Battle of Lepanto or the capture of Paschendaele (a half million lives lost in three months), all made realistic by mathematical modeling. Virtual casualties in their last agonies can be strewn about a 3D landscape by random number generators. The line between real and virtual war becomes more and more blurred. The Great Computer Game becomes a military training ground or a Grand Guignol Theatre with audience participation.

Could the virtual ever replace the real? If simulated and networked military engagements, played over and over again, resulted in the virtual destruction of most players, then some small measure of sense might seep into the collective brain of humanity. Could we achieve through mathematical simulation what the humane philosopher William James sought for and despaired of finding: a moral equivalent of war. Could we achieve the dream of Leibniz: that if a dispute arose we would standardly say: let us compute. This would indeed be the glory of mathematics.

References

Michael Zyda and Jerry Sheehan (Eds.): *Modeling and Simulation: Linking Entertainment & Defense*, National Academy Press, 1997.

Michael Capps, Perry McDowell & Michael Zyda: A Future for Entertainment-Defense Collaboration, *IEEE Computer Graphics & Applications*, January/February 2001, pp. 2–8.

Michael Zyda: *The Naval Postgraduate School MOVES PROGRAM – Entertainment*

Research Directions," Proceedings of the Summer Computer Simulation Conference, Vancouver, 16.–20. July 2000.

See also websites for Moves Institute (U.S. Naval Postgraduate School)

James Der Derian: *Virtuous War – Mapping the Military-Industrial-Media-Entertainment Network*, Chap. 7, Westview Press, 2001.

II
Perspectives from the Military

War Cannot Be Calculated

Svend Bergstein*

"Cry Havoc! And let slip the dogs of war."
Shakespeare, *Julius Caesar* III.i.270

War is throughout influenced by policy, but who can predict political interventions? War is about breaking soldiers' will to fight, but who can predict when it happens? In war, for a number of reasons, small deviations and misunderstandings occur, and they start avalanches of misfortune which only the iron will of a commander can attenuate, but never stop completely. It is called friction. For all these reasons war cannot be calculated.

1 Introduction

1.1 Setting the stage

"When the dogs of war slip loose accurate prognosis is almost impossible; the surreal becomes the typical."[1] Shakespeare's havoc and the quoted surreal represent the semantics of most principal statements one can find in the so-called Western thinking on this particular phenomenon – war. The "fog of war" is another all embracing expression, which has been widely accepted as a dictum. To go to war means to enter a fog of uncertainty.

In order to cope with this, specified authority to react according to circumstances is delegated down the chain of command. The origin of this particular approach to warfare is found in the Prussian/German General Staff, which existed 1657–1945. So far, this system has proven itself effective.

It should be noted, however, that the Marxist-Leninist thinking on war was – and in some parts of the World still is – based on the appreciation that war can be waged in a scientific manner, i.e., within the framework of the dialectics of historical materialism. It meant that the armed forces had to fight in accordance with rigid rules.

* Lieutenant Colonel (ret.), Former Minster for Science and Technology, Danmark.
Email: svend.bergstein@mail.dk

[1] Both quotations from *The Dogs of War*.

Especially during the Stalinist era, the Soviet Union had to pay dearly for this. In the winter war (1939) against Finland the Red Army initially suffered terrible losses to a seemingly very inferior, but determined Finnish army, which fought Western style.

Equally astonishing – on the East front, after having suffered terrible losses at Moscow (December 1941) and Stalingrad (December 1942), the German army succeeded in fighting a series of successful delaying battles against an – on paper – increasingly superior enemy. Against all odds, the German army succeeded in maintaining a cohesive defense until January 1945.

It should be noted, too, that in the so-called Eastern thinking on war, pragmatism always has been predominant:

> To the great Chinese military writer Sun Tzu, war was neither a duel nor a sport like contest. Rather, it represented one way, and a rather undesirable way, of settling disputes between social and political groups. For him it was a question, not of honorably meeting an opponent face to face, but of settling the dispute with the least possible disturbance of cosmic harmony, or dao. [...] In marked contrast to the West, the East has always considered the ambush not only perfectly respectable but much the most effective means of waging war. This leads to an emphasis on traps of all kinds, from the feigned retreat that was the specialty of the Mongols to the excreta-smeared bamboo stakes that so angered the Americans in Vietnam. When war involved a struggle between civilizations whose patterns of thought differed in this way, the results were often markedly savage. Unable to agree as to what constituted acceptable behavior and what did not, each side took the enemy‘s action as a pretext for reprisals.[2]

In brief, we see three archetypical approaches to what war is really about. There are, of course, other variations of organized violence: Symbolic tribal wars, for example, which involved very little bloodshed, have been used as an entry to manhood, and clan and family feuds have lasted for generations. But these activities can hardly be defined as wars in this connection.

Large scale terrorism is excluded as well as an act of war. The current so-called[3] humanitarian laws of war forbid it. These laws define – broadly speaking – that it is states only that can wage war, with one exception. Rebels, who are in control of a certain area and are able to live up to a number of requirements, are protected by the same laws.

Wars fought with an extensive and continuous use of weapons of mass destruction (nuclear warheads, biological or chemical agents) against large population centers have not been seen yet. However, there is no reason to believe that such wars would be less chaotic than so-called conventional war. This type of war will not be discussed in the paper, as we – fortunately – have no experiences to draw on.

[2] Creveld, *Technology and War*, p. 73.

[3] So-called, because it is a set of international conventions which are not fully accepted by all countries. Add to this that law enforcement seems to be the domain of the winners only.

Figure 1. Carl von Clausewitz at 50 years of age in a Prussian general's uniform. Lithography: Franz Michelis jr. after a painting by Wilhelm Wach. [Courtesy: Chris Bassford, National War College, Washington]

1.2 The main sources

In order to provide a common point of departure for a discussion of the topic in question I have chosen to use Karl A. Popper's harsh critique of what he calls "historicism" in his main work: *The Open Societies and its Enemies*. This is, of course, an *ideological choice*, and as such disputable.

My particular arguments concerning the unpredictability of war rest on the Prussian officer Carl von Clausewitz's book *On War.*[4] This choice is obvious, as Clausewitz's analysis of the relationship between politics and war, and of war itself, has had, and still has, a marked impact on the central European and American perception of war. Lenin studied On War as well, but failed to draw a lesson from it.

2 True Science?

Can the phenomenon "war" be approached in a truly scientific manner? It is obvious that the equipment of war (weapon systems, communications systems etc.) is to be regarded as a product of mathematics, natural sciences and technological development.

But how do we handle the imponderables of war? I return to this, but at first it is necessary to define what war is. It is obvious that war is evil, but war is also a

[4] Since this work is quoted copiously in the following it is referred to by the abbreviation OW, meaning the edition [Clausewitz 1982].

matter of fact. It is also a matter of fact that it is people who fight wars. The international society has in a number of conventions agreed on the legal definitions of which situations can be regarded as war and what is acceptable behavior in war.

Clausewitz invested much effort into uncovering how war can be defined as a phenomenon. Doing so, he realized that the relationship between policy and war is the key to uncovering the mysteries of war.

The following quotations from *On War* lead to a definition of war:

> War is the continuation of policy by other means. We see, therefore, that war is not merely a political act, but also a real political instrument, a continuation of *political intercourse,*[5] a carrying out of the same by other means.[6]

In a note from 1827 about plans for a revision, Clausewitz called attention to "...that *war is nothing but the continuation of state policy with other means.*"[7] And he stressed that by keeping this in mind, the reading of *On War* would be easier.

Clausewitz's initial definition of war in *On War* was "*War therefore is an act of violence intended to compel our opponent to fulfill our will.*"[8] (Clausewitz underlined this sentence.)

Elsewhere in *On War,* Clausewitz several times returned to the theme that war is an instrument for (foreign) policy, and he distinguished between the political purpose of war and the manifest goals "inside" the war.

In connection with a discussion over whether or not war is a science or an art, Clausewitz concluded:

> War is part of the intercourse of the human race. We say therefore war belongs not to the province of arts and sciences, but to the province of social life. It is a conflict of great interests which is settled by bloodshed, and only in that it is different from others. It would be better, instead of comparing it with any art, to liken it to business competition, which also is a conflict of human interests and activities; and it is still more like state policy, which again, on its part, may be looked upon as a kind of business competition on a great scale..."[9]

It is obvious that organized violence with a purpose has existed as far back in human existence as archaeology makes sense. We also find an abundant number of historical indicators pointing in the direction that war is a social phenomenon both within and among the warring parties. War *is* a conflict between people, and wars have always been decided upon and fought by people using whatever technol-

[5] In the source, the word (political) "commerce" is used. In another translation form 1976 by Michael Howard and Peter Paret, the word "activity" is used. I find a Danish translation "samkvem" (intercourse) more correct. (See list of sources)
OW, p. 119. (Bk. I, ch. 1: *War is a mere continuation of policy by other means*).

[6] ibid.

[7] My translation from the Danish text.

[8] OW, p. 101. (Bk I, ch. 1: *Definition*).

[9] ibid., p. 202. (Bk. II, ch. 3: *War is part of the intercourse of the human race*).

ogy is at hand. Wars have always been started because one or more parties at the moment of decision have found war either beneficial or unavoidable.

The ability to use organized violence (the military capabilities) can in itself be a motivating factor, if one of the parties finds itself militarily superior. But it does not exclude the fact that it is people who decide to go to war, plan and fight wars, campaigns and battles, and perform the combat, and it is people at all levels of war who make the final decision when enough is enough on their part.

I have chosen to consolidate the above observations in the following definition of war:

> War is the continuation of policy with other means, a political, social and human intercourse, which includes the use of organized violence for a political purpose.

It should be noted that this definition takes into consideration that diplomatic activities take place during war.

The decision makers at all levels of war have, since the beginning of the industrialization with an ever accelerating momentum, been supported by technology. Today this support ranges from off-the-shelf commercial communication and information systems to government controlled, secret complex simulations supported by extremely powerful computers.

But we have to bear in mind that war is a "dialectic clash", where the decision makers on both sides and at all levels try to establish a number of suitable options, estimating or calculating ends and means, causes and effects. But would it be chimerical to expect any estimate or calculation to provide a lasting solution? The answer to that question is what this paper is all about.

In view of the chosen definition of war, it seems to be an obvious choice to turn to social science in order to find a way to provide a deeper insight into the dynamics of war.

3 Popper

Popper's views on the social sciences are found in the following quotations:

> The future depends on ourselves, and we do not depend on any historical necessity. There are, however, influential social philosophies which hold the opposite view. They claim that everybody tries to use his brain to predict impending events; that it is certainly legitimate for a strategist to try to foresee the outcome of a battle; and that the boundaries between such a prediction and more sweeping historical prophecies are fluid. [...] They also believe that they have discovered laws of history which enable them to prophecy the course of historical events.[10]

[10] *The Open Society and its Enemies (OSE)*, volume I, p. 3.

> [...] in spite of their plausibility they are based on a gross misunderstanding of the method of science, and especially they neglect the distinction between scientific prediction and historical prophecy.[11]
>
> [...] in science, we never have sufficient reason for the belief that we have attained the truth. What we usually call "scientific knowledge" is, as a rule, not knowledge in this sense, but rather information regarding the various competing hypotheses and the way in which they have stood up to various tests [...] This view means, furthermore, that we have no proofs in science (excepting, of course pure mathematics and logic). In the empirical sciences, which alone can furnish us with information about the world we live in, proofs do not occur, if we mean by "proof" an argument which establishes once and forever the truth of a theory. (What may occur, however, are refutations of scientific theories). On the other hand, pure mathematics and logic, which permit of proofs, give us no information about the world, but only develop the means of describing it. Thus we could say: "In so far as scientific statements refer to the world of experience, they must be refutable; and, in so far as they are irrefutable, they do not refer to the world of experience." But although proof does not play any part in the empirical sciences, arguments still does; indeed, its part is at least as important as that played by observation and experiment.[12]

But Popper's cure for historicism is at hand:

> The only course open to the social sciences is to forget the verbal fireworks and to tackle the practical problem of our time with the help of the theoretical methods which are fundamentally the same in all sciences. I mean the methods of trial and error, of inventing hypotheses which can be practically tested, and of submitting them to practical tests. *A social technology is needed whose results can be tested by piecemeal social engineering.*[13]

What did Popper mean by "social technology"? I consulted a comprehensive dictionary[14] and found that the adjective "social" has a number of applications, ranging from
- "of or pertaining to society or its organization" to
- "grouping compactly".

Concerning the noun "technology", I found it can be defined as
- "theoretical knowledge of industry and the industrial arts" and as
- "the application of science to the arts" and as
- "that branch of ethnology which treats the development of the arts".

Concerning the noun "arts", I found it can be defined as
- "the skilful, systematic arrangement or adaptation of means for the attainment of some end, especially by human endeavor as opposed to natural forces" and as

[11] ibid.

[12] OSE, volume II, p. 13.

[13] ibid., p. 222.

[14] Webster's comprehensive Dictionary of the English Language, encyclopedic edition.

- "the practical application of knowledge or natural ability; skilled workmanship; mastery; dexterity" and as
- "a set of system or rules, principles etc., devised for procuring some scientific, esthetic, or practical result, as by exercise; a branch of learning to be studied in order to be applied."

The following example illustrates my interpretation of Popper's requirement for a scientific approach, based on the above specifications.

Within this particular field of social and human intercourse,[15] war, Popper's cure for historicism has been used for as far back in history as we have credible historical facts at hand. Take the ancient Greek phalanx, the very structure of the city state army. In a psychological perspective it can be regarded as a product of social technology, the purpose of which was to maintain social coherence[16] during battle. The main components of this particular social technology were the following:

- Knowledge of, and conviction in, the value of ones own rights and duties as a free citizen.
- Knowledge of, and conviction in, the necessity of discipline (in the sense of voluntary obedience).
- Knowledge of, an ability to handle, and possession of up to date weapons and armor.
- Knowledge of, and an ability to execute proven effective tactics.
- Conviction in the proven leadership qualities of commanding officers.
- A deliberately amplified, culturally induced sense of honor and dishonor.

The need for this social technology was extreme. It was a matter of life or death to maintain social coherence within the phalanx during battle. If the coherence broke, the phalanx would decompose and the individual foot soldier, the hoplite, would be an easy prey for the pursuing enemy cavalry.

The result of the social technology, the coherence, was tested by piecemeal social engineering, for example adjustments in tactics and change of leader, in order to improve the coherence in the next battle.

This process still goes on. We may use expressions as "team building" or "indoctrination" to classify the process, but social coherence is still extremely necessary within political crisis management groups, military staffs, military formations, tanks, aircraft, submarines etc. – not to forget infantry squads.

But how do we – in the mathematical context we deal with here – quantify the components in this particular social technology and measure the effects of any social engineering? Add to this that it is only to a minor degree that we can test the result, the coherence, through questionnaires, simulations, and peacetime

[15] One can speculate if there is a distinction at all between social and human intercourse?

[16] I use the word "coherence" in order to indicate the product of a logical proces. "Cohesion" is used as a more "loose" concept.

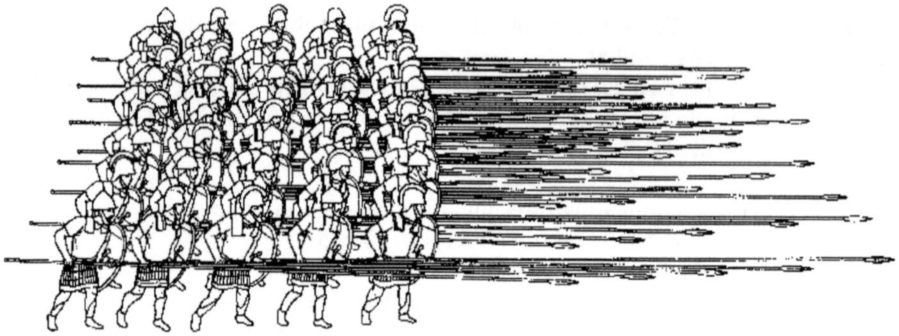

Figure 2. The first archeological proofs of disciplined attacks with massed infantry formations using pointed weapons are dated to app. 650 B.C. and are accredited to the Greek hoplite phalanx. Some of the last attempts to carry massed charges home with the bayonet as a primary weapon were done 1914–1915. The picture shows a fraction of a Macedonian phalanx as it was used by the emperor Alexander the Great (356–323 B.C.). [Illustration from *The World Atlas of Warfare*, Mitchell Beazley Publishers 1988]

exercises. The final test is combat, and only then we will know for sure. Here we find the first major source of uncertainty:

1) The social coherence of military units cannot be tested precisely during peacetime.

Social sciences may provide deeper insight in the dynamics of war and warfare. But we have not solved the crucial problem: It is possible to predict the outcome of war and battles.

Could mathematics provide the answer? Let us turn to Clausewitz for a more particular analysis of *war as a phenomenon* and *warfare as an activity.*

4 Clausewitz

4.1 How do we read Clausewitz?

Clausewitz was a child of the period of enlightenment in the last half of the 18th century, where philosophy and sciences flourished in Central Europe. His war experiences were the Napoleonic wars. He tried to establish an all-embracing theory on war as a phenomenon.

He died before he could finish his work. In a note on revision, probably from 1830, he stated that only the first chapter of the first book was fully finished. In this chapter he answers the question: What is war? He regarded the rest of *On War* as a collection of materials, where only the main lines can be used to recognize what war and warfare are all about.

When reading Clausewitz we have to keep a few preconditions in mind:

- We must try to understand the mindset of a well educated and intellectual officer who lived 1780–1831 in Central Europe.

- We also have not just to translate his contemporary German into modern German but we also, if desired, have to translate the modern German into any other modern language we master. These are no easy tasks at all. Important nuances may disappear and concepts be twisted.
- And most important – we have to be very careful not to accept concepts, information and conclusions which simply are outdated. Clausewitz's introduction to *On War* offers a clue where to be cautious.

> We propose to consider first the single elements of our subject, then each branch or part, and, last of all, the whole, in all its relations – therefore to advance from the simple to the complex. But it is necessary for us to commence with a glance at the nature of the whole, because it is particularly necessary that in the consideration of any of the parts their relation to the whole should be kept constantly in view.[17]

In the introduction to an unpublished manuscript on the theory of war, written between 1816 and 1818, he wrote:

> Thinking and observation, philosophy and experience must never despise or exclude each other – they mutually stand surety for each other.[18]

Obviously Clausewitz's ambition was to develop an all-embracing coherent theory from a conceptual structure – and to test it against reality. Though it is not mentioned in the introduction to chapter 1, in *On War* he tried to test both the conceptual framework and his theory against the reality he had at hand from historical sources and own experiences.

But we have to consider, if some – or even many – of the single elements or branches or parts in his conceptual structure may be irrelevant or plain obsolete.

Some guidance can be found in the following discourse:

> The description of phenomena and the communication of experiences are two principal and interdependent functions of language. The unambiguous description of a particular phenomenon excludes from consideration all other similar phenomena, but if the experience of the particular phenomenon is to be communicated to another person who has not observed the phenomenon, the description must in some way transcend the particular case. Otherwise, of course, communication is not possible. *Thus the interaction between particularity and generality must be fundamental in linguistic communication.* As an example one may take any noun from ordinary language. A noun has a dual function: first, it denotes a particular phenomenon or experience; second, it is a concept, that is, it also denotes a particular class of phenomena or experiences. The context in which the noun occurs may focus the attention on the particular significance, but the conceptual significance is the knowledge by means of which the particular case is understood.[19]

[17] OW, p. 101. (Bk. I, ch. 1: "*Introduktion*").

[18] My translation from a Danish translation of *On War*.

[19] T. Bergstein, *Quantum Physics and Ordinary Language*, p. 1.

A few remarks on the levels of war are needed, too. In brief, Clausewitz's perception of war and warfare were the following:

1. The political level generates the purpose of war.
2. The strategic level is responsible for the deployment of the armies for the battle.
3. The tactical level is responsible for the maneuvers, which bring the army units in contact with the enemy.
4. The combat level is where the actual use of violence takes place.

However, Clausewitz indicated that the above stratification was insufficient as far as both strategy[20] and tactics[21] are concerned. Later the German general staff introduced an operational level, the purpose of which was to establish a domain of their own, where the realities of warfare could be handled by professionals.

This construction was much disputed in the rest of Europe, because the German General Staff, besides its legal activities, also sought to influence policy making and committed war crimes during both world wars. But the construction proved itself effective, and today the following stratification is generally accepted:

1. The political level generates the purpose of war.
2. The strategic level is responsible for providing the overall idea of how to fight the war (the strategic concept).
3. The operational level is responsible for the deployment of the armies, navies, and air forces for the war.
4. The tactical level is responsible for the maneuvers, which bring the military units in contact with the enemy.
5. The combat level is where the actual use of violence takes place.

These levels of command are, of course, not sharply separated, but represent a continuum. Statesmen at Clausewitz's time often were mostly occupied with preparing for or fighting wars and thus were supposed to have insight in military matters. Napoleon was the last head of state who was able to manage war between states down to and including the tactical level. Stalin and Hitler tried too, but failed. Stalin realized his failure, when the German army threatened Moscow. Hitler continued to the very end.

Clausewitz had a somewhat dubious attitude to the concept "the art of war". In book II, "On the Theory of War" he initially used it to define all activities in war,[22] including the formation of units, logistics etc. Later in book II he used a whole chapter[23] to discuss "Art or Science of War", where he concluded: "We say therefore war belongs not to the province of art and science, but to the province of social life." (As I have quoted earlier.)

[20] OW, p. 243. (Bk. III, ch. 1, section beginning with: *But let us admit: there is no question here* [...] (ending with) [...], *the moral quantities are already reduced to a very small number.*)

[21] OW, p. 173. (Bk. II, ch. 1, section beginning with: *That doubtful cases may occur,* [...] (ending with) [...] *for the passage of a river at several points, etc.*)

[22] OW, p. 172. (Bk. II, ch. 1, section beginning with: *The Art of War is therefore* [...] (ending with) [...], *belong to the Art of War.*)

[23] OW, pp. 201-203. (Bk. II, ch. 3).

Today we use two expressions, more or less at random; "art of war" or "operational art". Both are really the same concept, and refer to the operational level, where the strategic intentions (the strategic concept) are transformed into operational plans; i.e., orders specifying, which units are going where, doing what, and when.

Let us turn to one question which is rather important in this connection: What did Clausewitz mean by "laws of probability" and "calculation of probability" - concepts he refers to in *On War*?

4.2 Probability

It is basic knowledge that all probability calculation in the end must rest on empirical data. And it should be noted that when attempting to predict the *outcome* of a war and the *overall progress* of the warfare "within" it, probability calculation must *begin* with gathering empirical data in order to be able to calculate the end state.

Now - land warfare in the Napoleonic era was a rather cumbersome affair measured with nowadays' standards. All armies moved by foot or horse, and alone the problem with providing fodder for the horses caused predictable stops in movements. All armies were much alike, too. There were three main branches, infantry, artillery, and cavalry. Weapons were by and large the same: Muzzle loaded guns and cannons, bayonets, sabers and lances.

Thus - if a commanding general of an army had reliable intelligence concerning the quantity, composition and whereabouts of an enemy army and equally reliable information concerning the terrain (including the possibility of providing fodder and food), it could not have been too difficult to estimate or calculate for example the probability of the position of that particular army at a particular time if it marched along a particular route. The following quotation supports this interpretation:

> But the slower the action proceeds in War, the more frequent and longer the periods of inaction, so much more easy can an error be repaired; therefore so much bolder a General will be in his calculations, so much more readily will he keep them below the line of the absolute, and build everything upon probabilities and conjecture. Thus, according as the course of the War is more or less slow, more or less time will be allowed for that *which the nature of a concrete case particularly requires, calculation of probability based on given circumstances.*[24]

In other words - Clausewitz pointed out that the premises must be specific. But the problem with providing reliable in-time intelligence was significant:

> Great part of the information (i.e. intelligence reports, my comment) obtained in War is contradictory, a still greater part is false, and by far the greatest part is of a doubtful character. What is required of an officer is a cer-

[24] OW, p. 116. (Bk. I, ch. 3: "*Frequent periods of interaction in war...*")

tain power of discrimination, which only knowledge of men and things and good judgment can give. The law of probability must be his guide. This is not a trifling difficulty even in respect of the first plans, which can be formed in the chamber outside the real sphere of War, but it is enormously increased when in the thick of War itself one report follows hard upon the heels of another; *it is then fortunate if these reports in contradicting each other show a certain balance of probability, and thus themselves call forth a scrutiny.*[25]

Clausewitz most likely meant that beside the information overload (a stress factor in itself) reliable but contradicting intelligence reports for example could indicate a diversionary attack, the purpose of which was to lure the defender to concentrate forces in the wrong direction.

After having reflected on danger, courage, and the relationship between courage and prudent calculation Clausewitz concluded

And we see therefore, how, from the commencement, the absolute, the mathematical as it is called, nowhere finds any sure basis in the calculations in the Art of War; *and that from the outset there is a play of possibilities, probabilities, good and bad luck, which spreads about with all the coarse and fine threads of its web, and make War of all branches of human activity the most like a gambling game.*[26]

Had Clausewitz conceptual problems here? He integrated the concepts of possibility, probability and luck in one sentence. He seemingly had troubles in constructing a conceptual structure. On the other hand – he did not state that war is a gambling game.

Having only sporadic knowledge of mathematics myself – not to mention a complete ignorance of its state at the beginning of the 19th century – I had the above quotations and arguments reviewed by a mathematician, who suggested inserting the following:

It is worthwhile to observe, that the mathematical probability concept as well as the elementary rules of probability calculation were founded through the study of games as a means to calculate the likelihood of winnings and losses during a game (Fermat, Pascal, Huygens). The exact rules defining a game make the calculated probabilities indisputable. More advanced applications of mathematical probability theory also rest on a number of exactly defined assumptions, which make the calculated probabilities indisputable within the conceptual framework in question. In the case of warfare Clausewitz points out that probability calculations always rest on an insufficient number of real world information, which are, furthermore, often highly uncertain. If Clausewitz is right, any calculated probability of the outcome of a warfare operation must therefore be highly disputable.[27]

[25] ibid., p. 162. (Bk. I, ch. 6, second section).

[26] OW, p. 117. (Bk. I, ch. 1: *War is a game both objectively and subjectively*).

[27] T. Bergstein, author of *Quantum Physics and Ordinary Language.*

I find it reasonable to claim that Clausewitz's use of the words "calculation" and "probability" is not to be understood in a strictly mathematical sense. They are "loose" ordinary language concepts. His reflections over the concept "friction" [28] support this interpretation.

But we have to remember, too, that military forces of today are both very different from and extremely agile and maneuverable compared to Clausewitz's armies. A clarification of the current conditions, under which armies, navies, and air forces operate during war may be useful.

4.3 Current conditions for operations

Assuming that we in war are not able to gain exact knowledge of the enemy's intentions, but only of his whereabouts and force compositions, we can sketch the conditions as follows:

Ground operations
Though the dividing line between air and land operations is rather fluid (armies do transport troops with helicopters, and use helicopters as weapon systems), the overarching problem in land operations is the terrain. The decisive components of an army, the tanks and other combat vehicles, the artillery (be it rockets or cannons) and not least the immense amounts of supplies needed, can in reality not be transported with sufficient speed by air in an operations area. The number of transport aircraft needed would be enormous.

Therefore the terrain dominates army operations. Especially the so-called passability factor is important. Mountains and swamps are insurmountable obstacles for heavy army units, and other terrain types, rivers, lakes, urbanized areas etc. can hamper movements considerably. Thus it is possible to deduce the routes heavy units are able to use. This is a decisive factor in planning an army operation.

But though we may have solid intelligence on the enemy's force composition, position and previous behavior, we should only estimate his *possible* intentions by analyzing the overall situation, where the available routes of approach are a dominating factor. When doing so, in a defensive operation we will be able to identify a number of our own *possible* options. The next step in the process is to identify the option, which is best suitable to counter all of the enemy's possible intentions. (In an offensive operation, the process is reversed.)

It would be plain stupid to select the enemy's most probable intention and plan on that alone, as the enemy will do his very best to deceive us with false information. If we did so, we would run a risk to find ourselves trapped, as swift movements can only be done with airborne light forces.

[28] Discussed later under the headline **Friction**

Naval operations

In confined waters (archipelagos, river deltas etc.) we may be able to identify possible routes of approach. But the *major* naval battles are fought in open waters over long distances using aircraft from carriers and long range missiles from surface ships or submarines, including torpedoes. The primary mission will be to gain control over the sea lines of communications, in order to insure that military transports and commerce can continue even during periods of war. Other missions, such as support of land or air operations, are likely, but as a principle, they must not hamper the primary mission.

Looking at the overall strategic and operational situation, we may be able to guess the enemy's most likely intentions (there are no mountains or swamps at sea). But good intelligence on the range and effectiveness of the enemy's weapon systems, and effective real time surveillance technologies (satellites, submarines, reconnaissance aircraft and drones) are the decisive factors.

But the surveillance technologies will be targeted by the enemy's counter-technologies be it in the form of so-called hard kills by interceptors or so-called soft kills using the electromagnetic spectrum to disturb transmissions. Counter-counter technologies and tactics are developed to cope with this problem, but we will never know for certain who has the upper hand here, until the systems are used during war, or are compromised one way or another.

Thus modern open water naval warfare will have the character of probing for possibilities, that is trying to get close enough to be in-range of the enemy with one's own weapon systems without exposing one's own forces. This probing for possibilities can be done for a considerable time (if the mission allows for it) because of the endurance of the naval vessels.

Air operations

The primary mission in air warfare is to gain air superiority. That is denying the enemy the use of the airspace. As a principle this is done primarily by targeting the enemy's ground facilities, air-stations and support facilities in order to reduce the enemy's capacity to launch aircraft or air defense missiles. A part of one's own air assets have to be kept in reserve to counter enemy attacks. In reality this principle is often turned upside down, because the weaker part is forced into the defensive.

Though the technology factor is very important in air warfare, there is no probing for possibilities to be done due to the relatively short endurance of aircraft compared to ships. The situation in air combat changes rapidly due to the speed of the aircraft, and the ability to maintain so-called situational awareness is of crucial importance in order to survive. In modern aircraft, the crew‘s efforts to maintain situational awareness is supported with computerized warning devices, which can literally speak to and warn the crew and recommend evasive maneuvers and other survival measures.

Especially when it comes to attacks on enemy installations on ground by several aircraft, there is not much room for improvisations. Detailed planning is mandatory in order to ensure timing and coordination, which are vital in order to hit the target with maximum concentration and minimum exposure to the defenses on ground. If attacked en route to the target, the timing and coordination can be completely spoiled and cause serious losses to enemy defenses in the actual attack phase.

On top of this, the technological counter-counter problematic also dominates in air warfare.

Having come so far in my explanations, I will specify my hypothesis.

5 The Hypothesis

The outcome of a war and the overall progress of the activities in it, the warfare, cannot be calculated with mathematics.

It should be noted that the adjective "overall" has been used for the following reason:

At the lowest level of warfare, where for instance a single air defense missile system engages an enemy aircraft, it is of course possible to describe an idealized situation, make a proper mathematical model and feed it with empirical data. The stochastic output of such a simulation can be used to develop a *procedure*. A procedure which states how many missiles the defenders have to launch against an approaching aircraft in order to ensure a sufficiently high probability of destruction. But when we approach the higher levels of warfare, we very rapidly realize that procedures become meaningless.

In the following I have used quotations from *On War* in order to identify additional major sources of uncertainty which can support the hypothesis.

6 On War

6.1 The quotations

I did my best to keep all of the previous observations in mind when I selected the quotations from *On War.* Doing so, I especially hope that I have been able to avoid the parts of *On War* that are too particular for Clausewitz's timeframe to be valid today.

The purpose of the quotations is to support my hypothesis. This means that I have been very selective in my choice of quotations. This was no easy task, of course, not least because *On War* is an unfinished work. Chapter 1, book I, is consistent, and is the only chapter Clausewitz himself found finished. The rest of *On War* is somewhat erratic.

Book I, "On the Nature of War", is so general in arguments and conclusion, that it by and large can be regarded as "durable". To a certain extent this can be said as well about book II, "On the Theory of War". Book III–VII are in general too specific for Clausewitz's timeframe to be of use in this context. Book VIII, "Plan of War", is partly "durable". (It is also difficult, but recommendable reading everybody who occupies themselves with arguing for the use of military forces).

For each quotation arguments will be provided concerning why this specific quotation points out a major source of uncertainty, which has to be observed during any attempt to establish a scientific prediction of war and warfare.

Let us begin with war as a phenomenon and continue with the praxis of warfare.

7 War

7.1 War as a phenomenon

After a lengthy analysis of what war is all about, Clausewitz came to the following conceptual conclusion on war as a phenomenon (I have made a minor editing of the text):

> War is therefore, not only chameleon-like in character, because it changes its color in some degree in each particular case, but it is also, as a whole, in relation to the predominant tendencies which are in it, a remarkable[29] trinity, composed
> - of the original violence of its elements, hatred and animosity, which may be looked upon as blind instinct;
> - of the play of probabilities and chance, which make it a free activity of the soul; and
> - of the subordinate nature of a political instrument, by which it belongs purely to the reason.
>
> The first of these three phases concerns more the people; the second, more the General and his army; the third, more the Government. The passions which break forth in War must already have latent existence in the peoples. The range which the display of courage and talents shall get in the realm of probabilities and of chance depends on the *particular characteristics* of the General and his Army, but the political objects belong to the government alone [...] The problem is, therefore, that theory shall keep itself poised in a manner between these three tendencies, as between three points of attraction.[30]

[29] It is disputed, whether "strange" would be a more correct translation.

[30] OW, pp. 121–122. (Bk. I, ch. 1: "*Result for theory*").

Arguments
That war is "chameleon-like" and "changes its color in some degree in each particular case" is obvious: History shows that even wars which were fought repeatedly over the same stretch of land progressed differently. This was not necessarily only because of a factor of unpredictability, but could as well be caused by for example changed strategy or tactics. After a war we would know the end state and might be able to reproduce the courses of events from war diaries etc. and establish a model of what happened.

Especially the possibility of acquiring knowledge of the probability of effectiveness of individual weapon systems dueling with other individual weapon systems could provide valuable empirical data to use in probability calculations of battles. The problem is that the validity of these data would quickly erode as major weapon systems (tanks, aircraft, naval vessels etc) are continuously modified in order to improve their effectiveness and survivability.

Summing up
2) The validity of empirical data gathered in war quickly erodes.

Let us look more closely on t*he passions, the particular characteristics*, and the *political objects* as they are described in the above quotation.

7.2 Passions

> The passions which break forth in War must already have latent existence in the peoples.

Arguments
Passions are without doubts important for the will to fight and the social cohesion of the people in general, including the armed forces. But how can we measure the degree of hatred and animosity, which break forth in war, with sufficient precision to be of value in a probability calculation? Opinion polls could be an option, but could we really rely on answers given in a much-stressed situation? Would it be possible at all for practical reasons? Would the polling itself push passions? And is the government interested in "honest" opinion polls at all? The results could be devastating for the policy.

Another matter is that opinion polls could disclose the latent passions in the people before war was even threatening. This could be useful for a government which for one reason or another wanted to go to war. But this would only be one of many deciding factors in the decision making process, whether or not to go to war.

But experience shows that passions can change rapidly in an information open society. The Vietnam War is a gloomy example. At the beginning of the war, the American people in general had a neutral or positive attitude towards the war. The positive attitude was only steeled when the American casualties began to show. But as the casualties increased, a breaking point was reached and the general attitude became more and more negative. An amplifying factor was the TV-transmissions of filmed and rather horrible events in the war. In the end it was public reaction which made the war policy impossible.

Summing up:
3) We cannot predict the impact of passions on the war effort.

7.3 Particular characteristics

> The range which the display of courage and talents shall get in the realm of probabilities and of chance depends on the particular characteristics of the General and his Army [...].

Arguments
This is without doubt true, but how should we be able to measure the courage and talents of the Commanding Officers in the chains of command in the armed forces? And if it was possible – how could we provide exact quantification for the whole of the armed forces? Especially when we take into consideration the metaphor that "a chain is not stronger than its weakest link."

Thus:
4) We cannot quantify the courage and talent of the armed forces as a whole.

7.4 Political objects

War is "[...] *of the subordinate nature of a political instrument, by which it belongs purely to the reason.*"[31] and "[...] *but the political objects belong to the government alone.*"

Arguments
This raises the question, whether politics during war is reasonable. Apart from the fact that governments in democracies do exercise political control of the war efforts within the limitations of a political system of checks and balances, the leaders of states at war are under heavy psychological pressures – in brief, stress.

A long period of continuous excessive stress can do strange things to some people. The cognition processes can be disturbed to such a degree that the person in question is seriously hampered in his efforts to perceive, understand, and learn from events or experiences.

[31] OW, p. 121. (Bk. I, ch. 1: *"Result for theory"*).

On top of this, there is a significant risk that the political and military top leaders simply do not understand the messages sent to them by their enemy counterpart, be it by diplomacy or via television, because the cultural gap is too wide.

Some of us may remember how Saddam Hussein on television asked a hostage, a little boy, whether he had had cornflakes and milk for breakfast. By doing so, he sent a message to the majority of Arab people in the Middle East, that the hostages were not hostages at all, but guests, as it is an Arab tradition, that when you have served a stranger bread and milk, he is accepted as a guest and thus protected by the host. To most of the world the remark seemed so absurd that it raised questions on the man's sanity.

The combined effects of stress and misperception do not promote reason. And there is a considerable risk of a reciprocal escalation of misperceptions. Therefore we cannot rely on a reasonable (in the sense "objective") political decision making process during war.

The following quotations from *On War* paint an even more complex picture of the influence of policy making on war and warfare:

> Policy, therefore, is interwoven with the whole action of War, and must exercise a continuous influence upon it, as far as the nature of the forces liberated by it will permit.[32]
>
> If the aim of the military action is an equivalent for the political object; that action will in general diminish as the political object diminishes, and in a greater degree the more the political object dominates. Thus it is the explanation how, without any contradiction in itself, there may be Wars of all degrees of importance and energy, from a War of extermination down to the mere use of an army of observation.[33]

(More) arguments

The continuous political influence upon warfare ("the whole action of War") can have both positive and negative effects:

During the Cuban Missile Crisis in 1962, President J. F. Kennedy prevented that the cold war (a war of mutual military observation) exploded into a war of extermination. Among other measures taken, he did that by direct control of the individual U.S. warships which confronted the Soviet warships in the waters off Cuba.

During the Second World War, on the East Front, NAZI Germany fought a war of extermination. Before the German onslaught, Stalin for his part was unable to perceive what was coming (he may have suffered from cognition disturbances). He forbid his generals to make preparations, and his political commissars through a terror regime made sure that his orders were followed.

The history of war is filled with evidence of both successes and disasters caused by political influence upon warfare, down to minute details.

Summing up

5) The effects of political influence upon warfare are unpredictable.

[32] ibid., p. 119. (Bk. I, ch. 1: "*War is always a serious means for a serious object*").

[33] ibid., p. 110. (Bk. I, ch. 1: "*The political object now reappears*").

7.5 Context for a theory

Clausewitz also found he here had established the context for a theory of war:

> These three tendencies, [the remarkable trinity of war, Bergstein] which appear like so many different law-givers, are deeply rooted in the nature of the subject, and at the same time variable in degree. A theory which would leave any of them out of account, or set up any arbitrary relation between them, would immediately become involved in such a contradiction with reality, that it might be regarded as destroyed at once by that alone. The problem is, therefore, that theory shall keep itself poised in a manner between these three tendencies, as between three points of attraction.[34]

That the three tendencies are variable in degree, I have just pointed to. But what Clausewitz meant with "poised between points of attraction" is a little mysterious.

But in Book II,[35] chapter 2, he after a lengthy argumentation identified the premise for a theory for the conduct of war (i.e., warfare); a headline stating that:

> Theory must be of the nature of observation, not of doctrine.[36]

Under this headline he analyzed the premise and came to a conclusion written in a rather florid language:

> If maxims and rules result themselves from the considerations which theory institutes, if the truth accretes itself into that form of crystal, then theory will not oppose this natural law of the mind; it will rather, if the arch ends in such a keystone, bring it prominently out; but so does this, only in order to satisfy the philosophical law of reason, in order to show distinctly to the point to which the lines all converge, not in order to form out in an algebraical formula *for use upon the battlefield*; for even these maxims and rules serve more to determine in the reflecting mind the leading outline of its habitual movements than as landmarks indicating to it the way in the act of execution.[37]

With the reference to the battlefield we now leave the concept of war as a phenomenon and enter the concept of warfare.

[34] OW, p. 122. (Bk. I, ch. 1: "*Result for a theory*")

[35] ibid., pp. 180–190

[36] ibid., p. 190.

[37] OW, pp. 191–192. (Bk II, ch. 2: "*Theory must be of the nature of observation, not of doctrine*").

8 Warfare

8.1 Doctrines

Before we go further, a few comments on the word "doctrine" are needed. I find that the word "learning" would have been a better translation of the original text, as it indicates that the aim of theory-making cannot be a textbook in itself, but rather reflections over reality. In "Western" military language of today the concept "doctrine" means "broad guidelines for waging war" at all levels of warfare, i.e., the operational, and tactical.

Arguments
What Clausewitz wrote in the last three lines in the last quotation is a definition of military doctrines, as they have existed for at least a hundred years under names as "Field Manual" or "Field Regulation" or "Air force Doctrine" etc. Doctrines of today are continuously reviewed by study groups, in order to keep them up to date with the latest developments within threats, technology, organization etc. These reviews can be supported by operations analysis, for example in order to study the probable effects of new weapon systems in a number of scenarios.

As said, doctrines are found at the operational and tactical levels in the armed forces. The army, the air force, and the navy, each branch and each type of military units have their own doctrines. There is an obvious need for that, as the guidelines for operational art and tactics must be different for tank units and infantry, aircraft carriers and submarines, bombers and fighter planes and so on. But cruel experiences in warfare show that even the most well thought out doctrines can fail, if for example ones own tanks or tank crews show themselves inferior to the enemy's for one reason or another.

Summing up
6) Even broad guidelines for waging war can fail.

A natural question to ask here is: Are we able to measure or estimate – by and large – the relative strength of the opposing military forces?

8.2 Relative strength, the problem

In case one part is overwhelmingly strong, is it then the sure winner? In order to answer this question, we have to make a detour into reality. Thereafter the question will be answered under the heading: Relative Strength, the Answer.

Our detour into reality begins with two questions:
1. What does it mean to "win" a battle?
2. What is military power really?

Let us turn to *On War* for an answer:

> The military power must be destroyed, that is, reduced to such a state as not to be able to prosecute the war. This is the sense in which we wish to be understood hereafter, whenever we use the expression "destruction of the enemy's military power".[38]

"Military power" is in itself as general a concept as it can be. The military power is the armed forces of a country, which are composed of army, air and naval forces, which again are divided in a number of branches, each with a number of military units (which often have different characteristics) and their logistical back up.

But what does it mean to "[...] reduce to such a state as not to be able to prosecute the war"? Does it mean that it is necessary to annihilate the enemy formation in order to win a battle?

8.3 To win a battle

In *On War* Clausewitz often pointed to the fact that it is the decisions in the individual engagements (combats) which add up to the overall decision in war.

Now we have three concepts in warfare to consider: *Destruction, decision* and *win* (or lose, for that matter).

The *destruction* can take place in a number of different ways: In air and sea warfare the word "destruction" is more often than not to be taken literally. When aircraft, ships and submarines are hit by modern weapons, the damage is often fatal. The vessels themselves are destroyed and their crews suffer heavy casualties. But they can also be "destroyed" in a more indirect way by denying them access to their bases or by destroying the bases. A remarkable thing is that the crews very seldom break down psychologically or rebel. Social coherence is normally very strong within these units. Another factor could be that they engage their enemies at long distances – the horrors are not right in their faces. A third factor could be an irrational sense of safety, as they fight their battles from cockpits, operation centers and other enclosures. The same irrational sense of safety that many car drivers have, when they (we) race the highways without realizing that they (we) sit in a tin can.

Much land warfare is fought from enclosures, too, i.e., tanks, armored personnel carriers, etc. But the horrors are right in the face of tank and carrier crews, when they see their own and the enemy's tanks and carriers blow up. Even when not in direct combat there are threats. Minefields, commando raids, artillery fire, air raids etc. And the urge to run away is always present, simply because there is a possibility to run, which does not exist in the air or at sea.

What makes soldiers run and what makes them fight on? It is a very difficult question, to which there are no straightforward answers. Here we will look at two examples only, and leave it with that.

[38] OW, p. 123. (Bk. I, ch. 2).

At the battle of Waterloo, June 18th, 1815, *the decisive moment* occurred when the best of the French infantry twice, *in column*, marched into the British *line* of infantry, which the French believed to be seriously shaken by artillery fire. This was proven tactics, and the idea was that the mere weight of the column should split the infantry line whereafter the French cavalry could chase the broken British infantry.

What happened was that the British infantry, rather unusually for the period, had been ordered to lay down in protection behind a low slope. When the front of the French column was within musket range, the British infantry rose, and began to fire salvo after salvo into the foremost part of the column. Something odd then happened inside the French columns:

> But the deliverance from seemingly certain death of so many Frenchmen at the head of the column also draws attention to another and more significant phenomenon. Although it was they who had suffered most from the British fire, it was also they who did what little was done in order to counter or return it effectively. The men at the rear did nothing, or did nothing useful. Indeed, it seems safe to go further. It was at the back of the columns, not at the front, that the collapse began, and the men in the rear who ran before those in the front.[39]

Immediately before the cease-fire in the Finnish-Soviet winter war (1939–40), a Finnish Army unit at a location named Kollaa fought a defensive battle, which is almost incomprehensible:

> Now as the Russian regiments advanced towards Tsumeikka and Kontro to attack the rear of the Kollaa defenders, other Russian tank divisions were pressuring the Finns at the Ulismainen swamps, trying to slash the road between Saarijärvi and Uomaa. The threat of being surrounded increased by the minute. Kollaa was weakening. [...] Even when the situation had seemed hopeless so many times in the past, the Finns had somehow managed to get some sort of help, so by that evening the lines were still intact. But by now the men were few; reinforcements were older troops who were unable to hold out the way experienced defenders had. And under the murderous artillery fire from the Reds, these older men were swallowed up almost daily, sometimes even before they reached the line. [...] But when the Ulismainen battles were over, there were only a few men left in the company and these remnants joined the small forces who still believed in the miracle at Kollaa. [...] Suddenly rumors of a cease-fire came from the front lines; now for the first time, the Kollaa defenders realized that foreign help was not coming after all. But whatever happened to the big Allied army? The world wants us to fight alone, they decided, and there would be now applause for the brave Finns.
>
> For the first time the Finnish counterassault did not succeed. The infantry was finished. The few men that were left never made it to the proper distance

[39] J. Keegan, *The Face of Battle*, p. 171.

> for throwing their hand grenades. They died of exhaustion and by Russian fire. The nerves of those who survived were gone.[40]

What is it that makes soldiers run? Week social coherence combined with a clear and present danger and a possibility to escape could be an answer.

What is it that makes soldiers counterattack against all odds? Strong social coherence combined with a belief in the cause could be an answer.

The classic example of a surprising outcome of air warfare is the Battle of Britain (1940, August 8 – October 30): The numerously very inferior Royal Air Force, thanks to the social coherence within the ranks and files and shrewd use of a new technology, radar, managed to stall the onslaught from the Luftwaffe. "Overall losses were 1.733 German planes shot down, to 915 British."[41]

Another example of the inferior beating the superior is the naval battle off Midway Island in the Pacific (1942, June 4–6). Three American aircraft carriers fought and won a strategically decisive battle against overwhelmingly strong Japanese naval forces.

> Two factors lead to the result: first and foremost, American knowledge of the Japanese secret codes, which presented Nimitz (the U.S. Admiral in command) with an accurate picture of the Japanese intentions and dispositions; second Yamamoto's (the Japanese Admiral in command) original dispersion of his tremendous armada to fit his own estimate of probable American intentions and reactions.[42]

In brief – superior intelligence on the American side combined with an ability to exploit it carried the day for the Americans. The Japanese fleet never recovered from the losses: four major aircraft carriers against one American. "Japan had lost her entire carrier force in being; the U.S. still had two in commission."[43]

So far we can sum up:

- If you destroy an enemy unit in combat, it means that this particular unit cannot fight for the rest of the war. This is certainly a win-situation, which often occurs in *air and sea combat*, where destruction often is to be taken literally. But sometimes a unit may slip away, unable to complete its mission for one reason or another.
- In *ground combat* we find a somewhat different situation. If you want to make sure, that the unit is destroyed, you will have to kill, seriously invalidate or capture the bulk of its soldiers. Otherwise they will withdraw and come back to fight another day.

[40] E. Engle and L. Paananen, *The Winter War*, pp. 137–138.

[41] *The Encyclopedia of Military History* (EMH), p. 1066.

[42] EMH, p. 1148.

[43] ibid., p. 1148.

But what does it really mean to win? It means to force through a *decision* to one's own advantage.

Now – both parties in a battle, be it big or small, have a mission to complete. At Waterloo the French columns' mission was to break through. The British infantry's mission was to hold the line. The British infantry won, as they completed their mission and the French pulled back. This was the *decisive moment* of the battle at Waterloo. The French Army began to loose its coherence and disintegrated, chased by the cavalry of the allied.

In the Battle of Britain, the mission of the Royal Air Force was to maintain air superiority[44] over the British Isles during daylight. The *decisive moment* occurred during the 15th of September 1940. "On that day Luftwaffe sent more than 1,000 bombers and app. 700 fighters against London. By nightfall, 56 assaulting planes had been downed at the expense of 26 British. The tide had turned and at the end of September Luftwaffe had to stop daylight operations".[45]

The Royal Air Force had won. *But they could not know on the 15th of September, that the decisive moment had occurred.* This is an important point to remember.

In the battle of the Midway Island both parties tried to gain control over the sea lines of communication in the central Pacific. The Japanese mission was to establish a naval base on the island in order to be able to maintain continuous naval presence. The American mission was to prevent it for the very same reason.

The so-called "center of gravity" (a metaphor for a vital force concentration) for the U.S. Navy was the Japanese carrier force. It had to be destroyed. This done, the Japanese ordinary battleships would be very vulnerable due to the lack of air cover. The Japanese thought the center of gravity was the land based air forces on the island. They were not aware of the presence of the U.S. carriers. The *decisive moment* of the battle occurred when the U.S. carrier aircraft caught the Japanese carriers with their flight decks crowded with aircraft in the process of being rearmed for another attack on the island. The Americans won.

But history tells us that sometimes a probable winner – be it in the air, at sea, or on the ground – does not appreciate that the decisive moment has come, and the chance of winning slips away. It is called an indecisive battle.

Lesson learned is the following:
7) Sometimes we cannot know for sure, whether we have won or lost.

8.4 Military power

What is military power? A more precise concept is "military capability". The following model – which like all models covers only parts of reality – provides a conceptual answer.

[44] In brief – Air Superiority means that you are able to press the enemy away from a certain airspace. Air Supremacy means that the enemy air forces are destroyed (in the Clausewitzian sense).

[45] EMH, p. 1065.

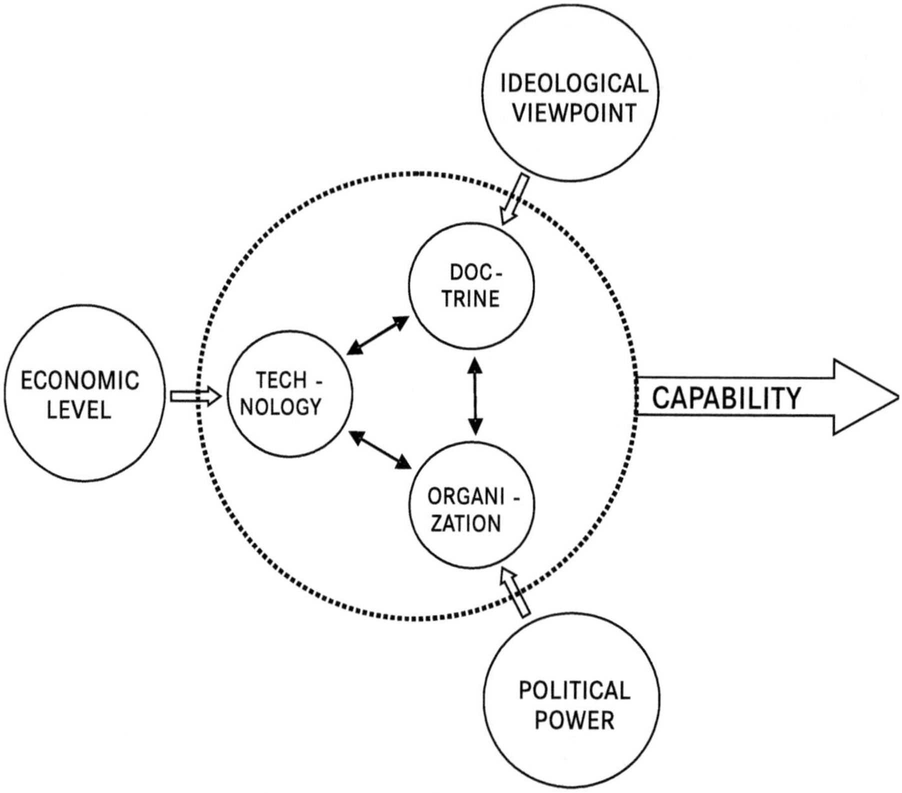

I have discussed the concept "doctrine" earlier. But who or what is it that in the end determines what a state finds to be acceptable guidelines for waging war. The mutual violence in warfare clearly has a tendency to escalate, so most civilized states today adhere to the international law of war, the set of conventions which restrains warfare: Excessive use of violence is forbidden, cultural heritage must not be damaged or used as cover, international red cross or crescent must not be used to deceive the enemy, prisoners must not be tortured or killed etc.

In essence doctrines are restricted by the ideological viewpoint on war by the state concerned. This ideological viewpoint has its roots in history and culture.

The concept "technology" refers to all the equipment of war, including the knowledge of how to handle and maintain it. The more advanced the economy of the state is, the more advanced equipment it can produce or acquire, handle and maintain.

Finally, the concept "organization" in this connection refers to the fact that military units are people equipped with, and organized in relation to, weapon systems, communication systems etc. Military organizations cost money, much money, and the funding must be provided through a political process, be it democratic or not.

The model can be used as well for a general description of a state's armed forces as for a detailed description of a specific unit.

Now – technology, doctrine and organization have to be balanced against each other in order to maximize the capability. (It is of little use to have a doctrine for fighting in swamps, if the unit in question is equipped with heavy tanks).

A well trained, properly balanced unit, supported by effective logistics *should* have maximum capability, i.e., be the stronger part if it confronted a less balanced unit. But it is not that simple. Let us turn to *On War*:

> But action in War is divided into two forms, attack and defense, which, as we shall hereafter explain more particularly, are very different and of unequal strength. Polarity therefore lies in that to which both bear a relation, in the decision, but not in the attack or defense itself.[46]

Read isolated from the context, the quotation makes little sense. It deals with phenomena at a conceptual level. What it means is the following:

- Attack is the weaker form, but it brings the decision.
- Defense is the stronger form, but it brings no decision.
- We cannot compare attack and defense directly, but only in relation to the decision.

Let us assume that we have two identical army units, A and B. A has dug in on the top of a slope in order to defend the position. B will attack in order to take the position. Both units want to complete their missions. All other parameters equal – on paper the two units have equal strengths. But B is doomed to lose, because the decisive moment will occur when B, on its way uphill, is shot to pieces by the protected and prepared defenders.

What Clausewitz wrote about attack and defense seems to be correct. B, for obvious reasons, was the weaker, but it tried to force through a decision. A, for the same reasons, was the stronger, but if B had not attacked, any discussion about a decision would have been irrelevant.

Let us assume that the officer who tasked unit B to attack, had read Clausewitz. He would then have started his planning process by looking at the desired decision, the conquest of the top of the slope. This would have led to reflections over which kind of unit was needed to traverse the open slope uphill, where the unit would be sure to take intensive fire from the defenders. A tank unit with support from artillery would be the ideal choice. At last he would investigate the size and composition of the defending force in order to estimate the optimal size of the tank unit and the amount of artillery fire needed. Thereafter orders would be issued and the attack would be carried through.

Or would it? Assume the defenders were of the same breed as the Finnish defenders at Kollaa or as the aircrews of the Royal Air Force. The attack could then have failed.

[46] OW, p. 113. (Bk. I, ch. 1: *"Attack and defence are things differing in kind and of unequal force. Polarity is, therefore, not applicable to them"*).

What if the British infantry at Waterloo had not been ordered to take shelter during the artillery bombardment? What if the Japanese had forced the U.S. secret codes and not vice versa?

There is a lesson to be learned from all this:
7) Relative strength can only be estimated, not calculated.

8.5 Relative strength, the answer

We can, of course, imagine situations where the capability of the attackers (or the defenders) obviously is overwhelmingly superior. Why then bother to calculate? And still, the calculation would have to take into consideration all the sources of uncertainty.

It gets worse.

8.6 Friction

> As long as we have no personal knowledge of war, we cannot conceive where the difficulties lie of which so much is said, and what the genius and those extraordinary mental powers required in a general have really to do. All appears so simple, all the requisite branches of knowledge appear so plain, all the combinations so unimportant, that in comparison with them the easiest problem in higher mathematics impresses us with a certain scientific dignity. But if we have seen war, all becomes intelligible; and still, after all, it is extremely difficult to describe what it is which brings about this change, to specify this invisible and completely efficient factor.
>
> Everything is very simple in war, but the simplest thing is difficult. These difficulties accumulate and produce a friction which no man can imagine exactly who has not seen war [and] in war, through the influence of an infinity of petty circumstances, which cannot properly be described on paper, things disappoint us, and we fall short of the mark. A powerful iron will overcome this friction; it crushes the obstacles, but certainly the machine along with them. We shall often meet with this result. Like an obelisk towards which the principal streets of a town converge, the strong will of a proud spirit stands prominent and commanding in the middle of the art of war.[47]

Clausewitz's friction and the cure for it, the iron will of a commander, is the underlying foundation of all basic training of cadets in military academies. People who want to command soldiers in war must be prepared to force their will through under the worst of conditions, as well as they have to prepare for the unexpected, no matter how minute their task may be.

[47] OW, p. 164. (Bk. I, ch. 7).

We have a proverb in Denmark which explains why: *"Little mound can overturn big load"*.

A mathematician would probably turn to chaos theory in order to explain the phenomenon. But this interesting theory carries a sinister message, too, if I have understood it right: The butterfly that flutters in Hong Kong could cause a number of disasters in the rest of the world.

Lets us look at a current example[48] of friction: On March 4, 2002, everything went wrong when a team of U.S. Navy Seals, esteemed commandos, were ambushed on the ridge of Mount Takur Ghar in Afghanistan. Seven soldiers died in a rescue effort that revealed both *mistakes* and *determination*. Let us have a look at the mistakes first.

The initial plan was to helicopter lift a U.S. Navy Seal team, well trained commandoes, to a safe landing zone at a distance from the top of the mountain. From there they, under cover of darkness, cautiously should deploy to the top of the ridge in order to establish a much needed observation post. But a number of "simple things" delayed the helicopter lift so much that is was decided to drop the team directly on the ridge top in order to be there before daylight.

But other "simple things" caused that the U.S. intelligence sources failed to detect enemy fighters on the ridge.

The helicopter approached the landing zone and lowered its ramp. The Navy Seal team was lined up to leave in a hurry. A few feet over the ground the helicopter suddenly took enemy fire and was seriously damaged. When it turned violently to get away, one from the team slid over the ramp. Social coherence made one of the helicopter crew members try to grasp him, but he, too, slid over the ramp and was only saved by his safety line which left him dangling under the helicopter. He was eventually hauled back as the helicopter flew away and eventually was forced to make an emergency landing four miles from the ridge top.

One team member was left on the ridge top. But well trained commandos have a very developed sense of honor. It is a dictum that they never leave a buddy behind.

Therefore another helicopter with a Navy Seal team later arrived on the scene. Because of communications failures, "simple things" you may call them, the helicopter crew knew nothing about the conditions on the ridge top, and when they arrived, they took heavy fire and crash-landed. The Navy Seal team and the crew got out and a fire fight erupted. Two of them were wounded. They withdrew to another position down the mountain.

Due to other communications malfunctions, the next two rescue helicopters with a rescue team of 20 Rangers (well trained troopers, too, and specialized in rescue operations) believed they had to land on the top of the ridge to extract the first team. When they arrived, the first of the helicopters took heavy fire, had casualties and crash-landed. The other helicopter escaped and landed at a considerable distance from the first.

Let us interrupt the tale here.

[48] "Bravery and Breakdowns in a Ridgetop Battle – A Wintry Ordeal at 10,000 Feet" by Bradley Graham, Washington Post Staff Writer, May 24/25, 2002.

What happened was "friction" in the most high tech force in the world. The "simple things" caused seven deaths and a number of wounded.

But on the ridge top, something remarkable happened, too: Our particular variant of Popperian social technology[49] seemed to work within both the helicopter crews, the Navy Seal teams, and the Ranger team, and the Clausewitzian iron will of leaders was demonstrated as well. Factors which were decisive, too, when the ancient Greek phalanx clashed with the enemy.

On the ridge top the ultimate disaster was averted by a devoted and professional force of U.S. Navy Seals and Rangers.

Were there anybody to blame for serious neglects? Probably not. "Little things" happened here and there, and they were most likely corrected as fast as possible – if they were observed at all.

But let us close this piece of current military history by telling that infrared films taken from an observation drone proved that the unfortunate Navy Seal, who slipped over the ramp in the first place, afterwards moved around on the ridge top and fought his enemies to the bitter end. I believe he had no doubts that his buddies were coming back to get him out.

The history of war is filled with examples of friction. Friction is to be regarded as a permanent condition in war, as Clausewitz stated.

Clausewitz's language was filled with metaphors. Why he used the word "friction" we cannot know. But the following quotation offers a hint what friction is all about:

> Perhaps Clausewitz's mention of friction and its effects on warfare operations has some resemblance with the increase of entropy (disorder) in many thermodynamic systems as well as with the concomitant decrease of informational content in such systems.[50]

The conclusion must be the following:

9) Even a small number of unavoidable minor deviations or accidents can cause disastrous effects.

9 Closing Remarks

I have identified a number of major sources of uncertainty in war and warfare. There are more to be found, for example the constantly diminishing so-called teeth-to-tail ratio: The number of fighting soldiers of all kinds is constantly dwindling in relation to the rest; administrators, logisticians, people manning the vast network of all kinds of communications means, intelligence personnel, specialists

[49] The components of this "military" social technology have of course been modified since the days of the Greek phalanx. A major new component is the dramatically improved probability to survive wounds as the military medical services have become more and more effective. See *The Face of Battle*, pp.269–279 for a more profound analysis of the will to combat.

[50] T. Bergstein, author of *Quantum Physics and Ordinary Language.*

in international law of war, welfare personnel etc. All of them are necessary for the high tech military machine to work.

But there are still low tech military forces around the world with very high teeth-to-tail ratios. In the future these forces most likely will try to fight a war on their own terms, not ours; i.e., what we today call asymmetric wars, for example in the enormous slum areas of mega cities with say 75 million or more inhabitants.

How would the Western democratic liberal states cope with a situation where the future al Quaeda fighters sought shelter, not in a rather desolated Afghanistan, but in the mega slum? Vast numbers of infantry soldiers would be needed (read extensive conscription). Could the Western democratic liberal states fight "Vietnam wars" in the "jungles" of the mega cities against people who would have nothing to lose but their lives?

There are other future horror scenarios to consider, but none of these appear to be more calculable than wars of today.

Anyway – for now we have the following nine major sources of uncertainty to consider:

1. The social coherence of military units cannot be tested sufficiently during peacetime.
2. The validity of empirical data gathered in war quickly erodes.
3. We cannot predict the impact of the peoples' passions on the war effort.
4. We cannot quantify the courage and talent of the armed forces as a whole.
5. The effects of political influence upon warfare are unpredictable.
6. Even broad guidelines for waging war can fail.
7. Relative strength can only be estimated, not calculated.
8. Sometimes we cannot know for sure, whether we have won or lost.
9. Even a small number of unavoidable minor deviations or accidents can cause disastrous effects.

Do these nine major sources of uncertainty support the hypothesis? The hypothesis is:

The outcome of a war and the overall progress of the activities in it, the warfare, cannot be calculated with mathematics.

Is it possible to construct a mathematical model which both takes into consideration the individual effects of the sources of uncertainty and the way those effects interact? I cannot provide an answer, as I myself have only limited knowledge of mathematics.

But I find the above nine sources of uncertainty to be strong indicators pointing in the direction that my hypothesis is valid.

If you do not agree – then allow me an unfair remark as a last resort:

How comes, that nobody so far has been able to develop a "durable" mathematical model for stock trading which could make the inventor rich in a hurry?

This should be an easy task compared to the problems we face in war. Not least, because stock brokers normally do not try to kill each other.

Figure 3. Problems worthy of attack prove their worth by hitting back [from: *Grooks* by Piet Hein]. But who or what is the problem?

References

Bergstein, T.: *Quantum Physics and Ordinary Language.* The Macmillan Press 1972.

Clausewitz, Carl von: *On War.* Penguin Books, London 1982. (OW)

–, *Om Krig.* Danish translation of *On War,* Rhodos, Copenhagen 1986.

Creveld, Martin van: *Technology and War.* Macmillan, New York 1986.

Dupuy, R. E. and Dupuy, T. N.: *The Encyclopedia of Military History.* 2nd. rev. ed. Harper & Row, Publishers: New York 1986. (EMH)

Engle, Eloise and Paananen, Lauri: *The Winter War.* Stackpole Books, Harrisburg, USA 1973.

Graham, Bradley: "Bravery and Breakdown in a Ridgetop Battle"; *Washington Post,* May 24, 2002.

–, "A Wintry Ordeal at 10,000 Feet"; *Washington Post,* May 25, 2002.

Hanson, Victor Davis: *The Dogs of War,* in: http://www.nationalreview.com/hanson/hanson110601.shtml, November 6, 2001.

Keegan, John: *The Face of Battle.* Jonathan Cape, London 1976.

Popper, K. R.: *The Open Society and its Enemies.* 5th. rev. ed., Routledge & Kegan Poul, London 1966. (OSE)

Warfare Can Be Calculated

SVEND CLAUSEN*

Combat Modelling (CM) is calculating of combat outcomes based on mathematical models. During the cold war a lot of research was spent to develop CM, but still CM can hardly be considered a scientific undertaking. CM is rather a kind of practical tool for support to military decision makers. CM may be based on different types of mathematical models, traditionally deterministic models of the Lanchester type. These models are often used without due regard to underlying assumptions, and the outcomes may differ from the "true" mean outcomes usually wanted. CM based on stochastic models, especially time continuous and state discrete Markovian processes is in principle better than deterministic models. Stochastic CM has an innate potential for determining the "true" mean outcome usually wanted, but unfortunately stochastic CM often leads to very complex models.

The Danish Defence Research Establishment (DDRE) has developed a stochastic combat model, Defence Dynamics (DD) intended to describe tri-service combat. DD overcomes many of the weaknesses of the Lanchester type of models. DD is based on a universal combat model for a duel between two arbitrary military units. This model is a stochastic model able to describe almost any kind of duel. By use of a new concept: "A representative pair" the model has been generalized to arbitrary homogeneous and heterogeneous combat situations. Unfortunately the generalization requires an approximation, but it seems to be tolerable. DD determines outcomes much closer to the "true" mean outcomes than is possible with traditional CM. DD has supported practical problem solving both for the Danish defence and for NATO.

1 Introduction

Combat modelling is a kind of mathematical modelling intended to calculate combat outcomes. During the cold war combat modelling became a major tool to support NATO military decision makers. In those times very few people wondered if warfare might be calculated. The important problem was how warfare should be calculated. Generally speaking it was agreed that combat modelling is of no practical value with regard to reliable prediction of real life combat outcomes. The uncertainty related to this kind of prediction was considered too high. This conclusion is in no way surprising, because real life combat is influenced by a huge number of both qualitative and quantitative factors, e.g.,

* Danish Defence Research Establishment. Email: sc@ddre.dk

- Personnel
- Leadership
- Moral
- Training and education
- Weapon- and sensor systems
- Command, control and information systems
- Strategies and tactics
- Terrain
- Weather
- Light conditions

which may only be known to a limited and uncertain extent.

Nevertheless combat modelling was during the cold war still considered a useful tool, when military decision makers had to deal with problems like:
- Long term planning with regard to structuring of military forces
- Acquisition of weapon- and sensor systems
- Acquisition of CCIS systems
- Development of new military concepts and doctrines
- Optimisation of logistic systems

In relation to this kind of problems it was generally believed that it would be sufficient to consider mean combat outcomes. It was furthermore believed that combat modelling based on the Lanchester tradition would be sufficient for providing combat outcomes close to or equal to the "true" mean combat outcomes wanted.

But despite all the research and despite all the resources spent, combat modelling can hardly be considered a fully scientific undertaking. The reason for this statement is that combat models can hardly be exposed to empirical testing. Certainly it is not physicaly impossible to carry out adequate empirical testing on combat models, but in practice it will usually be considered unacceptable both for economic and moral reasons. Some scattered pieces of empirical evidence may be obtained from studies of war history, but this will hardly be sufficient to establish combat modelling as scientific.

Nevertheless combat modelling is generally accepted as a practical tool for military decision makers. Combat modelling may be useful for military decision makers to support summarizing, processing and reaching conclusions from relevant and available information. This information may be quite overwhelming and almost impossible to treat in any other way. Combat modelling may be considered a kind of thinking device enabling the military decision maker to a better understanding and solution of problems.

2 Possible Types of Combat Models

In fact there is a broad spectrum of possible types of mathematical models, which might provide the theoretical basis for combat modelling. Underlying each of these types of models is a set of specific assumptions, limitations and possibilities. Different mathematical criteria with regard to classification of combat models may be established. Based on a classification it might be easier to discuss, criticize and select a type of combat model relevant for a specific application.

The following set of mathematical criteria is considered sufficient for the present purpose:

- Descriptive models / optimisation models
- Deterministic models / stochastic models
- Time continuous models / time discrete models
- State continuous models / state discrete models.

In fact most military problems suited for combat modelling are decision problems. Consequently it should be obvious to select *optimisation models* (i.e., mathematical models, which include both a decision function and an objective function) as a basis for combat modelling. But unfortunately optimisation models will in most cases become overwhelming in mathematical complexity when applied to traditional military problems, so they will not be considered any further at this occasion. The only possibility left is *descriptive models*, i.e., mathematical models, which do not include any explicit decision function.

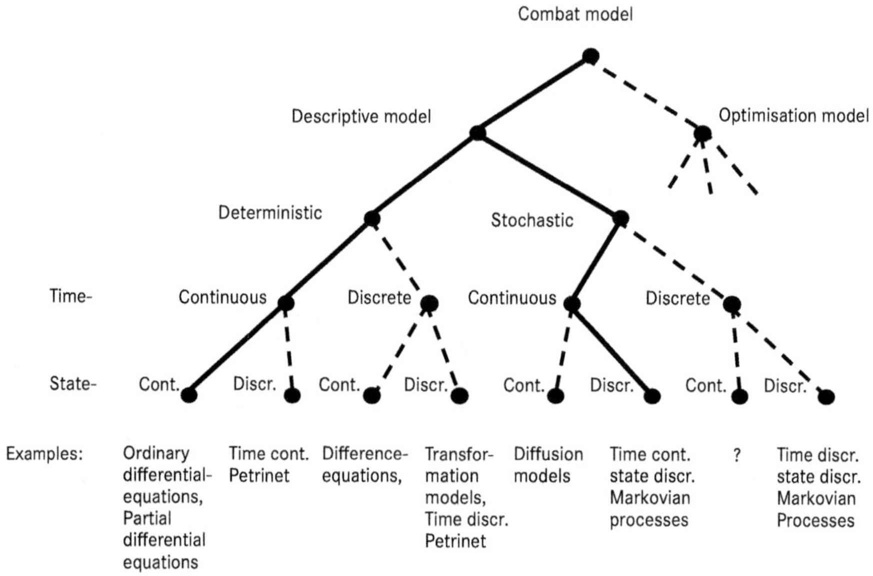

Figure 1. Possible mathematical types of combat models.

To keep it simple only two options are open to determine a combat outcome, especially a mean combat outcome by use of descriptive models: Deterministic and stochastic descriptive models. A *deterministic model* can usually only be justified if it is able to produce a combat outcome close to or equal to some "true" mean outcome. But this is meaningless because a mean outcome usually is not even defined in relation to a deterministic model. With a *stochastic model* it will be meaningful and in principle possible to determine a mean combat outcome, which is "true" at least within the framework specified by the stochastic model.

Of course time is a major factor for combat modelling and it has to be portrayed within the model. Again only two simple options need to be considered: Time discrete – and time continuous models. It is a basic fact of war that events relevant for combat modelling may occur at any time, so for that reason it should be obvious to apply *time continuous models.* Nevertheless time discrete models are quite frequently used in practice because they are especially well suited for discrete numerical computations. But in principle use of time discrete models introduces an error, because the events are forced within the model to occur only at the discrete points of time allowed by the model. This error will be reduced if the time steps chosen are diminished, but then the amount of work for solving the model and the cumulated numerical errors will increase. Time discrete models exclude the use of strong analytical tools such as differentiation and integration.

In combat modelling the number of operational military units at an arbitrary point of time is usually considered the state of the relevant military system. This state factor may also be portrayed in two different ways: Discrete or continuous. A *state discrete model* is the natural choice, because the real life number of units is usually considered an integer. But unfortunately a state discrete model may be too demanding to operate if the number of units on each side is high. With a *state continuous model* the real life number of units will be considered a real number and this may be a gross approximation, especially when the number of units are small, especially smaller than one.

Based on this kind of considerations it should in principle be obvious to choose *descriptive, stochastic, time continuous and state discrete types of models* as the best-suited type of mathematical model with regard to combat modelling. It might for instance be a time continuous and state discrete Markovian process model. Unfortunately this type of combat model has only been implemented a few times, because it has a tendency to become unacceptably complex.

In common practice combat modelling is quite often based on *descriptive, deterministic, time continuous and state continuous types of models.* Usually it is a set of ordinary differential equations of the Lanchester type. The argument for using differential equations is that they are sufficient to determine the mean combat outcome. But this argument is doubtful, when the mean outcome is not defined within a deterministic model.

3 NATO Combat Modelling

Today combat modelling within NATO is considered to be a fairly well established and almost concluded research area. But this assessment is highly doubtful. Most combat modelling within NATO is still to a high extent based on deterministic descriptive models, i.e., ordinary differential or difference equations of the Lanchester type often used without due respect to underlying assumptions and limitations.

To exemplify this criticism a few combat models of the Lanchester type based on ordinary differential equations are considered. The simplicity of the examples chosen is necessary for purposes of presentation and computation, but the criticism will also be valid with regard to more advanced models of the Lanchester type.

3.1 Lanchester square model

The Lanchester square model is the best known of the classical Lanchester models. Lanchester considered this model especially relevant for modern warfare (1917):

$$\frac{\mathrm{d}b(t)}{\mathrm{d}t} = -\kappa_r \cdot r(t) \qquad b(0) = B$$

$$\frac{\mathrm{d}r(t)}{\mathrm{d}t} = -\kappa_b \cdot b(t) \qquad r(0) = R$$

B initial number of blue units
$b(t)$ number of operational blue units at time t
κ_b kill rate for one operational blue unit against operational red units

R initial number of red units
$r(t)$ number of operational red units at time t
κ_r kill rate for one operational red unit against operational blue units

This model describes combat between two homogeneous forces (a homogeneous force is a force including only one type of military units, e.g., tanks), both fighting under the assumption about full tactical information.

Full tactical information means that an arbitrary operational unit is at any time able to detect at least that many hostile operational units as it is capable of killing. Furthermore it is assumed that all operational units on each side are able to fully share their information and coordinate their firepower among the operational hostile units. In fact firepower is the only limiting factor. The capabilities for detection and coordination are assumed sufficient for firing.

This assumption may be quite unrealistic, because it is difficult to identify many modern combat situations, where the assumption is fulfilled on both sides. For example very few army operations are carried out with full tactical information on both sides. Within NATO the square model is often used without sufficient consideration to this assumption.

The square model is simple and it may be solved by analytical means. From the analytical solution it is possible to derive the following criterion for victory:

$$R^2 \cdot \kappa_r > B^2 \cdot \kappa_b \Rightarrow \text{red victory}$$

$$R^2 \cdot \kappa_r < B^2 \cdot \kappa_b \Rightarrow \text{blue victory}$$

$$R^2 \cdot \kappa_r = B^2 \cdot \kappa_b \Rightarrow \text{mutual destruction}$$

The saying, that "Quantity is more important than quality", presumably originated from this criterion.

Numerical example:

$$B = 200, \; R = 400$$
$$\kappa_b = 0.40, \; \kappa_r = 0.15$$

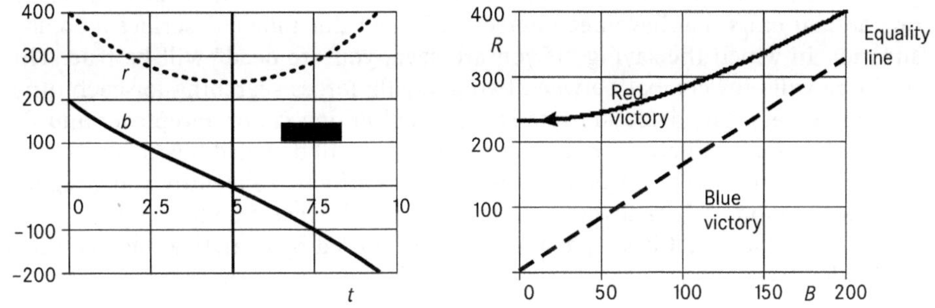

Figure 2. Combat outcomes derived from the Lanchester square model.

The Lanchester square model is a time and state continuous model in practice often assumed to be a kind of mean value model. The logic behind the model deteriorates with diminishing numbers of operational units and it certainly collapses when the number of operational units becomes less than one or even negative. The model is a pair of ordinary non-linear differential equations and it exhibits chaotic behaviour in combat situations close to the equality line (equality leads to mutual destruction).

3.2 Guerrilla model

Another quite popular combat model of the Lanchester type is a model that might be called "the guerrilla model":

$$\frac{\mathrm{d}r(t)}{\mathrm{d}t} = -\delta_b \cdot r(t) \cdot b(t) \; ; \quad r(0) = R$$

$$\frac{\mathrm{d}b(t)}{\mathrm{d}t} = -\delta_r \cdot r(t) \cdot b(t) \; ; \quad b(0) = B$$

R initial number of red units
$r(t)$ number of operational red units at time t
δ_r detection rate for one operational red unit against an arbitrary operational blue unit

B initial number of blue units
$b(t)$ number of operational blue units at time t
δ_b detection rate for one operational blue unit against an arbitrary operational red unit

This model also describes combat between two homogeneous forces. One possible interpretation of the model is that each force is able to kill all hostile operational units that have been detected. In fact this model describes a combat situation in which the saying "If you are seen, you are dead" will be true. This might be valid for combat between two guerrilla forces searching for each other in a wilderness. The detection rate is the only limiting factor. Firepower and the capability for coordination are sufficient for killing all detected hostile units. This assumption may also be unrealistic, because it is difficult to identify many regular combat situations, where detection is the only crucial factor on both sides.

The guerrilla model is simple and it may be solved by analytical means. From the analytical solution it is possible to derive the following criterion for victory:

$$R \cdot \delta_r > B \cdot \delta_b \Rightarrow \text{ red victory}$$

$$R \cdot \delta_r < B \cdot \delta_b \Rightarrow \text{ blue victory}$$

$$R \cdot \delta_r = B \cdot \delta_b \Rightarrow \text{ mutual destruction}$$

With the assumption valid for "the guerrilla model" quality and quantity obviously are of equal importance.

This type of model has also been used to describe outcomes resulting from indirect fire combat, i.e., outcomes relevant when each of two artillery forces at random shoot against the territory occupied by the other.

Numerical example:

$$B = 200, \; R = 400$$
$$\delta_b = 0.40, \; \delta_r = 0.15$$

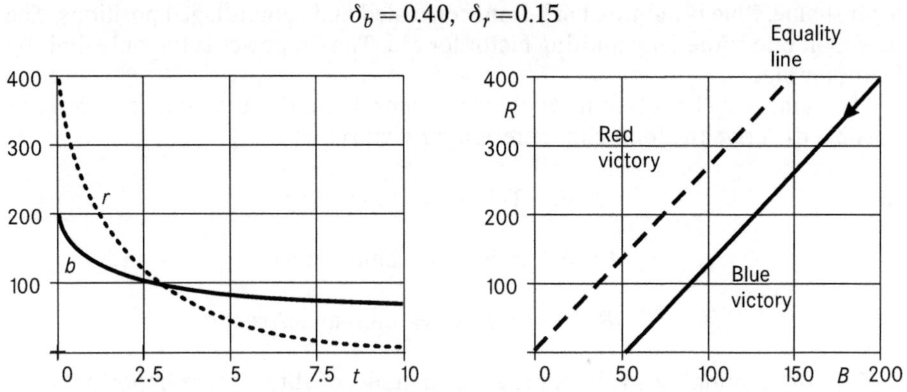

Figure 3. Combat outcomes derived from the Guerilla model.

"The guerrilla model" also is a time and state continuous model in practice often assumed to be a kind of mean value model. The logic behind the model deteriorates with diminishing numbers of operational units and it certainly collapses when the number of operational units becomes less than one. With this model the number of operational units cannot become negative.

This model is a pair of ordinary non-linear differential equations, but it does not exhibit any chaotic behaviour.

3.3 Mixed combat model

The last example of a combat model of the Lanchester type treated here is a mixed combat model (a mixture of the square model and "the guerrilla model"):

$$\frac{dr(t)}{dt} = -\kappa_b \cdot b(t) \qquad r(0) = R$$

$$\frac{db(t)}{dt} = -\delta_r \cdot b(t) \cdot r(t) \quad b(0) = B$$

R initial number of red units
$r(t)$ number of operational red units at time t
δ_r detection rate for one operational red unit against an arbitrary operational blue unit

B initial number of blue units
$b(t)$ number of operational blue units at time t
κ_b kill rate for one operational blue unit against operational red units

The model describes combat between two homogeneous forces. The interpretation might be that red is advancing over an open terrain looking for and fighting against blue. Blue is fighting back from concealed and camouflaged positions. The detection rate is the only limiting factor for red. The firepower is the only limiting factor for blue.

The model may be solved by analytical means. From the analytical solution it is possible to derive the following criterion for victory:

$$R^2 \cdot \delta_r > 2 \cdot B \cdot \kappa_b \Rightarrow \text{ red victory}$$

$$R^2 \cdot \delta_r < 2 \cdot B \cdot \kappa_b \Rightarrow \text{ blue victory}$$

$$R^2 \cdot \delta_r = 2 \cdot B \cdot \kappa_b \Rightarrow \text{ mutual destruction}$$

With the assumptions valid for the mixed model quality is more important than quantity for red, while quality and quantity are equally important to blue.

Numerical example:

$$B = 200, \; R = 400$$
$$\kappa_b = 0.40, \; \delta_r = 0.15$$

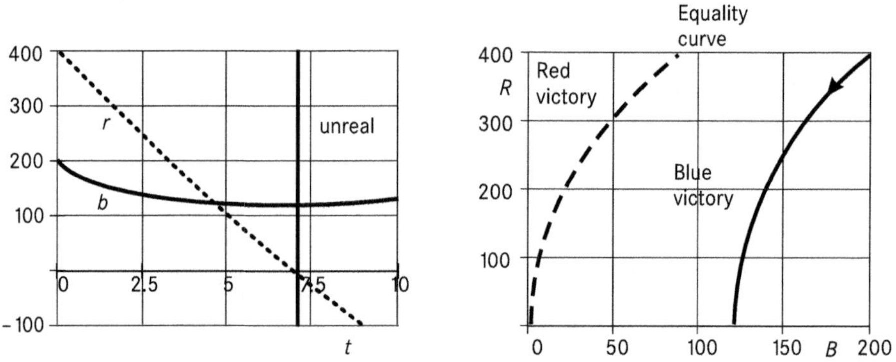

Figure 4. Combat outcomes derived from the mixed combat model.

This is a time and state continuous model, in practice often assumed to be a kind of mean value model. The logic behind the model deteriorates with diminishing numbers of operational units and it certainly collapses when the number of operational units becomes less than one or even negative.

The model is a pair of ordinary non-linear differential equations, but it does not exhibit any chaotic behaviour.

The three simple combat models presented and discussed exemplify deterministic combat models of the Lanchester type, which are frequently used within NATO. The examples have demonstrated that the models are based on certain crucial assumptions embedded within the models. Quite often these types of models are used without sufficient awareness of and regard to the assumptions.

3.4 Comparisons of deterministic and stochastic combat models

Another point of criticism against deterministic combat models of the Lanchester type is that the concept of a mean combat outcome is not even defined within these models. To define and determine a mean combat outcome it is necessary to use a stochastic model.

To illustrate this point the three deterministic combat models already considered are compared to three corresponding continuous and state discrete Markovian process models. The stochastic models are based on the same assumptions and parameters as the deterministic models. For each of the three relevant situations a numerical comparison between the combat outcome determined with the deterministic model of the Lanchester type and the "true" mean outcome determined by the corresponding Markovian model has been carried out. To keep it simple and manageable the initial numbers of the two military forces considered have been kept low: 1 blue and 2 red military units. (Note that in the figure below "EB" and "ER" signify the "true" mean outcomes determined by Markovian models for blue and red respectively. "B" and "R" are the outcomes determined by deterministic models.)

Numerical example:
Comparison between the outcome determined by the square Lanchester model and the "true" mean outcome determined by a corresponding time continuous and state discrete Markovian process combat model.

$$B = 1, \; R = 2$$
$$\kappa_b = 0.40, \; \kappa_r = 0.15$$

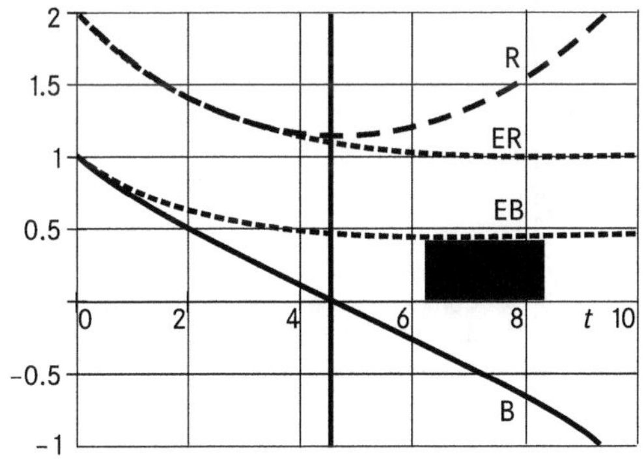

Figure 5.
Comparison between combat outcomes derived from the Lanchester square model and mean outcomes derived from a corresponding stochastic model.

Numerical example:
Comparison between the outcome determined by "the Guerrilla model" and a "true" mean outcome determined by a corresponding time continuous and state discrete Markovian process combat model.

$$B = 1, \; R = 2$$
$$\delta_b = 0.40, \; \delta_r = 0.15$$

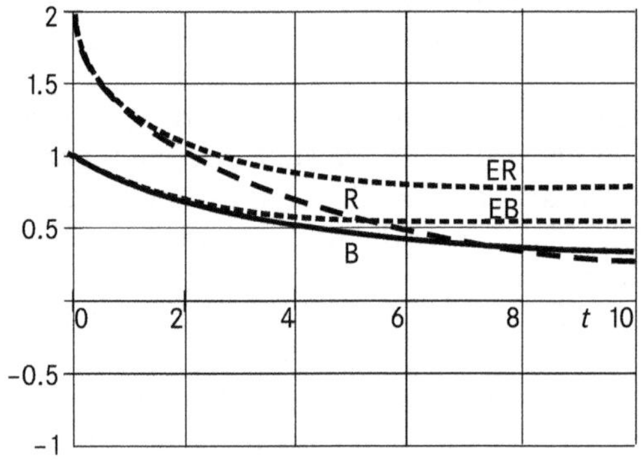

Figure 6.
Comparison between combat outcomes derived from the Guerilla model and mean outcomes derived from a corresponding stochastic model.

Numerical example:
Comparison between the outcome determined by the mixed model and a "true" mean outcome determined by a corresponding time continuous and state discrete Markovian process combat model.

$$B = 1, \; R = 2$$
$$\kappa_b = 0.40, \; \delta_r = 0.15$$

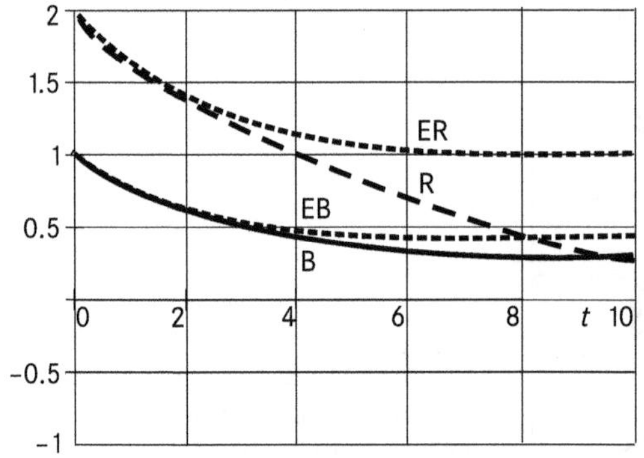

Figure 7.
Comparison between combat outcomes derived from the mixed model and mean outcomes derived from a corresponding stochastic model.

The three comparisons reveal considerable differences between the combat outcome determined by a deterministic model of the Lanchester type and the "true" mean combat outcome determined by a corresponding Markovian process model. The differences will not be large for high numbers of units, but they will increase with diminishing numbers of units left. So with regard to determining the "true" mean outcome of a combat a stochastic model obviously is preferable to a deterministic model especially for small numbers of units. In fact the mean combat outcome is only defined in relation to a stochastic model.

Furthermore this stochastic combat model does have other advantages. A state discrete Markovian model only operates with integer numbers of operational units. The integer number can never become negative and the "smaller than one" problem has completely disappeared.

For all these reasons stochastic models are more attractive than corresponding deterministic models. But unfortunately stochastic models also suffer from serious drawbacks compared to the deterministic models. These drawbacks may be most prohibitive in practice:
- The mathematical complexity has increased dramatically.
- The number of equations blows up with factors of maybe millions.
- The time for solution of the models becomes unacceptably lengthy.

4 Defence Dynamics

In 1980 The Danish Defence Command asked the Danish Defence Establishment (DDRE) to develop a tri-service combat model as a tool for long term defence planning. Later this model became known as *Defence Dynamics* (DD). The idea was that DD should be able to predict a mean combat outcome if a possible Danish tri-service defence structure was attacked by WAPA. The Danish Defence Command had set up a list with 10 - 20 possible Danish defence structures and 5 - 10 possible WAPA attack structures against Denmark.

After prolonged and difficult discussions concerning combat modelling DDRE decided to base DD on a so-called *"Universal combat model"*. This model was intended to predict stochastic outcomes for a duel between two military units independent of their types and services. Furthermore this model was intended to be most flexible with regard to different kinds of relevant assumptions concerning the actual combat situation. It was decided to apply time continuous and state discrete Markovian processes as the relevant mathematical framework. This type of mathematical model was chosen to ensure that it would be possible to determine the "true" mean combat outcome defined within the model framework. As exemplified earlier this type of mathematical model would avoid both the "smaller than zero" and "smaller than one" problem, which otherwise might be most frustrating in combat situations involving only a few military units. The universal combat model was intended to serve as a basis for development of a full-scale tri-service model.

DDRE succeeded in developing the so-called "universal combat model", which with very few exceptions is applicable to duel situations in general. One exception is a duel between a sea mine and a landing ship. For the few exceptional cases it was necessary to develop specific models.

4.1 One-on-one combat

The "Universal Combat Model" was originally based on a quite limited number of basic factors, which it was considered important to take into consideration:

- *Tactical manoeuvres*
  During combat a military unit can usually change between a passive and an active state. Within the passive state the unit is not able to detect and fight the hostile unit and the hostile unit is not able to detect and fight the unit. Within the active state the unit may be able to detect and fight the hostile unit and the hostile unit may be able to detect and fight the unit.

- *Stand off advantage*
  If the unit has a stand off advantage due to long-range sensor- and weapon systems it is able to detect and fire upon the hostile unit without any risk of retaliation.

- *Detection*
  Detection of the hostile unit depends on the unit's sensor systems and the signature of the hostile unit compared to its background. Detection may be highly influenced by weather, light conditions, decoys and means of concealment and camouflage.

- *Firing*
  Conditioned by detection of the hostile unit the unit is able to fire upon it. Firing depends on the actual weapon systems and types of ammunition available to the unit.

- *Kill*
  Conditioned by detection of and firing the unit may be able to kill the hostile unit. Kill depends on the actual weapon systems, the types of ammunition available to the unit and the physical protection of the hostile unit.

Later on further features have been added to "the universal combat model", for example:
- coordination among units within a group,
- decoy effects due to killed hostile units,
- limited ammunition supplies,
- limited sensor- and weapon ranges (front sections),

- initial detection, when the unit first appears from the passive state,
- Suppression.

To keep the presentation simple these further features will not be considered here.

The basic factors are reflected within the universal combat model partly by the following possible states for a military unit:

s The unit is passive.
k The unit has a stand off advantage compared to the hostile unit.
n The unit is active, but it has not been detected by the hostile unit.
d The unit is active and it has been detected by the hostile unit.
e The unit is active (and the hostile unit is passive). Note "e = n + d".
u The unit is killed.

Of course the hostile unit may independently of the unit occupy the same kind of states with a single exception; it is not possible for both units simultaneously to have a stand off advantage (state k).

Taken together a pair of two opposing units may consequently occupy and change between the following 16 combined states: ss, se, su, ks, kn, kd, ku, es, nn, nd, dn, dd, eu, us, ue and uu. Note the combined states where the hostile unit is in state k have been omitted. The 16 combined states are considered to be the possible states for the universal combat model, i.e., a time continuous and state discrete Markovian process.

To make this mathematical model operational the possible transitions between the states are characterized by the following (transition) rates:

Rates	Blue	Red
Activation	η_b	η_r
Passivation	σ_b	σ_r
Loss of stand off advantage	ε_b	ε_r
Detection	δ_b	δ_r
Loss of detection	γ_b	γ_r
Kill	κ_b	κ_r

Table 1. Transition rates relevant for one-on-one combat.

The information necessary for setting up a time continuous and state discrete Markovian process may be summarized within the transition rate matrix characteristic for this type of stochastic model (Table 2).

The diagonal elements within the matrix are, as a consequence of mathematical technicalities, defined as the negative sum of all other elements within the relevant row. Note furthermore the state uu will be impossible with this type of model, because it does not allow two events (two killings) to occur simultaneously.

The corresponding Chapman-Kolmogorov equation for determination of the time dependent probability distribution for the actual combined state of the two units is described by 14 first order linear differential equations, which may

A	ss	se	su	ks	kn	kd	ku	es	nn	nd	dn	dd	eu	us	ue	uu
ss	-	σ_r		σ_b				σ_b								
se	η_r	-			σ_b	σ_b			σ_b	σ_b	σ_b	σ_b				
su			-				σ_b						σ_b			
ks	η_b			-	σ_r	σ_r										
kn		η_b		η_r	-	γ_b										
kd					δ_b	-										
ku			η_b			κ_b	-									
es				ε_b				-	σ_r	σ_r	σ_r	σ_r				
nn					ε_b			η_r	-	γ_b	γ_r					
nd						ε_b			δ_b	-		γ_r				
dn									δ_r		-	γ_b				
dd										δ_r	δ_b	-				
eu							ε_b			κ_b		κ_b	-			
us														-	σ_r	
ue											κ_r	κ_r		η_r	-	
uu																-

Table 2. Transition rate matrix relevant for one-on-one combat.

be solved relatively easily by use of either analytical or numerical methods. The information included in the transition rate matrix may also be demonstrated in a more conspicuous graphical way by use of a transition diagram. The combined states are represented by nodes and the possible transitions between the states are represented by arrows connecting the relevant nodes.

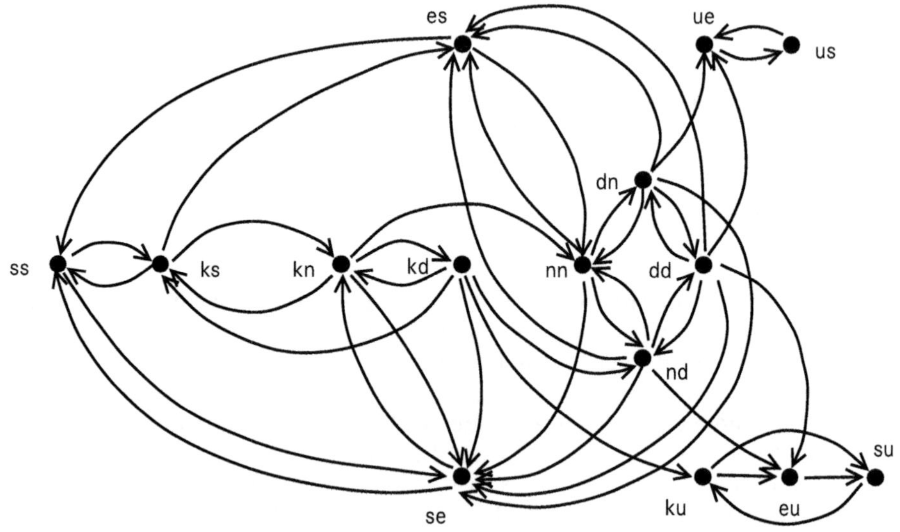

Figure 8. Transition diagram for one-on-one combat.

4.2 Many-on-many homogeneous combat

To generalize one-on-one combat to many-on-many homogeneous combat it is necessary to introduce further military units on each side:

B Initial number of blue homogeneous units
R Initial number of red homogeneous units.

But first and foremost it is necessary to have an idea that makes it possible to avoid the overwhelming number of combined states due to possible combinations of blue and red operational units. Note, that already 200 blue units and 400 red units immediately imply $(200 + 1) \cdot (400 + 1) = 80\,601$ possible combined states, when the only states allowed for a unit is operational or killed. This might be considered an immediate threat of a set of 80 600 ordinary differential equations.

To escape from this major complication the idea of *a representative pair of opposing units* has been introduced:

In principle all units on each side fight under identical conditions. For that reason it should be sufficient for modelling purposes to select an arbitrary pair of opposing representative units fighting a one-on-one combat and then superimpose the influences from all the other non-representative units.

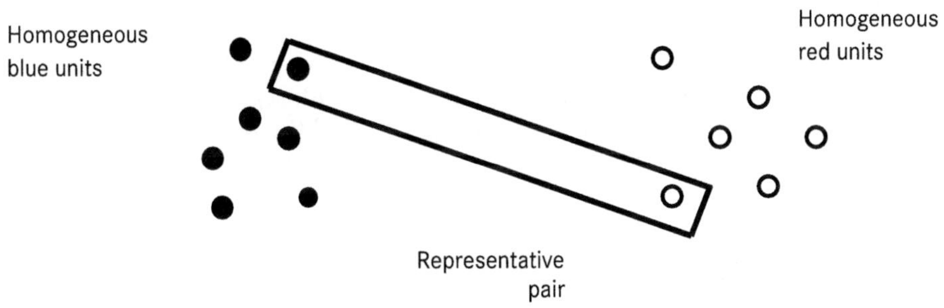

Figure 9. The concept of a representative pair.

Based on this idea it might be sufficient to elaborate on the Markovian model for one-on-one combat, hoping to avoid a dramatic increase in the number of possible states. Of course a number of conditions related to the original one-on-one universal combat model has to be reconsidered to develop a model for a representative pair within a many-on-many homogeneous combat situation.

States/transitions related to detection/non-detection of a representative unit have to be subjective, i.e., only relevant to the opposing unit within the representative pair. Other opposing units may have a different point of view with regard to detection/non-detection of the representative unit. It means for instance that a unit may be detected by the opposing unit within the representative pair, but not by any other opposing unit and vice versa.

States/transitions related to everything else are (in most cases) considered objective, i.e., all opposing units have the same point of view concerning the representative unit. It means for instance that if a unit is killed, it is killed from the point of view of all opposing units.

If the representative pair is in *state kd*, it means that the representative blue unit has a stand off advantage (state k) with regard to all opposing units. Furthermore the representative blue unit has detected the representative red unit (state d) with certainty and an arbitrary non-representative red unit with a probability p_k:

$$p_k = \frac{p_{kd}}{p_k} \; ; \quad p_{k\bullet} = p_{ks} + p_{kn} + p_{kd} + p_{ku}$$

The expected kill rate from the representative blue unit in state k against an arbitrary detected red unit becomes:

$$\alpha_{bk} = E\left\{\frac{\kappa_b}{1+X}\right\} = \sum_{x=0}^{R-1} \frac{\kappa_b}{1+x}\binom{R-1}{x} p_k^x (1-p_k)^{R-1-x} = \frac{\kappa_b}{Rp_k}\left[1-(1-p_k)^R\right]$$

X is the stochastic number of non-representative red units detected by the representative blue unit in state k. X is binomially distributed with the parameters $R - 1$ and p_k.

Something quite similar is valid for the representative pair in *state ed*. The representative blue unit is active and exposed (state e) with regard to opposing units. Furthermore the representative blue unit has detected the representative red unit (state d) with certainty and an arbitrary non-representative red unit with a probability p_e.

$$p_e = \frac{p_{ed}}{p_{e\bullet}} \; ; \quad p_{e\bullet} = p_{es} + p_{nn} + p_{nd} + p_{dn} + p_{dd} + p_{eu}$$

The expected kill rate from the representative blue unit in state e against an arbitrary detected red unit becomes:

$$\alpha_{be} = E\left\{\frac{\kappa_b}{1+X}\right\} = \sum_{x=0}^{R-1} \frac{\kappa_b}{1+x}\binom{R-1}{x} p_e^x (1-p_e)^{R-1-x} = \frac{\kappa_b}{Rp_e}\left[1-(1-p_e)^R\right]$$

Here X is the stochastic number of non-representative red units detected by the representative blue unit in state e. X is binomially distributed with the parameters $R - 1$ and p_e.

Now the total expected kill rate from all non-representative blue units against the representative red unit in state e may be determined by:

$$K_b = (B-1)\left(\frac{p_{kd}}{p_{\bullet e}}\alpha_{bk} + \frac{p_{ed}}{p_{\bullet e}}\alpha_{be}\right)$$

α_{bk}, α_{be} and K_b is state dependent kill rates, which shall be introduced into the transition rate matrix for one-on-one combat to describe the full threat of a kill on the representative red unit due to all blue units.

Corresponding α_{re} and K_r are state dependent transition rates, which shall be introduced into the transition rate matrix for one-on-one combat to describe the full threat of a kill on the representative blue unit due to all red units. But unfortunately this derivation required an *approximation*. The kill rates α_{bk}, α_{be} and α_{re} for a representative unit against an arbitrary opposing unit are determined by averaging over all non-representative opposing units instead of using the actual (but unknown) number of detected non-representative opposing units.

The great advantage with this approximation is that it reduces the number of possible states within the relevant Markovian process from maybe millions to 16, but never the less it still is an approximation. A number of sensitivity analyses have been carried out for modest combat situations. The mean number of operational units determined based on this approximation deviated in no case more than 9% from the "true" mean number of operational units determined with a correct Markovian process model.

Based on this approximation the transition rate matrix valid for the representative pair within a many-on-many homogeneous combat becomes:

A	ss	se	su	ks	kn	kd	ku	es	nn	nd	dn	dd	eu	us	ue	uu
ss	-	σ_r		σ_b				σ_b								
se	η_r	-			σ_b	σ_b			σ_b	σ_b	σ_b	σ_b				
su		K_b	-				σ_b						σ_b			
ks	η_b			-	σ_r	σ_r										
kn		η_b		η_r	-	γ_b										
kd					δ_b	-										
ku			η_b		K_b	$\alpha_{bk}+K_b$	-									
es				ε_b				-	σ_r	σ_r	σ_r	σ_r				
nn					ε_b			η_r	-	γ_b	γ_r					
nd						ε_b			δ_b	-		γ_r				
dn									δ_r		-	γ_b				
dd										δ_r	δ_b	-				
eu							ε_b		K_b	$\alpha_{be}+K_b$	K_b	$\alpha_{be}+K_b$	-			
us								K_r						-	σ_r	
ue									K_r	K_r	$\alpha_{re}+K_r$	$\alpha_{re}+K_r$		η_r	-	
uu													K_r		K_b	-

Table 3. Transition rate matrix relevant for one-on-many homogenerous combat.

So in this way many-on-many homogeneous combat is still represented by a time continuous and state discrete Markovian process with only 16 possible states. It means that the corresponding Chapman-Kolmogorov equation for determination of the probability distribution for the actual state of the representative pair is described by 15 first order non-linear differential equations, which may be solved relatively easily by numerical methods of integration.

Note the representative pair is now allowed to be in the state uu due to the influence from the non-representative units. Note furthermore all the transition rates, which have been modified or added because of the influence from the non-representative units. The diagonal elements within the matrix are, as a consequence of mathematical technicalities, still defined as the negative sum of all other elements within the relevant row.

The information included in the transition rate matrix for many-on-many homogeneous combat may be demonstrated in a more conspicuous graphical way by use of a corresponding transition diagram. Note, the dashed arrows indicate rates, which have been modified or added compared to the transition rate matrix for one-on-one combat.

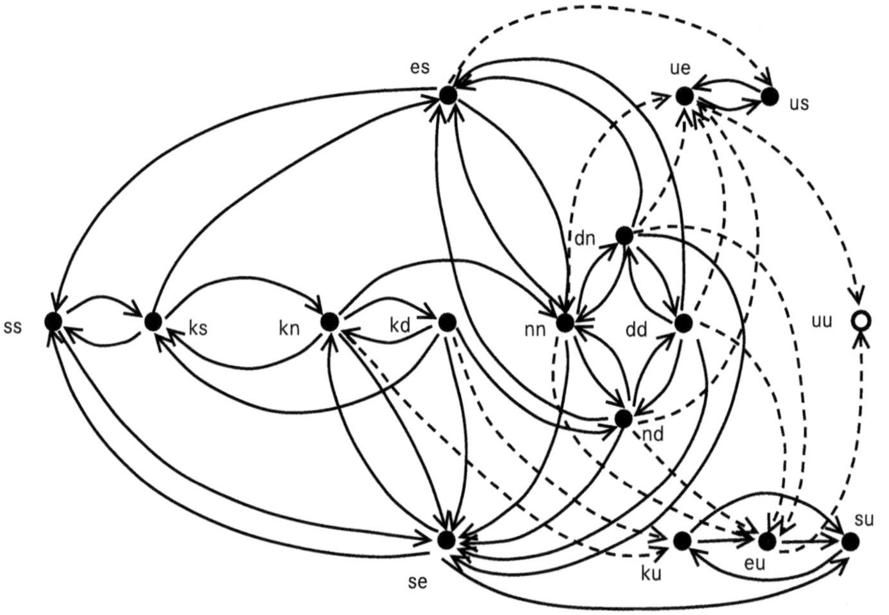

Figure 10. Transition diagram for many-on-many homogeneous combat.

4.3 Many-on-many heterogeneous combat

A heterogeneous battle may be split into a number of parallel and serial heterogeneous sub battles. To describe a heterogeneous sub battle the many-on-many homogeneous combat model has to be generalized to a many-on-many heterogeneous combat model. To achieve this it is necessary to take into account both the type and the size of each blue and red homogeneous group participating within the heterogeneous sub battle.

It will be necessary to define a representative pair of opposing units for each relevant combination of homogeneous forces fighting each other within each heterogeneous sub battle.

The principles and the approximation necessary for this generalization are quite similar to those applied for generalization from one-on-one combat to many-on-many homogeneous combat. For each relevant representative pair of opposing units it is now necessary to superimpose not only the influence from all non-representative units within their own homogeneous force. Now it is also necessary to superimpose the influence from all units within all the other homogeneous forces taking part in the heterogeneous sub battle.

When DD was first applied to support long term planning for the Danish Defence Command a few thousands of representative pairs were defined. The maximum number of non-linear differential equations, which were set up and integrated simultaneously, amounted to approximately 20.000, when a heterogeneous sub battle going on in Schleswig-Holstein was really intense.

Within the heterogeneous battle each sub heterogeneous battle is described with a many - on - many heterogeneous combat model, which determines the time dependent probability distributions for each relevant representative pair. From these probability distributions the mean number of blue and red units surviving the sub battle may easily be determined

Due to the averaging approximation it has not up till now been possible to determine the "true" outcome of a heterogeneous battle exactly. But the battle outcome determined by Defence Dynamics is still considered to be a magnitude of size closer to the "true" battle outcome than outcomes generated with deterministic combat models of the Lanchester type.

For the time being DDRE hopes to avoid introducing the basic approximation into Defence Dynamics by use of a special blending of numerical and Monte Carlo simulation methods for solving the model. In this way it should be possible to generate combat outcomes very close to or identical to "true" mean combat outcomes. But this still requires a lot of research to be seen.

4.4 Practical considerations to use of Defence Dynamics

DD is considered a tool for summarizing, processing and reaching conclusions with relevant and available information. This tool is intended to provide support to decision makers operating within a broad spectrum of military problems. In any case use of DD requires the following type of input information:

- A blue force structure, including types and numbers of weapon systems, ammunition and sensor systems
- A red force structure, including types and numbers of weapon systems, ammunition and sensor systems
- Strategies and concepts for both blue and red force
- CCIS, Command, Control and Information Systems
- The weather situation
- Light conditions
- Topographical information
- A technological database concerning weapon- and sensor systems and parameters for all relevant one-on-one combat models.

It certainly is a most demanding job to provide and maintain all this information necessary for using DD.

Based on the input information DD generates a combat outcome. This combat outcome is an approximation to the "true" mean outcome. Based on this outcome the decision maker may find support to make one or another decision.

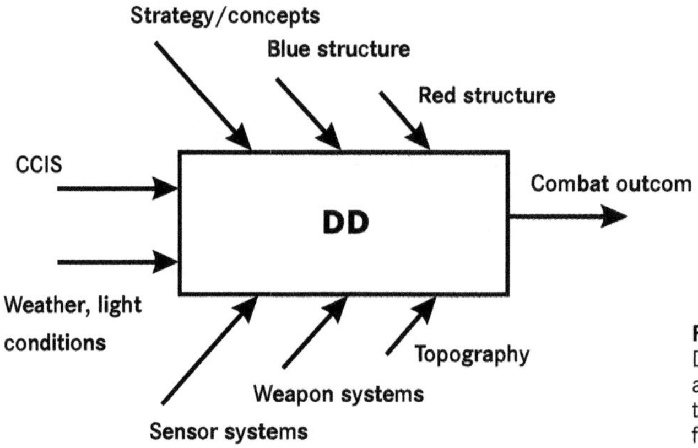

Figure 11. Defence Dynamics generates a combat outcome based on the type of information specified.

The practical procedure for using DD may be illustrated by the following flow chart:

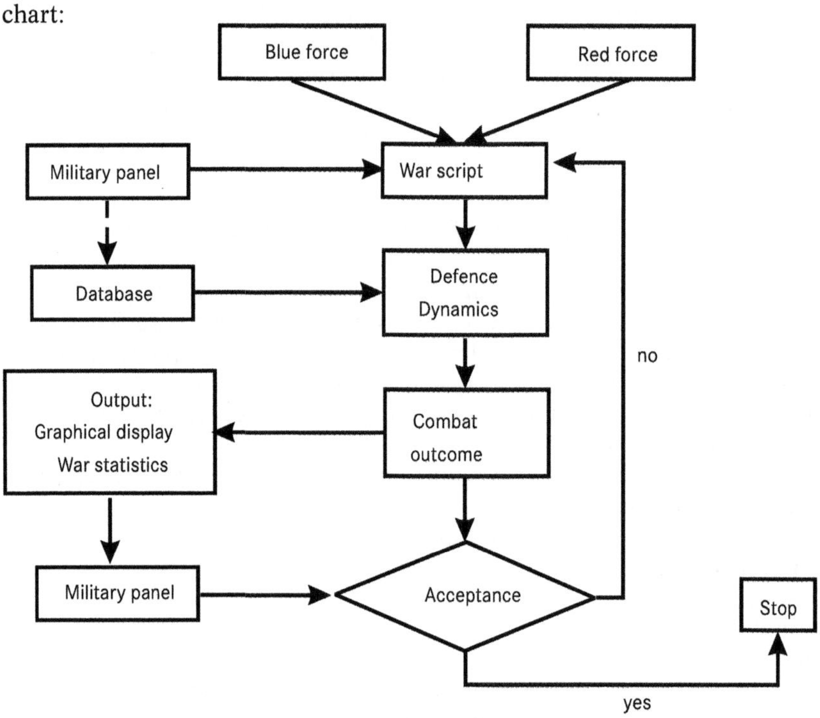

Figure 12. Practical procedure for using Defence Dynamics.

A military panel defines the relevant blue and red force. Afterwards the military panel works out a so-called war script for the heterogeneous battle situation considered. The war script specifies all movements and engagements for blue and red forces based on actual strategies and concepts. DD exploits the war script to control the setting up and removal of the relevant many-on-many heterogeneous combat models necessary to describe parallel and serial sub battle situations. A team of military and analytical people has prepared the database with information concerning all relevant weapon-sensor systems. This team furthermore has to analyze all relevant representative one-on one combat situations to provide the parameters necessary. DD draws the relevant parameters directly from the database when it establishes the combat models. Based on all this input information an approximated mean battle outcome is determined. This outcome is presented to the military panel by use of graphical displays and war statistics. The panel decides if the outcome is in agreement with the strategies and concepts intended. If so the outcome is accepted, otherwise the war script is revised and a new computer run is implemented. This will continue until the panel finally accepts the combat outcome. From the concluding outcome the conclusions relevant to the actual practical application will be drawn.

The total process is time consuming and difficult to carry out. Furthermore the process will quite often provoke disagreements among the members of the military panel. Such disagreements may delay the process further.

Despite the problems and the heavy amount of work involved DD has been used on a number of occasions. The most important applications have been:

1982 "The October War". The Danish Long Term Planning Group. The Danish Defence Command.

1989 Evaluation of possible new Air Defence Systems. The Danish Advisory and Analysis Group. The Danish Defence Ministry.

1989 Division Jutland defensive combat in Schleswig-Holstein. The Danish Advisory and Analysis Group. The Danish Ministry of Defence.

1989 Comparison of the TOW system and Leopard tanks used for defensive purposes. (European disarmament negotiations.) ACAG/NATO.

1992 Development of a new doctrine for the Danish army mobile combat. The Danish Defence College.

1994 Analyses intended to investigate the concept of Stable Defence. Defence Research Group, NATO.

1994 Development of an optimal structure and size for a Danish armoured battalion. The Danish Advisory and Analysis Group. The Danish Ministry of Defence.

References

Bonder, S. (1981): Mathematical modelling of military conflict situations, in: *Proceedings of Symposia in Applied Mathematics*, Volume 25, 1981, pp. 1–51.

Christensen, T., Clausen, S. (1984): Defence dynamics – predictions of war outcomes and evaluation of force structures, in: R. K. Huber (ed.), *Systems Analysis and Modeling in Defense*, Plenum Press, New York 1984, pp. 93–112.

Christensen, T., Clausen, S. (1988): *Stochastic combat modelling based on a representative pair of opposing units*. Danish Defence Research Establishment 1988.

Clausen, S. (1987): *Krig på formler*, Danish Defence Research Establishment 1987.

Clausen, S. (2002): *Matematisk Modellering*, IMM, Technical University of Denmark 2002.

Deitchman, S. J. (1962): A Lanchester model of guerrilla warfare, *Operations Research*, Vol. 10 (1962), pp. 818–827.

Dockery, J. T., Santoro, R. T. (1988): Lanchester revisited: progress in modelling C2 in combat, *Signal*, July 1988, pp. 41–48.

Koopman, B. O. (1970): A study of the logical basis of combat simulation, *Operations Research*, Vol. 18 (1970), No. 5, pp. 855–882.

Lanchester, F. W. (1916): *Aircraft in Warfare: The Dawn of the Fourth Arm*, Constable and Company, London 1916.

Lepingwell, John W. R. (1987): The laws of combat. Lanchester reexamined, *International Security*, Vol. 12 (1987), No. 1, pp. 89–134.

Morse, P. M., Kimball, G. E. (1951): *Methods of Operations Research*, John Wiley, New York 1951.

Taylor, James G. (1983): *Lanchester Models of Warfare*, 2 vols., Naval Postgrade Scholl Monterey, Calif., U.S.A. 1983.

Duels of Systems and Forces

HELGE LÖFSTEDT*

This paper begins with a brief summary of the vital changes in warfare during the last decades as seen by the author: Mass armies are no longer common tools for warfare. The "big" war has become unusual, while limited (and controlled) use of military forces for political purpose still is in use. However the ability to maneuver and fight with weapons still remains a basic military skill and modeling of combat remains important for understanding war. A formulation of a basic problem for single weapon systems is made and the effects of technology development are discussed in that context. Further comments are given on problems in information processing in modern weapons and in Command, Control, Communication and Intelligence, where the technological changes are expected to be revolutionary. A short comment is also given on transforming single weapons data to force characteristics, which is important for connecting analyses on low levels with those on high levels. The author also gives some comments on Operations Other Than War and Asymmetric War – which are new challenges for military analysts as well as officers. Finally a critique is given on the state of the art in military modeling.

1 Introduction

This paper focuses on duels of systems and forces in war and will give some thoughts on technology changes and the use of quantitative modeling in that context. These thoughts are based on my experience from more than thirty years as a military analyst (operations research) at the Swedish Defence Research Agency.

The central part of my paper gives some different background views to help understand the present development in warfare. The technological change is of course a fascinating and terrifying subject, which also will be central in this presentation. However, it is also vital to mention the development of attitudes to violence, which gives a framework for warfare activities.

2 Characteristics in Present and Future Warfare

Before going into the duels of systems and forces, it will be useful to mention some characteristics in present and future warfare. The first characteristic is that we have been able to avoid all-out-war with mass armies during the last half century.

* Swedish Defence Research Agency. Email: helge.lofstedt@foi.se

There may be several reasons for that. One can be the technological development of weapons and war systems that have made all-out-war with mass armies out of date. A phrase that characterizes the consequences of technological development during recent years and will be used in coming years is "Revolution in Military Affairs" – RMA. I will come back to that concept.

However, there is another explanation to why the time of mass armies and all-out-war is over and that is "Revolution in Attitudes to the Military" – RAM. That concept is not as well known as RMA, but I think it is even more valid. The concept RAM is from a war historian and philosopher – Jeremy Black.

The change in attitudes to the military has many causes. Underlying all these causes is the change in society due to industrial and technological revolutions. Risk of violent death for people in ordinary civilian life has been substantially reduced and many people have a rather decent living. Large casualties among ones own forces are no longer accepted. There also exists a high moral uneasiness in causing collateral damage and even casualties among enemy forces. In parallel to that, most societies have abandoned military triumphalism – a victory may not be worth the prize. A reluctance to serve in the military has in many societies contributed to a radical decrease in number of personnel serving in military forces.

The media is very critical toward war and the CNN effect is important. There is also a heavier emphasis on how to get fast political results out of military activities. And the political result is less than ever a simple function of destroying the enemy.

However, even if there has been no all-out-war in the last half century, there have been wars! These could be characterized as low-level conflicts – asymmetric wars between conventional forces and guerillas and terrorists and some low technology wars – and there have even been some conventional wars on medium levels. Even if the words low and medium conflicts are used, those wars have caused much suffering. And it is difficult to see an end to conflicts like these.

The nuclear threat has been, and will be, a continuous heavy factor in many security policy considerations.

My conclusion is that it is still meaningful to try to understand warfare in all its shapes.

3 A Basic Duel

In all above conflicts the basic phenomena are duels between lethal weapons. The development of non-lethal weapons during the last decades remains only a marginal phenomenon and mostly used for riot control or special circumstances in war operations. Before elaborating on the technological changes I will define what I think is a basic duel between lethal weapons.

Fundamental in many combat situations is a weapon firing at a target that has ability to hide or to move out of the dangerous area. The target tries to hide as long as possible and will make the shortest possible exposure. The weapons operator will then make the time for aiming as short as possible in order to deliver a shot with enough precision to hit the target.

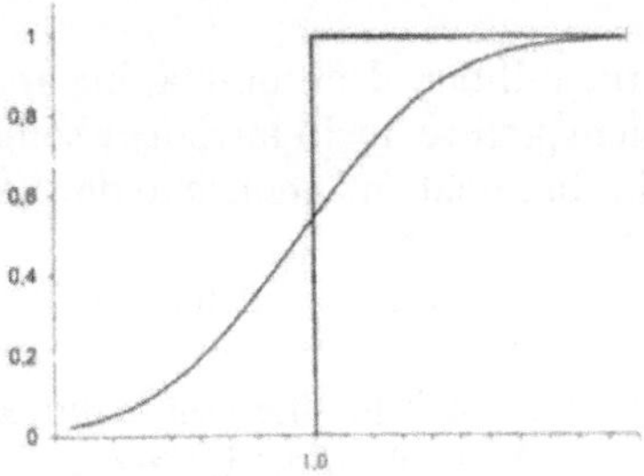

Figure 1. A basic duel.

A single weapon aiming at a target:
t_d Time to aim and deliver at shot
t_e Target exposure time

Deterministic:
$P(\text{hit}) = 1$ if $t_e/t_d > 1{,}0$

$P(\text{hit}) = 0$ if $t_e/t_d < 1{,}0$

Or stochastic:
$P(\text{hit}) = F(t_e, t_d)$

Even if this is a basic situation in military activities, it is not the most frequent. Most of the time in military activities in combat is spent *maneuvering* weapons to achieve favorable positions for such duels. Most military activities are preparations for such duels.

A characteristic of a duel is a struggle in time between two opponents. The value of better performance in t_d must be related to data in t_e and vice versa. There is no meaning in good performance in t_d if the target is even better in t_e. Many activities are aimed at being in the right place at the right time and achieving a time advantage. A simple weapon can be efficient against a more advanced weapon by firing some hasty shots and scooting out of the opponents lethal area before he is able to return the fire – that tactic is called shoot and scoot.

Another characteristic is that the result is stochastic. The actors are human and cannot be trained to deterministic behavior. Even if it often is necessary or convenient to consider duels deterministically, as a part of a more complicated analysis, one must remember that the end result would be stochastic.

Notice that neither t_d nor t_e are directly represented in the Lanchester equation that was mentioned in the previous paper. However, the phenomena are there in indirect ways.

A very important characteristic is that the duel concerns information processing. An example is a situation where one weapon outranges an enemy and is able to deliver a shot without the enemy being able to shoot back. Information processing is vital to be able to outrange an enemy. One modern way to do this is to use guided weapons, which are dependent on information processing.

4 Technological Change

Information processing is a vital part of most military activities. Information technology is also the field where the projected advances are likely to be in a true sense revolutionary. There will also be advances in many other fields, but I cite also a conservative judgement. O'Hanlon states that information technology is the only aera where the advances will be truly revolutionary. However, in that field there may be technological advances that have not yet been implemented by military applications. (Krepinevich)

Here it seems suitable to elaborate a little on the military decision-making and management of operations. To give a more complete picture, I will introduce some of the words used in the military organizations in that field and choose to do that by explaining a stream of acronyms – CCCCIIWS!

The first C stands for command, which means that orders and instructions should be delivered to their destinations in time.

The second C stands for control, which means feedback for the commanders being aware of the situation in their unit of command. Much effort is devoted to give the commanders an *enhanced situational awareness*. The means are better information processing to reduce delays and to find better ways of presenting information for commanders and staff officers.

Before the information technology era the usual acronym for management in military activities was Command and Control. Sometimes that acronym or others are still used.

The third C stands for communication and communications systems moving information between commanders and staffs. Modern communications systems are developed towards *networking*, which means that information is transmitted also sideways and not only up and down in a hierarchy as in the old military way. A modern acronym in the military world is Network Centric Warfare (NCW). An important problem is how staff officers and others will be able to avoid being blocked by too much unnecessary information. There is a need for a flexible way to evaluate information and to let through only what is wanted. The amount of information that can be carried by a single radio channel is however limited. This and other limitations highlight the importance of developing software capable of information processing close to the source of information.

The fourth C stands for *computers*, which are more and more introduced in

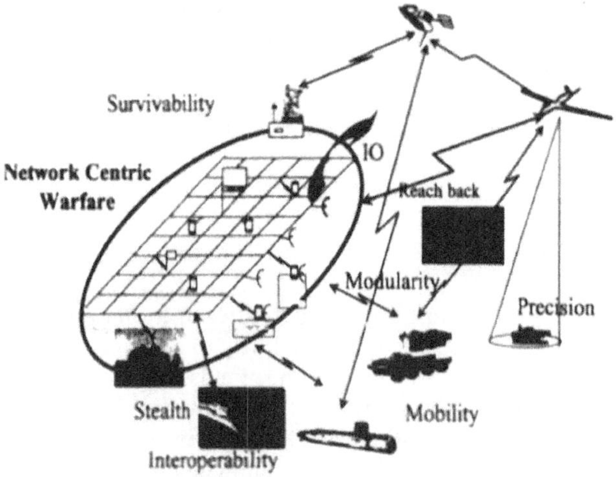

Figure 2.
Network Centric Warfare.

military management and for a wide range of missions. The trend of remarkable progress in computer technology shows no immediate signs of slowing down. The benefits of improved computing speed will be devices such as handheld computers and wrist telephones. These types of technology will continue to tie together the future battlefield and make it possible to equip many more soldiers with high performance computational gear. The capability could be used in scout teams to call in long-range firepower. In addition, automatic target recognition will make autonomous munitions that are resilient to countermeasures. Automatic target recognition will also permit substantially increased use of unmanned flying platforms (UAVs).

The first I stands for *intelligence* or information, which it is all about. The IW stands for *information warfare*. The definition of IW is wide and covers for instance electronic jamming, camouflage and also deception. It is a field where much effort is already spent. The technology part in this field started with electronic warfare that has been present from the beginning of electronic communication. It has expanded much during the last decades and will continue to expand in the future. Information operations may even be a new way of conducting war.

Characteristic in information warfare are the very complex duels that may take place and the long chain of counter-counter-counter loops. A first step may be a simple camouflage against a sensor. Then the sensors may get information processing that will see through that simple camouflage. The next step may then be better camouflage or a tactical change that reduces the exposure.

I will come back to IW after some word on sensors. Further on there will be a presentation on information warfare in the following paper of this volume.

S stands for sensors and I gave an introduction to that in the beginning of this paper. Sensors are vital for giving information to guided weapons, for aiding operators and conventional gunners, but also getting tactical information that is vital to maneuvering on the many levels of hierarchy that exist in military organizations. In connection with sensors there will be more and more automatic information processing, for instance for target recognition. Further, miniaturization of electronic components is improving the performance data of radar on satellites, aircraft, unmanned flying platforms, missiles and guided munitions.

A modern joint surveillance and target attack radar system (JSTARS) aircraft can form detailed images of the ground while simultaneously scanning for mobile targets in a search mode. Among the most important areas of radar development are attempts to penetrate below the forest canopy and to a degree certain types of soil. Other types – bistatic and multifrequence radars – will be able to challenge stealth technology.

Infrared sensors will also be improved with further increases in detection range. Improved information processing will also be able to detect weak signals passing through light rain. An infrared-system under development (LANTIRN) will allow use from aircraft in flight-levels more than 10000 meters. Other systems may be able to pick up heat signals from snipers camouflaged from visual observation.

Night-vision light-imaging detectors for infantrymen work over a range of one to two kilometers, meaning that soldiers now can see as far as their weapons can accurately fire. More advanced devices use infrared detectors, which will give information even in perfectly dark conditions.

In brief, advances in sensor technology have been huge and development will continue. However, two main principles constrain their potential. They must all function according to the laws of physics. And the everlasting information war – for every new sensor developed, countervailing efforts will be made to defeat it.

Radar and radio waves can penetrate clouds and rain and to a lesser extent foliage and soil. But they cannot physically penetrate deeply into soil or water. Nor can they reach inside sealed metal containers. These limitations open up possibilities for an adversary to develop countermeasures that explore those limitations.

Another fundamental constraint on radar is that it is active and energy emitting, which makes it susceptible to countermeasures. A counter-counter action is to minimize the use of radar. For example, highly directional wide-bandwidth radar beams minimize the chances of detection. And then the counter-counter-counter action is detection networks with even more advanced information processing.

A target that possesses a radar receiver will generally notice that it is being examined before weapons can be launched against it. It will then have the opportunity to take defensive actions: shooting back on the enemy, turning on radar jammers or simply ducking for cover. One possible result is that large aircraft like JSTARS will fly further and further away from the battlefield in order to stay out of range of antiair missiles. However, beyond a certain distance, the curvature of the earth will set the limit.

Military forces that are maneuvering on the ground, flying, or sailing will give a high technology adversary important advantages in a symmetric warfare. The advantages will be less notable in other combat settings.

5 Future Warfare Due to Technological Change

What do the trends in key military technologies portend for future warfare? And what about the validity of the most common hypothesis that a revolution in military affairs (RMA) is on the way?

The official opinions in the United States and many other countries reflect grandiose and ambitious interpretations of what a revolution in military affairs should entail. There are information systems, sensors, new weapons concepts, much lighter and more deployable military vehicles, missile defenses and other capabilities. They also include the terms dominant maneuver, precision engagement, full dimension protection and focused logistics.

However, some fields show few signs of major breakthroughs. The speed and weight of most means of transportation are not likely to change drastically. Neither are battlefield-armored vehicles likely to improve drastically. The aspirations for dominant maneuver and focused logistics may also be too optimistic.

A field with great expectations that O'Hanlon confirms is autonomous and homing munitions, which will improve in absolute terms over the next two decades. Whether they improve as dramatically in relative terms remains to be seen. Enemies may learn to reduce the effectiveness of these munitions substantially through the use of sophisticated countermeasures. The potential of these weapons is considerable but it will be realized depending on the opponent and other circumstances. O'Hanlon says:

> [...] the popular notation of information dominance simply goes too far. It is one thing to determine that a silicon-based modernization is the best way to modernize U.S. defense capabilities. It is another to claim that warfare will radically transform in the years ahead. The former conclusion seems a sober analytical judgment. The latter is a leap of faith upon which a sober scientific survey casts considerable doubt.

The following characteristics are also from O'Hanlon:

Likely technological accomplishments between 2000 and 2020:

- Computers – there will be a tenfold increase in capacity.
- Networking of data between various platforms will be operative.
- Autonomous munitions and unmanned sensors will be increasingly effective.
- Satellite imagery will become much more widely available.
- Long range radar sensors will be able to detect large vehicles under forest covers.

Challenges likely to remain elusive through 2020:

- Aircraft, ground vehicles and ships will not become radically faster or lighter.
- The large oceans will not become "transparent" to any sensor.
- Soldiers and small arms in urban environment and to some extent also in forests will be very difficult to track from long distances.
- Soldiers operating in small groups will often be able to get the first shots against any intervening forces.
- Sensors and munitions will not improve enough to hold targets in buildings and underground at risk.
- Things transported inside well-sealed metal boxes will not be detectable by long range sensors.

To evaluate the development and to reach and further elaborate answers like the above, it is important to systematically survey the potential of a wide range of defense technologies and systems. For developing military applications there is a need for analyses, experiments and evaluations. There will be a need for military analysis from many aspects.

6 Diffuse Conflicts

So far analysis of duels of systems in combat have been the common subject. And the technological development has been commented on. The capacity to perform in combat is still the ultimate reason for military forces. However, most missions that modern military forces conduct are in Operations Other Than War (the military acronym is OOTW). Further, most wars have been *asymmetric* in some aspect. And also most other conflicts have changed to a more diffuse pattern, where the connection between military means and political goals are much more complex than before! Operations other than war can be seen as the extreme opposite to war in this context. So I start there.

Operations other than war are mainly: peace operations and aid to the civilian authorities in catastrophes of some kind. Peace operations could be defined as (D. Davis 1998):

- Peace Making primarily conducted by the diplomatic community – the negotiation, mediation, treaty and accord building that remains central to all Peace operations.
- Peace Building primarily conducted by governmental agencies, non-governmental organizations and the UN – the attempt to address the root causes and exacerbating circumstances that hinder the Peace Making effort. These usually include Human Rights functions, humanitarian Relief, Refugee Activities, Good Governance, Infrastructure, Economics and Demilitarization.
- Peace Support primarily conducted by military organizations – The use of Command and Control, Force, Security, Presence, Observation, Liaison and Logistics to provide the space necessary for Peace Making and Peace Building to occur. Often it is convenient to use the words *Peace Enforcement and Peace Keeping.*

When military organizations are used for operations other than war the main effort is often to provide mobile logistics with an efficient management. I will not go further into that complex but will instead concentrate on duels with weapons and forces in Peace Support Operations and diffuse conflicts.

The use of weapons in operations other than war can be characterized by:

- An early step is to establish conditions allowing for *negotiations* from a position of strength.
- Military force should then *assist* in the resolution of the conflict and any attempts to alter the conditions.
- As the conflict de-escalates, the military role decreases.
- Troops require conventional combat skill tempered by stringent *rules of engagement.*
- The commanders need also effective use of intelligence, civic action and psychological operations.

Notice the rank order of these characterizations. Military means are used to establish conditions for *negotiations*. Notice also that duels between weapons and forces are much more restricted in this context than in war. The *rules of engagement* are very important. In general the military must exhibit a much higher regard for those rules than in conventional war situations. That implies that technological features of weapons could become secondary. Further: use of weapons is more and more controlled from higher levels in the chain of command due to their responsibility to act with diplomacy in most situations. In peace keeping operations common soldiers may use weapons only in self-defense and to save lives in evident cases.

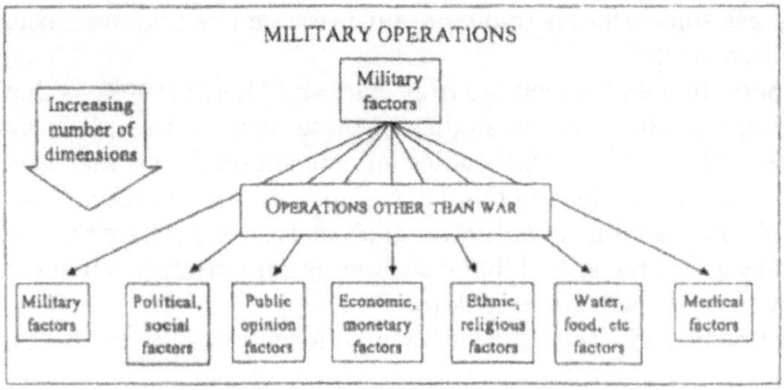

Figure 3. Multiple dimensions of conflict.

In brief: military operations other than war and related humanitarian activities take place in environments involving the interplay of many different types of social factors and processes (Woodcock), see Fig. 3.

I would like to add that these comments are valid also for many of the diffuse conflicts that we see at present and probably will see in the future.

The Vietnam War is a good example of warfare where the old rules for achieving a victory were not valid. That war was asymmetric in many ways and so will other wars between two parties develop, where one combatant use high technology means and the other combatant use guerilla or terror means.

In future diffuse conflicts the range of problems might still be expanding. Highly complicated behaviors can occur. Even well intentioned and well-thought actions may have very negative unintended consequences. Providing food and water to starving people can be interpreted as giving significant advantages to adversaries. Events can be stage-managed for the benefit of international media in order to change the perception of apparent realities and influence opinion. Under such circumstances it is evident that the planning and execution of many operations should be based on analyses that take into account the complexity of the human, societal and political environment within which they take place.

Empirical evidence indicates that quantitative analysis of operations other than war and the "diffused new wars" are less frequent than in analyses of conventional war. There are some good reasons for that but also some not so good.

To support this kind of operations there is a need for systematic studies of behavioral patterns that can be identified and measured. Data from which those patterns can be identified are the historical, social and other records of interstate conflicts. Here the difficulties have been rising. The relevant specialists often do not dwell on trying to pinpoint similarities and differences among a number of different events. There is a trend to direct more attention to the uniqueness of the particular event under examination rather than its similarities to other events. Most conflict data sets provide extensive detail on a selected number of different types of conflict episodes, rather than a brief description of a broader set of conflicts. However there are some schools of quantitative researchers that are closer to meeting some of these needs.

My comment on these thoughts, which are from Barr and Sherill (1996), is that we must realize that operations research studies in these matters are still in an early stage, compared with mathematical studies on conventional war, that now have more than fifty years of development and experience. An observation, however, suggests that resistance against quantitative analysis is stronger here than in the more conventional war analyses. And that resistance is quite strong even in the conventional military field as I will come back to.

I cannot resist the temptation to end this part with a formulation by Dr. Karen Toombs Parsons:

> Most analysts in the policy community [...] rely on more traditional qualitative research techniques. Many have an almost *theological aversion* to trying code information about social/relevant phenomena. They proudly proclaim that they do real analysis, not "crunch numbers".

7 Levels in Combat Dueling

We now continue to discuss duels between systems and forces and explore some aspects of modeling such duels.

To evaluate the implications and the possibilities of technological changes, the analyses must be conducted in different levels of resolution. In military analyses the natural breakdown in levels is:

- strategic,
- operational,
- tactical,
- military technical.

At each level models are needed for alternative representations of dueling, to reflect different facets of a problem or to serve different purposes.

On a military technical level there are one-on-one models. They are of course a simplified description for pure weapons analysis. Such duels can be symmetric – for instance tank versus tank – or asymmetric – tank versus anti-tank missiles. Such a model will not permit any analysis of interaction between friendly systems and units or for analyzing management of military forces in combat. To allow this the models must represent *few-on-few* or *many-on-many*, which is another class of models.

Even if modern computers can handle enormous amounts of information it will be too impractical with simulations of type many-on-many when large amounts of weapons and equipment must be analyzed. Then it could be useful with a model of type *force-on-force*. In such a model clusters of weapons and equipment are combined and represented as units with a few characteristics. Thereby the volume of descriptions of the total mass to be modeled can be reduced compared with what has to be done in a many-on-many model. It appears convenient but another problem arises and that is to make the descriptions in the force-on-force model *compatible* with results from simulation models. Here is a field for interesting mathematical applications, which I will elaborate on later.

Then you have a choice. When you are analyzing a problem on high or medium level a choice has to be made: to use a low resolution model with approximations or a high resolution model that explicitly gets you down to a lower level. This deserves of course a lengthier discussion than what is possible to give here. The principal reasons are as follows (P Davis):

There is a need for low resolution modeling on a high level for:
- comprehension and understanding of interaction between forces (seeing the forest rather than the trees),
- analysis for structure decisions and strategic or operational level decisions,
- low cost and rapid analysis,
- making use of low resolution knowledge and data (frictions of war!).

There is a need for high-resolution models on a low level for:
- understanding tactical and technological phenomena,
- representing systems knowledge and simulating reality,
- calibrating or informing lower-resolution models,

Notice the complication that important *data* on "frictions of war" mostly are available at low resolution on a high level. And a fundamental *understanding* of tactical or technological phenomena in war is mostly connected with high-resolution analysis on a low level.

To conduct analyses on different levels, the most common way is to use alternative models for each level. Often those models and databases are inconsistent. At best such a collection of models is combined with "good engineering practice" which means there are some procedures for using the more detailed models to calibrate selected input parameters in the lower resolution models.

However, during the 1990s there has been a development of methods for systematic and consistent variable resolution modeling with consistent mathematics. Lanchester differential equations are central in that development. In general they create an opportunity to reach a mathematically consistent way of transmitting information from low-level models (simulation models) to higher-level models.

Even with Lanchester models the amounts of information will be impractical when large amounts of weapons and equipment shall be analyzed. However, the Lanchester formulation gives the opportunity to model *force-on-force.* In such a model clusters of weapons and equipment are combined and represented as units with a few characteristics.

The mathematics used is eigenvalue solutions to matrices. Here I give a general outline of the mathematics. Jaiswal gives a complete description. The basic Lanchester formulation is:

$$\frac{dB_i}{dt} = -\sum_{j=1}^{m} \beta_{ij} R_j \qquad i = 1, 2, \ldots n,$$

$$\frac{dR_j}{dt} = -\sum_{i=1}^{n} \gamma_{ij} B_i \qquad j = 1, 2, \ldots m,$$

where β_{ij} is the number of i-th type of Blue force weapons destroyed by one unit of allocated j-th type of Red force weapons per unit time and analogously for γ_{ij}.

With some good assumptions it is possible to derive eigenvalues to these matrices and reduce the voluminous differential equations to:

$$\frac{dV_b}{dt} = -C_r V_r \ ,$$

$$\frac{dV_r}{dt} = -C_b V_b \ ,$$

where C_r is the aggregated attrition rate coefficient for Red force and analogously for C_b.

The units of V_r and V_b are often chosen according to the ordinary military organization – i.e., company or division. That means a single combat value is used for a company or a division.

The method hinted at above will require theoretical efforts to develop good approximations to separate phenomena occurring on different levels. One will then recognize that relationships often are complex and not intuitive. Davis thinks that there is a great potential for developing powerful variable resolution models, but a great deal of work has yet to be done, if we are to achieve that.

I agree with Davis and will add that with good models on high levels that are consistent with data on lower levels, it might be possible to develop better analyses on problems like force structuring and operational planning. Aspin and Huber have given some examples. The mathematics they use in their models is very simple but they are dependent on aggregated data from lower levels.

8 Mathematics in Modeling Duels

In previous parts I have given some indications of mathematics used in analyses of duels of systems and forces. I will also give an overview of mathematical techniques that can be found in such analyses.

Let us start with *probability* modeling as given by my elementary introductory example. A more realistic but still generic case is a weapon system for defense against missile attack, for instance on a ship. That weapon system is characterized by its probability of acquiring for a specific target, acquisition time and lethality time, that is the time to achieve a kill on an attacking missile once it is acquired.

These parameters are design features. Trade-offs between these weapons characteristics exist. The simple probalistic model may provide insight in deciding the desirable parameter configuration. That kind of model can also be used to understand the defensive threat experienced by the attacker.

Such methods are used in a wide range of analyses – from infantry weapons to nuclear. There are of course many ways to conduct those analyses and I will not go further into that field. My example above is from Almeida.

My basic example indicates that time delay in information processing can be important to calculate. It is thus evident that *queuing* models can be useful.

One example can be a forward air control officer that shall assign combat air fighters to missions, taking into account the aircrafts armament, the requirements to get effects on a certain target and other factors. By running such a model under different scenarios for various combinations of systems parameters, the analysts can estimate effects in terms of expected targets destroyed. This example is from Samuelson and Sims (1995).

Lanchester differential equations have already been mentioned and I will not elaborate on that.

The idea of using information theory and entropy to measure effects of reconnaissance, intelligence and other activities related to preparing and conducting military operations seems also rather natural. Barr and Sherrel (1996) have demonstrated an application on a target detection process that makes it possible to derive a measure of information, which makes it possible to assess the value of new sensors in a way more general than the specific scenarios from which they are derived.

The examples above use conventional mathematics. There are huge amounts of models of that kind.

Dodd and Richardson (1996) gives an examination on some newer analytical and mathematical approaches, which could be useful for modeling information processes in the context of future conflict operations. The aim is to develop analytical methods for those processes that can be represented mathematically as transformations from inputs to outputs. They find that top-down detailed descriptions, for example traditional *artificial intelligence* (AI) knowledge-based methods, must be comprehensive so that every possible eventuality is covered. Maintenance of internal consistency within the rule set is often a problem.

A *Systems Dynamics* model can be developed quickly from an influence diagram using commercial tools. The main difficulty is the detailed differential equations that must be well specified.

Self-learning algorithms – for example neuronal networks – are able to generalize well if they are trained on large amounts of data, which span the range of relevant space. It is also desirable to have a well-designed network structure.

Military decisions by their very nature are often undertaken in conditions of extreme stress. Add to that the uncertainties in tactical and operational objectives in the chain of command, reliability and credibility of intelligence reports and fuzziness in the boundary regions and it is clear why military command and decision-making is due for further research.

There are other newer mathematical methods like *catastrophe theory, complexity theory* and *intelligent automata* approaches, which could be used for modeling combat and military decision making. A further description of new mathematics for modeling as support for decision-making is given in Woodcock (2001).

All analysts are not convinced of the usefulness of advanced mathematics. Carlsen and Hamrin (2002), who have tried to evaluate some statements of Woodcock and others, do not share their enthusiasm. They find that reading good military writers – for instance prophecies by Clausewitz – is an easier way to reach understanding of the complexity in military matters.

Moffat and Witty (2002) seems to be more successful in using catastrophe theory. They have considered a mathematical model – Bayesian Decision – of decision-making and military command. They also use elementary catastrophe theory. Their conclusion is that the application of this approach has led to insights into the command process, together with simple and elegant mathematical algorithms for the representation of these aspects of the command process in simulation of conflict. Thereby they think it will be possible to avoid large and cumbersome sets of decision rules.

A reflection is that Moffat and Witty have a more thorough representation of the dueling element than the other mathematical models, which I here have commented on. The core of their analysis is the commander's perception of the force ratio. The ordinary way in most models is that the dueling element per se is not modeled! Of course the enemy is represented in most models. However, often the opposing forces are behaving according to some fixed input. Even if that data is taken from a war game, the human dueling behavior is not modeled and thereby not analyzed as thoroughly as when it is modeled.

One reason for Moffat and Witty having a higher value could be that they go deeper into modeling of human behavior. My conclusion is that the newborn analysis in decision-making and command processes may open new fields for military analysis. That turns my attention on game theory. Look at the definition of *game theory* given by Myerson (1990):

> Game theory is about analyzing situations in which two or more individuals make decisions that will influence one another's welfare.

A shorter definition is "interactive decision theory". Theories that fit into those definitions ought to be useful also for analyses on dueling aspects of conflicts. And notice that here the dueling element is built into the core of the model. In my analytical experience I have often also thought that a problem could be analyzed in a better way if I had the time to formulate a game theoretic formulation and carry on from that.

However, I find no examples in the literature. Game theory has not often been used in analysis of modern warfare, with exceptions of nuclear warfare. And the literature on game seems to be dominated by economic problems. The military examples are few and elementary. Perhaps an explanation is the distinction between zero-sum games and nonzero-sum games. War fighting tends to be considered as zero-sum games, and theory in that domain is not as rich (it is even trivial in many cases) as nonzero-sum games. Operations other than war and a new perspective of war given by diffuse conflicts may open a new field also for game theory in war science. Further, the definitions given above focus on problems that would be relevant for commanders of dueling forces. I think also that the techniques of Recursive Games and Games with incomplete information (Kuhn) could be given interesting military applications. Thereby the maneuvering of dueling forces might be further developed in the same way as chess players develop their ability through studies of chess theory.

However, my conclusion so far is that there are not many good examples of new mathematical methods used in analysis of war. One reason for that could be that the state of the art in military science is not yet ready for more advanced quantitative theory building. So let us look at the state of the art in military science.

9 The State of the Art in Military Science

The defense community is becoming more and more dependent on combat models – including simulation and war gaming. Around 1990 some senior analysts in the USA – insiders of the national security model-development and analysis community – separately suspected that the use of combat models could be considerably developed. They therefore conducted a study on the state of military combat modeling. A report on that study was published by Paul Davis in 1990, which had the status of a white paper prepared for the Office of the Secretary of Defense Advanced Research Projects Agency. This report is summarized in Figs. 4 and 5.

The message in that report is that there was relatively too much enthusiasm about the advances in computer performance, communications and human interfaces, and much too little interest in the substance of the models. There was a *lack of military science.*

The report contains several aspects of self critique from military analysts. Their effort was mainly on building and using models as tools and conducting analyses on specific questions and tasks. Minimal attention was paid to "theory" per se as well as experiments and historical analysis.

They found that there is a fundamental need for improving the true *consistency* across levels. They suggest *variable resolution models* for making it easier to do cross-resolution work. A weakness was that individuals and organizations working with high-resolution models commonly look down upon aggregated models and vice versa. It is uncommon for analysts to move back and forth among levels of resolutions and perspectives. So limiting factors are more conceptual and organizational and less due to computational capacity.

Theory building and model calibration should use low-resolution data on a high level as well as high-resolution data. However, the *bottom-up paradigm* was dominating. Further, far greater emphasis should be placed on comprehensibility and appropriate documentation. A heavier emphasis was needed on long-term funding of military combat modeling as applied science, not mere tool building.

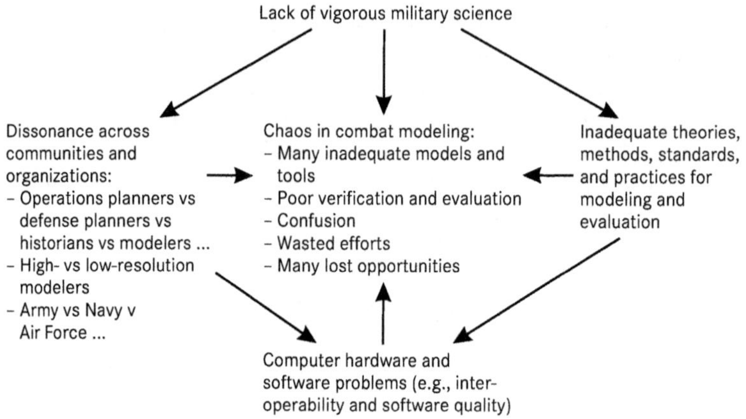

Figure 4. The state of the art in military modeling - the state 1990.

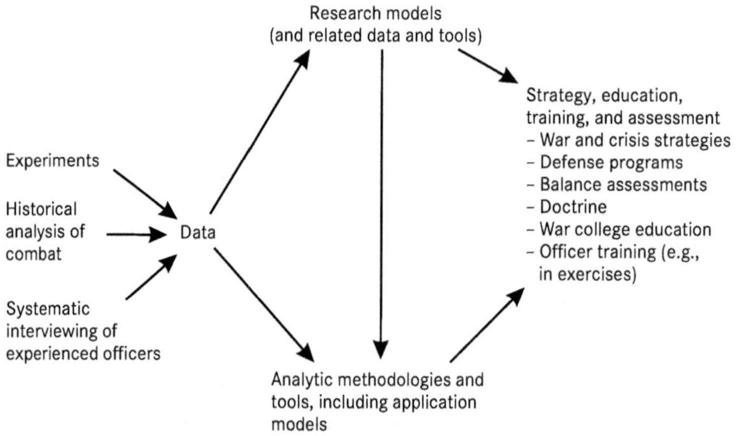

Figure 5. The state of the art in military modeling - the ideal.

There were few journals on military science. The closest was MORS journal. The academic journals of International Security and Defense analysis were of some value to military science.

Currently, the universities are of little help in developing first-rate analysts for military applications. Training centers are impressive. At the same time neither of those are appropriate places for basing scientific research, because their natural dispositions are to support here-and-now needs. Training centers give some attention to the collection of anecdotal and subjective information, but still there is much theoretical work that should be done.

As a conclusion, Davis and Blumenthal (1991) stated a need for the development of military science.

The white paper resulted in reforms in the USA. The National Defense University was reformed and strengthened to promote and nurture military science. The empirical base was strengthened and models seen as embodiment of knowledge. But the tradition of emphasizing intuitive art over science is still strong.

For some decades there has also been an international trend of academizing military officers' training. Among others, Sweden is following that trend. Lately an institution for war science is being built up within the Swedish Defense University. However, I don't think many people know the criticism and thoughts in the cited paper from 1991 in the USA. And the tradition of emphasizing intuitive art over science is still strong!

References

Almeida Rui, Gavier Donald P. and Jacobs Patricia A. (1995): Simple Probability Models for Assessing the Value of Information in Defense Against Missile Attack; in: Bracken: *Warfare Modeling*, John Wiley and Sons 1995, pp. 393–405.

Aspin, Les (2000): An Approach to Sizing American Conventional Forces for the Post-Soviet Era, in: *Strategy and Force Planning*, Naval War College Newport, USA 2000, pp. 448–468.

Barr, Donald R. and Sherill, E. Todd (1996): Entropy Modeling; in: *Analytic Approaches to the Study of Future Conflict*, The Canadian Peacekeeping Press 1996, pp. 114–139.

Black, Jeremy (2000): *War – Past, Present and Future*, St Martins Press, New York 2000.

Bracken, Jerome, Kress, Moshe and Rosenthal, Richard E. (1995): *Warfare Modeling*, MORS and John Wiley and Sons 1995.

Carlsen, Henrik and Hamrin, Martin (2002): *En litteraturstudie kring ledning och beslutsfattning (FoRMA)*, FOI memo 02-283.

Davis, David (1996): *Introduction in Analysis for Peace Operations*, The Canadian Peacekeeping Press 1996.

Davis, Paul K. and Blumenthal, Donald (1991): *The Base of Sand Problem – A White Paper on the State of Military Combat Modeling*, RAND California 1991.

Davis, Paul K. (ed) (1994): *New Challenges for Defense Planning – Rethinking How Much is Enough*, RAND 1994.

Davis, Paul K. (1995): An Introduction to Variable-Resolution Modeling; in: Bracken et al.: *Warfare Modeling*, John Wiley and Sons 1995, pp. 5–35.

Dodd, Lorraine and Richardson, Sean (1996): Adaptive C3I Systems Modelling; in: *Analytic Approaches to the Study of Future Conflict*, The Canadian Peacekeeping Press 1996, pp. 140–153.

Hillestad, Richard J. and Juncosa Mario L. (1995): Cutting some trees to see the forest: On Aggregation and Disaggregation in Combat Models; in: Bracken et al.: *Warfare Modeling*, John Wiley and Sons 1995, pp. 39–62.

Huber, Reiner K. and Avenhaus, Rudolf (1996): *Models for Security Policy in the Post-Cold War Era*, Nomos Verlagsgesellschaft, Baden-Baden 1996.

O'Hanlon, Michael (2000): *Technological Change and the Future of Warfare*, Brookings Institution Press, Washington 2000.

Jaiswal, N. K. (1997): *Military Operations Research: Quantitative Decision Making*, Kluwer 1997.

Krepinevich, Andrelw F. (2000): Cavalry to computer: The Pattern of Military Revolutions; in: *Strategy and Force Planning*, Naval War College, Newport, USA 2000, pp. 480. 496

Kuhn, Harlod W. (ed) (1997): *Classics in Game Theory*, Princeton University Press 1997.

Military Operations Research. The Journal of Military Operations Research Society, Alexandria, VA, USA.

Moffat, J. and Witty, S. (2002): Bayesian decision making and military command and control, *Journal of Operations Research Society* 2002, no. 7, pp. 709–718.

Myerson, Roger B. (1990): *Game Theory - Analysis of Conflict*, Harvard University Press, Cambridge, Massachusets and London 1990.

Samuelson, D. A. and Sims, R. L. (1995): Quick-Response Assessment of Forward Air Control and Communications Using a Queuing Model and Digital Simulation; in: Bracken et al.: *Warfare Modeling*, John Wiley and Sons 1995, pp. 407–415.

Söderqvist, Olof and Isacson, Tore (2001): *FORMA uppsummering 2001*, FOI Memo 02-349.

Toombs Parsons, Karen (1996): Predicting where the next OOTW will occur; in: *Analytic Approaches to the Study of Future Conflict*, The Canadian Peacekeeping Press 1996, pp. 14–28.

Woodcock, Ted (2001): Political and military leaders must understand societal dynamics, *Journal of Swedish Royal Academy of War Science*, 2001:4, pp. 21–63.

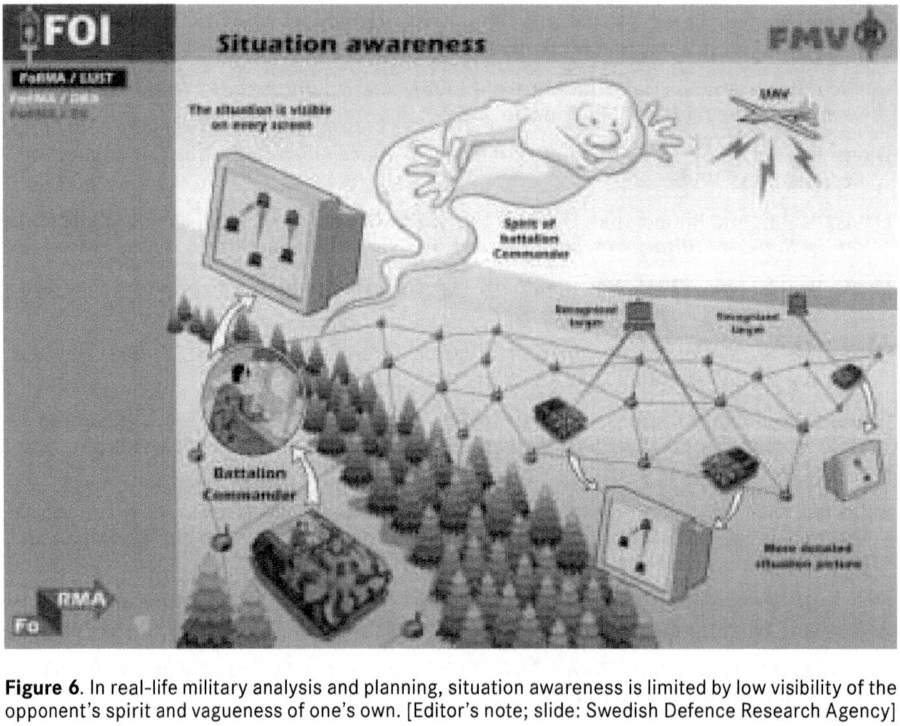

Figure 6. In real-life military analysis and planning, situation awareness is limited by low visibility of the opponent's spirit and vagueness of one's own. [Editor's note; slide: Swedish Defence Research Agency]

On Facts and Fiction of "Information Warfare"

Ute Bernhardt, Ingo Ruhmann*

Information Warfare has become a keyword for a revolution in military operations with far reaching political consequences, just as well as it is a phrase for pure military science fiction. It is thus necessary to separate the media hype from the development of information technology and its broad use by the military. It is shown how Information Warfare has developed from its origins in the use of computers for command and control, in weapons systems, and from electronical and psychological warfare into a new and comprehensive way of military operations. The political implications of this change are addressed together with the limits to conflict resolution by technical means.

1 Concepts, Meaning and the Media

Just as the preparations for a new war against Iraq are under way (work on this paper was finished in December 2002), the media once again put an emphasis on the use of Information Warfare as a new tool in armed conflict. U.S. officials are praising Information Warfare as "mining of data bases and lots of false targets generated. And, if most of the computers in the country immediately go down, that's not a bad way to start a war"[1].

In these very few words, taken at random from current publications, many of the expectations, misunderstandings, false leads and pure hype coupled with Information Warfare can be identified. A "mining of databases" under ordinary circumstances would rather be interpreted as "data mining", the concept of a complex search process in different data bases. The article, however, continues with descriptions of tools for hacking into adversaries' computers as described in the Department of the Army's Field Manual 100-6 "Information Operations"[2] thus suggesting an offensive meaning of "mining". If we assume that the official interviewed knows the terminology in his field, as a first point, the problems of explaining Information Warfare to the media become obvious.

* FIfF e.V. (Computer Professionals for Peace and Social Responsibility), Goetheplatz 4, D-28203 Bremen, Germany. Email: ute@kriton.bn.shuttle.de, ingo.ruhmann@acm.org

[1] David A. Fulghum: War planning for Iraq continues on target, in: *Aviation Week & Space Technology*, Sept. 23, 2002, pp. 22–23.

[2] U.S. Department of the Army: *Field Manual 100-6*, 16 June 1997, originally: http://www.atsc-army.org/cgi-bin/atdl.dll/query/download/FM/100-6/fm100-6.zip. Available today at: http://fas.org/irp/doddir/army/fm100-6/index.html.

The second assertion – generating false targets by computer – is a straightforward job for electronic warfare units since the days of World War II. What then was done by dispensing aluminum foils or emitting radio signals is now being done by computer-controlled frequency emitters that automatically fine-tune the emitted signal to the characteristics of an enemy's electronic equipment. As a second point, Information Warfare comes as a modernization of electronic warfare with profound effects on warfighting capabilities.

The third assertion, having most computers in an adversaries' country go down at the start of a war, obviously is a nice thing to have in modern combat. It most of all shows the central role computers play for warfighting in any modern army. There also is a method to achieve this goal: An atomic blast test in the upper atmosphere over the South Pacific proved at the beginning of the 1960s, that the electromagnetic pulse (EMP) generated by a blast at high altitude reliably destroys most electronic equipment in a wide diameter. An atomic bomb explosion over Iraq for exactly this purpose was publicly discussed before the first Gulf War 1991[3]. Unfortunately, the EMP affects the equipment of friendly forces just as well as that of an adversary. Today, directed energy weapons are reported to be ready for battlefield use that deliver an EMP on a desired target in a small area quite similar to a ray gun in science fiction movies. As a third point, Information Warfare blurs the line between fiction and reality through skillful combination of technology proven experimentally at laboratory scale and media reports of weapons developed from these experiments.

No matter how inconsistent yet, these three hints can be used to develop a first and basic understanding of Information Warfare. The basic idea of Information Warfare is the central value of *information* – or rather data – for coordinated and purposeful military activities. Since information today is acquired, communicated and manipulated by computer, any electronic communication and computing device becomes an element for Information Warfare activities, explicitly including civilian media infrastructures. Information Warfare combines formerly separated activities in psychological and electronic warfare with the use of new opportunities generated by computer technology.

Keeping in mind that Information Warfare is not a completely new invention, we can start with the definition of Information Warfare in the terms of the U.S. Commander in Chief:

> Actions taken to achieve information superiority by affecting adversary information, information-based processes, information systems, and computer-based networks while defending one's own information, information-based processes, information systems and computer-based networks.[4]

The Army interprets its military information environment (MIE) as a part of the global information environment (GIE) and defines Information Operations (IO) in its Information Warfare Doctrine Field Manual 100-6 as:

[3] John Barry: The nuclear option: thinking the unthinkable, *Newsweek*, 14.01.91, pp. 12–13.

[4] CJCSI 3210.01, *Information Warfare Policy*, Washington 1998.

> Continuous military operations within the MIE that enable, enhance, and protect the friendly force's ability to collect, process, and act on information to achieve an advantage across the full range of military operations; IO include interacting with the GIE and exploiting or denying an adversary's information and decision capabilities.[5]

If Information Warfare is understood as it is defined by its military proponents, there is a large number of Information Warfare methods, tools, and weapons. A press release can be just as valuable a tool in Information Warfare as a computer hack into an enemies data base. There is an even larger number of Information Warfare participants. As "significant players" in Information Warfare are seen by the U.S. Army "the media, think tanks, academic institutions, nongovernment organizations (NGOs), international agencies, and individuals with access to the information highway. [...] Their activities may cause an unanticipated or unintentional effect on military operations."[6]

When we note that information processes in this military view encompass the technological as well as the human elements of information-based processes, the global information environment – spanning the internet as well as the media and all users of these various information channels – becomes a military staging and operations area.

2 Levels of Information Warfare Reality

The scope of activities covered by Information Warfare principles is a key factor for the credibility of the concept. Influencing the media and of course psychological and electronic warfighting capabilities have for a long time been elements in modern combat. It are these elements of Information Warfare, developed and proven since more than 50 years, that make Information Warfare work on the battlefield.

Exploiting the weaknesses of information technology has become a new field of military activities with vast opportunities for destructive results. The large number of weaknesses of common computer systems – of which only a small number become evident to the general public in the form of computer viruses and the like – combined with the dependency of modern armies on computers, do not allow to ignore computer manipulations as tools in combat.

Information Warfare spans a vast array of different levels that range from nonviolent and preparatory means like the ones used in psychological warfare or the exploitation of information technology weaknesses up to a massive destruction of infrastructures necessary for command and control. At the core of Information Warfare thus lies a set of *obvious capabilities* in the use of electronics, information

[5] ibid.

[6] *Field Manual 100-6*, Chapter 1.

and communication technology as well as military force and a *concept*: to achieve a maximum effect.

First of all, it is necessary to assess the use of information technology by the military. The use of computers for military purposes nowadays is most obvious in "smart" bombs or highly complex weapons systems such as warplanes that fly "by wire". Historically, so-called "mission critical" computer systems were fielded for the first time in World War II. Analog computers appeared in British air defense systems that used radar data as an input for computers that in turn controlled anti-aircraft guns. According to Wiener's memories, the computers ran his linear prediction code algorithm[7] and delivered basic ideas for Wiener's work on cybernetics.

Electronics were used to equip the most expensive and complex weapons platforms. At the end of World War II, bombers in different air forces had so-called bomb sights as electronic steering devices for the final approach on a target, while the first fighters had cathode ray tubes for target acquisition by radar on board. Since the 1950s, more and more phases of flight control, communication and targeting have been supported by "mission critical computer systems"[8]. The Vietnam war saw terrain-following bomber types like the F-111 flying at altitudes below 50 meters towards their targets where the pilots directed the plane while the on-board computers actually flew the plane and kept it from crashing into obstacles. Today, aviation computer – avionics – systems are used to keep inherently unstable airplanes in the sky[9]. The success in the use of these systems was one factor in the integration of computer systems into land-based weapons platforms as well, leading to vehicle computer systems – so-called vetronics – for tanks and others.

Modern autonomous weapons systems make an even higher demand on computer power than manned vehicles. Their development also started with the control of airborne weapons systems. An early example is the analog flight-control computer on the German V-2 rocket, which was used to steer the rocket on its exhaust plume during liftoff as on any ballistic missile of a certain size. After engine shutdown, the V-2 flew on a ballistic trajectory[10]. The first integrated circuit chips were put onto Minuteman intercontinental ballistic missiles to severely reduce their target devia-

[7] He even noted that the program made use of the fact that the pilots mostly took evasive action as recommended by their Luftwaffe drill. Wiener gives a short, but very interesting description of his work in World War II in: Norbert Wiener: *Kybernetik. Regelung und Nachrichtenübertragung im Lebewesen und in der Maschine*, Düsseldorf 1963, pp. 30ff. (German translation of *Cybernetics or Control and Communication in the Animal and the Machine*, Cambridge, Mass. 1961).

[8] See an overview in: Ingo Ruhmann: Supercomputer mit Flügeln: Avionik; in: Ute Bernhardt, Ingo Ruhmann: *Ein sauberer Tod. Informatik und Krieg*, Marburg 1991, pp. 127–153.

[9] Examples are the Swedish JAS 39 "Gripen", and the U.S. F-117A stealth fighter and B-2 flying wing bomber.

[10] Helmut Hoelzer: 50 Jahre Analogcomputer; in: Norbert Bolz, Friedrich Kittler, Christoph Tholen (Hg.): *Computer als Medium*; München 1994, pp. 69–90. For additional accuracy, improved V-2 models until engine shutdown were supported from a ground station by a guiding radio beam ("Leitstrahl" – the steered V-2 models had an antenna on one of their rocket fins, for more details see: http://www.v2rocket.com/start/deployment/leitstrahl.html).

tion – or circular error probable[11]. With the dramatic cost reductions in integrated circuits came the integration of computer processors into fire-and-forget weapons and "smart" ammunition as well as all types of weapons platforms.

Computers in autonomous weapons

The technology that has evolved into modern autonomous weapons was developed from remotely controlled systems. An early example of one of the more prominent technologies are video-controlled platforms. The first experiments on video guidance of aerial vehicles date back to World War II when German troops experimented with planes that were controlled by radio and equipped with electronic television cameras to transmit pictures to a control station on the ground. The Maverick missile in the Vietnam marked the first use of video-guided missiles in combat where an operator had to keep a target in the center of a video picture transmitted by the missile warhead. Today, computer processors aboard video-controlled systems keep a target in their aim automatically once this is locked on by an operator.

Laser-guidance technology has been developed in the U.S. since 1962. Laser-guided bombs were first used 1968 in Vietnam, their first major success came in 1972 with the destruction of a bridge. Laser-guided systems are equipped with a detector that measures the intensity of a laser reflected from a target that in turn is illuminated by a laser beam. An on-board processor generates commands for the steering system to guide the bomb or missile towards the point of maximum laser reflection. The laser illuminators and warheads can be synchronized through the pulse modulation of the laser beam. Target illumination with lasers has been done from forces on the ground. Only the Gulf War saw the use of planes able to pinpoint targets with lasers exactly enough for bombs to reach the desired goal[12]. Laser- and video-guided autonomous weapons were only made possible through the miniaturization of computer equipment.

Autonomously flying vehicles are mainly used to monitor a target area or – to a lesser extent – as weapons. The first flying weapon was the V-1 flying bomb in World War II that navigated by keeping a straight course with the help of a simple gyro compass. The target was reached when the fuel had run out. The V1-concept was further developed after the war in the U.S., where in 1960 the MACE cruise missile was demonstrated that could compare radar data with preprogrammed images and the Regulus that could fly at supersonic speed[13]. Still, both were susceptible to interception and target deviation. The modern concept of cruise missiles became operational when computers could be put on board that not only could cruise the missile at low altitude, but also could compare a three-dimensional

[11] See Holger Iburg: *Abschreckung und Software. Computertechnologie als Instrument der amerikanischen Sicherheitspolitik*, Frankfurt 1991, p. 109.

[12] See Alberto Mondini: Laser für militärische Zwecke, *Umschau* **3**, 1985, pp. 172–178; Chris Cooper: Military use of lasers, *Miltronics*, August 1983, pp. 127–131; an overview on technology and bomb types: http://www.fas.org/man/dod-101/sys/smart/lgb.htm.

[13] Die Cruise Missiles, *Österr. Militärzeitschrift*, Nr. 1, 1979.

radar map of way points and the target area against incoming signals from the on board radar to achieve a precision of just a few meters on target. The accuracy of navigation allows for cruise missiles to be used against targets in urban areas. This sensor-based kind of navigation is supported today by the GPS satellite navigation system in order to reduce the emission of radar signals.

Smaller versions of cruise missiles are called drones; these are mostly used for intelligence gathering over a confined area of a battlefield. Unmanned small reconnaissance vehicles are used by many modern armies in combat. While many armies have used drones for some time – the Israeli Scout, the Canadian-French-German CL-289, the French Brevel or the U.S. Gnat – the U.S. forces managed to draw the interest of the media with the debut of the large Predator spy drone over Bosnia operated from Hungary[14]. The Predator is now the vehicle with the longest endurance over the battlefield, which allows it to be flown from greater distances. Newer designs call for micro drones with the size of a peanut and the ability to fly over a battlefield as well as within rooms in an urban warfare setting[15]. The development today combines these developments and aims to have autonomous missiles that dispense small weapons drones[16]. One early example is the tank-breaking drone Taifun of the German Army that can fly over a target area, scan the area below for radar signatures of tanks and propel itself onto the target, when such a target is recognized[17]. Predators are currently considered as weapons-carrying platforms for the Afghan theater.

Data communication – weaving platforms and weapons into a strike package

The accuracy of weapons delivery in the Gulf War resulted not as much from the development of any single technology – be it airplanes or steering mechanisms of weapons – as from the integration of formerly singular platforms and their weapons into a digital communication system. Data sharing between platform and weapons delivery mechanism – be it a bomb or a rocket – data sharing between different platforms on and over the battlefield and data sharing between the platform and intelligence gathering units as well as auxiliary information and data distribution networks improved the effectiveness of weapons delivery systems.

[14] David A. Fulghum: Predator to make debut over Bosnia, *Aviation Week & Space Technology*, July 10, 1995, pp. 47–48.

[15] See a discussion of beginnings, cf. Bruce D. Nordwall: Micro air vehicles hold great promise, challenges, *Aviation Week & Space Technology*, April 14, 1997, pp. 67–68. U.S. Special Forces are using UAVs today, cf. Michael Dornheim, Michael A. Taverna: War on Terrorism Boosts Deployment of Mini-UAVs, *Aviation Week & Space Technology*, July 8, 2002, pp. 48–50. The French Army today has an acquisition program for such a type of UAV, cf. Michael A. Taverna: French plan for miniature UAV demonstration, procurement, *Aviation Week & Space Technology*, June 17, 2002, p. 63.

[16] David Fulghum, Robert Wall: Small UAVs built for use from large UAV's, missiles, *Aviation Week & Space Technology*, July 22, 2002, pp. 192–195.

[17] The concept for the German Bundeswehr weapon dates back from 1988, see: Kampfdrohne Heer, *Wehrtechnik*, No 5, 1990, p. 52.

A precondition for any platform to deliver any explosive device with precision is accurate navigation. Navigation of piloted and autonomous airborne systems for a long time depended on gyroscopes that had to be corrected by radar measures. A change came with the U.S. Global Positioning System (GPS), developed after experiences in the Vietnam War. GPS consists of 24 satellites in low earth orbit that emit coded signals allowing for a measurement of one's location. The first 11 satellites were put into orbit between 1978 and 1985, the last of the original configuration in 1993[18]. GPS is controlled by five military ground stations around the world and coordinated by USAF Space Command at Schriever AFB, Colorado[19]. GPS signals were divided into, (1) the military "precision" code, allowing a position fix with less than 10 meter accuracy, and (2) the deliberately less precise "coarse/acquisition" code of GPS for civilian users, which had a 100 meter accuracy. Since different technologies using ground-based signal emitters (Differential GPS) or the Russian satellite navigation system "Glonass" as a corrective to distorted GPS signals have improved the precision of the civilian signal, it was switched to military precision in 1999 through a directive of U.S. President Clinton[20] in an unsuccessful move to discourage the build-up of the European counterpart "Galileo". After U.S. Congress mandated in 1994 that all new systems must make use of GPS navigation whenever necessary, it is by now the standard navigation system for U.S. Forces, giving precise information on their respective location to manned airplanes, land and sea vehicles, autonomous vehicles, intelligent munitions as well as individual soldiers on the battlefield.

Navigation, however, plays only a small part in the activities of the kind demonstrated since the Gulf War. Since mid-World War II, bombers entering hostile aerospace are accompanied not only by fighters to engage hostile fighter planes, but also by planes with capabilities for navigation and coordination. These command and auxiliary tasks today have led to the development of a triad of command and control systems that make up the decisive factor in the effectiveness of modern air forces.

In bombing campaigns, forward air controllers selected targets and observed the effect of air strikes from small planes. The Vietnam War showed the necessity for an airborne early warning and control aircraft that could coordinate bombing campaigns for an extended period of time. U.S. Forces equipped a Hercules C-130 transport with communications and electronics to coordinate the execution of the air tasking order. The result was the Airborne Battlefield Command Control

[18] Mark Tapscott: Extending GPS on land, sea and air, *Defense Electronics*, July 1993, pp. 42–47, p. 42. The last of the first round of GPS satellites were switched off in 1997 and have been replaced by updated Block-II and Block-III satellites. Until 1997, 37 GPS satellites were launched, cf. Craig Covault: New GPS broaden navigation accuracy, *Aviation Week & Space Technology*, Jan. 20, 1997, p. 22.

[19] GPS was initially controlled by the 14th U.S. Air Fleet at Falcon Air Force Base, Colorado. The Air base was renamed to Schriever AFB in 1998, the GPS part reorganized into the 50th Space Wing's 2nd Space Operations Squadron of the USAF Space Command. See: http://www.spacecom.af.mil/hqafspc/Library/FactSheets/FactsSheets.asp?FactChoice=9.

[20] Jeanne Rubner: USA beenden Störung von Satelliten-Signalen, *Süddeutsche Zeitung*, 3.5.2000, p. 8.

Center (ABCCC) that was substantially modified in the 1990s. The massive use of computers in the system today allows an ABCCC to command and control activities in the airspace over a target area, to share data with other command posts and is capable to give commanders a so-called "God's eye View" of the battlespace[21].

Since World War II the defense of carrier battle groups against air strikes and other attacks has led to the development of airborne early warning systems that loiter in the airspace over battle groups extending the range of radar sensors against incoming aircraft and of other sensors against hostile submarines. In the outgoing 1960s, attack planes flying below the horizon of ground-based radar stations made the use of downward-looking early warning radar systems necessary for defense of land forces as well. The result in 1972 was the development of the Airborne Warning and Control System (AWACS). AWACS systems characteristically have a large rotating radar radome situated on their back to give a wide-angle view of the airspace around and below the plane. The NATO E-3 AWACS fuse the data they receive with signatures from their data base and coordinate fighters to defend an airspace. AWACS need vast computing power for sensor data fusion, communication and control in real-time and data exchange with other platforms[22].

The third leg of the airborne command and control systems triad is the E8-C Joint Surveillance Target Attack Radar Systems (JSTARS). It was also developed in the 1970s, became operational at the beginning of the 1990s and is used to detect and monitor movements of vehicles and troops on the ground in real-time over an area of an Army corps. JSTARS uses a multi-mode side looking radar to detect, track, and classify moving objects on the ground deep behind enemy lines. Radar sensor input must be interpreted by computers and is matched against signatures of known objects[23]. The on-board computers store the tracks of possible targets as they move. The data are transmitted in real-time to ground stations[24] and between JSTARS and AWACS. JSTARS were the only platform able to track down Scud rocket launchers during the Gulf War. JSTARS are improved today to attack of

[21] Myron Struck: Airborne C^2 platform proves indispensable in Gulf War, *Defense Electronics*, May 1991, pp. 22–23; Robert Wall: New ABCCC tactics used in NATO air strikes, *Aviation Week & Space Technology*, April 26, 1999, pp. 32–35.

[22] See: David A. Fulghum: Scud hunting may drop under 10-minute mark, *Aviation Week & Space Technology*, Feb. 21, 1994, p. 90. David Hughes: AWACS data fusion under evaluation, *Aviation Week & Space Technology*, March 7, 1994, pp. 49–50. wt-Telex, *Wehrtechnik* 5/93, p. 18. AWACS must not be mistaken to exist only in the kind of the configured Boeing 707 known from NATO Forces. The E-2C Hawkeye is the turboprop maritime counterpart for carrier battle groups, the Brazil and the Swedish Air Forces use Saab business jets, Russian Forces modified transports like the IL-76, see John Fricker: Russian AWACS programs face funding problems, *Aviation Week & Space Technology*, Dec. 4, 1995, pp. 89–92.

[23] JSTARS, *Soldat und Technik*, No. 9, 1997, pp. 518–522; David Hughes: Mitre, Air Force Explore Data Fusion for Joint-STARS, *Aviation Week and Space Technology*, March 7, 1994, pp. 47–48.

[24] At the beginning, the basic configuration of JSTARS consisted of one VAX 11750 to control the radar, a Litton 85A for target movement measurement, four AN/AYK-14 processors with 625 million instructions per second each for signals data processing. Control is exercised by three VAX 860. The software measures 800.000 lines of code, see John Haystead: JSTARS – real-time warning and control for surface warfare, *Defense Electronics*, July 1990, pp. 31–39, pp. 34ff.

quickly moving and maneuvering targets with the GPS-guided Joint Direct Attack Munition released from a fighter plane at high altitude. This is achieved by triangulating the position of a target from two surveillance platforms like JSTARS – or a JSTARS with a Predator drone in the near future – and transmitting the data to the free-flight bombs[25].

The "smart" bomb hitting a target thus is only the result of a computer-supported communication between intelligence platforms ranging from GPS satellites to sensor-brimming planes like JSTARS. Bombers get the data for their flight management computers and their flight route from these and other sources indicating enemy positions on the way to the target area. When the ordnance officer in the bomber identifies a target, locks the sensors of the weapon on it and releases the load, only the final step of a complex set of preprogrammed activities is executed. The taped videos of the warhead or of the ordnance control pod of the fighter, the data from JSTARS and the ABCCC are again used for the so-called battle damage assessment and to restart the cycle, if necessary.

From the air to the ground

An obvious implication of the systems described is their use for ground forces as well. GPS is a result of orientation problems in the jungles of Vietnam. The civilian mode was deliberately built into the system to develop a civilian market for small hand-held GPS receivers driving down the prices for military orders as well. JSTARS explicitly was developed to aide ground commanders in battle through real-time data on hostile troop movements. Commanders can link up to JSTARS and see the same data as the on board crew.

The integration of the battlespace in the air and on the ground was the centerpiece of reform in military doctrine and tactics in the 1980s. The so-called AirLand-Battle doctrine – formulated in U.S. Army Field Manual 100-5 – is based on experiences in the Vietnam and Jom Kippur Wars and views a battlespace not along a front line, but as a three-dimensional maneuvering area of ground and air forces reaching deep into enemy territory. Prerequisite to the AirLand-Battle doctrine is the massing up of firepower on a small area against an enemy superior in numbers but inferior in command, control communications and – most of all – intelligence. Only systems as JSTARS and computerized C^3I – explicitly described in AirLand-Battle doctrine papers – give the advantage in speed, decisiveness, and force. AirLand-Battle is the doctrine for the U.S. Army since the 1980s and was exercised in Gulf War.

The development has gone on. The use of satellite navigation and the transmission of various intelligence data today has led to infantry units that are equipped with computers for orientation, communication and weapons control. In the 1990s, the U.S. Army published an ambitious plan called Force XXI to develop

25 David A. Fulghum: Moving targets vulnerable to radar / weapons mix, *Aviation Week & Space Technology*, Dec. 2, 2002, pp. 66–70.

the concepts for the soldier of the 21st century. To test new equipment, but most of all new tactics, the second armored division in Ft. Hood, Texas, was completely restructured and equipped with laptops, digital communication gear, and a totally interoperable C^3I-system[26]. The division exercised extensively and under realistic conditions[27]. Additionally, further means for enhanced "power projection" through improved weapons and cooperation were tested, to give small units greater firepower[28]. The idea is to equip the soldiers with GPS navigation to pinpoint their own location and to transmit it to other friendly forces. A miniature video camera is used to transmit target information, a handheld computer is used to communicate maps, target and other data. In exercises at least, a higher lethality of these units was demonstrated as a result of their improved coordination and accuracy in fight. Many of the systems tested have made their way onto the battlefield. Press pictures of special operations units on the battlefields in Afghanistan sometimes show troops equipped with GPS receivers, miniature video cameras or small video visors on some of the helmets.

This has shown that military organizations depend most heavily on computers when military command is concerned. Computer-controlled sensors collect the intelligence data that are first transmitted via computer networks to military headquarters where they are combined into a coherent military status report. What is going on at what place on the battlefield or on the whole globe is the central prerequisite for any commander to make a decision. Military communication networks propagate his decision and necessary additional data to the troops on the spot, thus allowing a commander to control his troops. Through sensors and communications links, the development of the initiated military maneuver is retransmitted to the commander. These four steps of *command, control, communications* and *intelligence* – called C^3I by its acronym – are the cornerstones of any purposeful military activity. The integration of all friendly military units on the battlefield into a C^3I system allows a large number of military units to act in coordination and with timely precision.

It should be noted that computerized command and control has intensified the trend toward decentralized command through improved communications and the distribution of intelligence data to commanders in the field giving them an improved "situational awareness" of their tactical situation and of that of hostile troops. Since they now can call upon the support of intelligence assets and highly precise aerial strikes, the obvious result is an intensification of firepower in the first place. This, on its part, has to be matched with an improved coordination between all friendly units. The data on exchanges on a local scale that are supported by surveillance units provide input for a situational overview on higher levels of the command chain. Even in the Gulf War, video imagery from special forces in Iraq was transmitted with almost no time lapse via satellite to central

[26] See also Paul E. Menoher, Jr.: *Force XXI: Redesigning the Army Through Warfighting Experiments*, at http://www.fas.org/irp/agency/army/tradoc/usaic/mipb/1996-2/menoher1.htm

[27] Army selects experimental force, *U.S. Army News*, Dec. 6, 1994.

[28] Robert K. Ackermann: Bytes transform Army, turn service roles upside down, *Signal*, May 1994, pp. 21–24.

command posts, connecting central decision making with the "warfighters" on the front lines. It thus would be an illusion to believe that decentralized warfare also decentralizes command and control - it rather brings the formerly loosely connected local commanders into the established chain of command.

The coordination of strikes and movements of air force, tanks and infantry, that started in World War II and was made possible especially through FM communication, has since been perfected into battlefield tactics - which fail, however, if they cannot be supported by computer. As with many kinds of computer-controlled activities, integrated C^3I can lead to an improved awareness of what is going on in the so-called "theater of war", to a reduction of time to reach a decision and to transmit necessary commands to the units in the theater, and an intense use of force, when better coordination allows more friendly units to operate effectively in any given area.

Just as an example, some years ago, aerial reconnaissance of the battlefield was hindered by the time it took a plane to return to base, to develop and interpret the film and retransmit the information to the commanders in the field. Today, JSTARS deliver real-time-data of a corps-sized battlefield complete with the classification of threats to any commander with the appropriate equipment. There is no time lag between target detection, threat assessment, the issuing of strike commands and the assessment of the battle damage inflicted: The commander can view the progress of an operation in real-time just as it unfolds and develops and can modify plans according to the battle situation.

C^3I on the battlefield today stands for a massive gain in speed and fire power that can be achieved by the same number of troops, leading to the notion of computers as a "force multiplier". The idea of integrated C^3I as a "force multiplier" is based on studies dating back to the early 1970s, when the U.S. Army noted an increase in military force by a factor of 2,7 to 2,9 through the use of computerized tactical command and control systems based on the appropriate communications infrastructure[29]. In the new strategic situation of military engagements of a solitary superpower with a technological lead in information technology, this idea now is a main reason for a shift in priorities for U.S. Forces. Although warplanes and aircraft carriers still amount for the largest sums in Pentagon acquisitions, the main focus of acquisitions is now on the communications networks that distribute the increasing amount of sensor data acquired by a growing number of platforms[30].

Information Warfare as a doctrine has developed from the combination of the two basic strands of development, namely computers in weapons systems and computers for command and control. Computers in weapons systems have increased their speed and accuracy. The growth in processor-controlled weapons, sensors and platforms allowed the connection of new terminal nodes to an expanding

[29] However accurate this may be, it was used to shape opinions also in Europe, see Eberhard Munk: Organisatorische und verfahrensmäßige Aspekte der Bedarfsdeckung bei Führungsinformationssystemen; in: H. W. Hofmann, R. K. Huber, P. Molzberger: *Führungs- und Informationssysteme*, München 1982, pp. 23–46, p. 30.

[30] David A. Fulghum: Pentagon priorities shift to data and networks, *Aviation Week & Space Technology*, April 22, 2002, pp. 22–23.

computer network which in turn made possible new tactics like AirLand-Battle. Securing computer systems, communication channels and sensors of one's own units or disrupting adversaries' systems by means of electronic warfare, physical destruction in a bombardment or just efficient psychological influencing of the decision-makers make up the goals and means of Information Warfare.

For these reasons, current Information Warfare units and their infrastructure often can be traced back to their descent from different origins. The U.S. Joint Electronic Warfare Center was renamed into Joint Command and Control Warfare Center in 1994 and was given responsibility for psychological warfare, operational security, and destruction of command, control, communications and intelligence (C^3I)[31]. What began in 1953 under the name of U.S. Air Force's Special Communications Center, then mutated in the 1970s into the Air Force Electronic Warfare Center, finally became the U.S. Air Force's Information Warfare Center (AFIWC) in the 1990s[32]. The same holds for networked computer systems. The World-Wide Military Command and Control System (WWMCCS), established in 1960 and leading to the development of Internet technology, mutated since 1971 into the Prototype WWMCCS Intercomputer Network (PWIN), and then again in 1993 into the current U.S. Global Command and Control System (GCCS), which provides U.S. forces with "an enhanced common operational picture, force status, intelligence support, enemy order of battle, related facility information, and air tasking orders"[33].

If the definition of Information Warfare and the necessary operative capabilities are taken seriously, one can compare operative elements and capabilities in the U.S. and similarly developed forces in the following overview.

- *Electronic warfare*
  Since the battle of Tannenberg August 1914, where the Germans fully exploited the Russian indiscriminate use of radio communications, interception of radio transmissions as a means of intelligence gathering has proved itself very effective. The methods have been direct tapping of information, if possible, or triangulation in order to pinpoint location of enemy forces. Together with the active disruption or misleading of the adversary's use of the electromagnetic spectrum, these activities are referred to as "electronic warfare" and have become a cornerstone in warfare. The advent of computers has only changed the means employed: As already described, aluminum foils (chaff) to blind enemy radar sites were replaced by computer generated emissions that specifically match the frequency and modulation of the radar sites to be foiled, while at the same time leaving the friendly systems operational. Technically thus, it is correct to state that enemy radar sites are fooled by computer-generated false images. It is by

[31] JEWC takes on new name to fit expanded duties, *Aviation Week & Space Technology*, Oct. 10, 1994, pp. 54–55.

[32] http://www.aia.af.mil/hqaia/afiwc.

[33] According to the Assistant U.S. Secretary of Defense: http://www.c3i.osd.mil/faq/.

contrast a gross exaggeration when the media imply that this is being done by means of hacking or other forms of computer intrusion.
Any kind of use of the electromagnetic spectrum by military organizations, as well as their use of the civilian communication infrastructure, makes up electronic warfare. EW explicitly includes the physical destruction of this infrastructure. A well published example for this destructive aspect of EW are HARM rockets that, when fired by fighter planes, lock their target seeker on the emissions of an anti-aircraft radar and destruct the site on impact. Belonging in this category are also non-nuclear EMP generators that have been developed and are on the verge of being fielded as a weapons system payload[34].
Since World War II, this electronic war of many different nations, even amongst allies, has been raging on without pause on a 24 hours-basis around the whole globe.

- *Psychological warfare*
Fooling an enemy is as old as war itself. The Chinese Warlord Sun Tsu is cited with the words: "Oh heavenly art of skillfulness and surreptitiousness", through which a warrior will try to fool an enemy. Deception is used to lead enemies to false beliefs of oneself: "The smart warrior will force his will upon his enemy, but he will never let an enemy force his will upon him."[35] From the stratagems of a warlord who lived 500 b.c., this tactic has developed into a broad method of influencing an adversary as well as one's own troops in many armies. Clausewitz defines war as an act to force an enemy into fulfilling one's will[36] and devotes a chapter to the ruse of war[37]. Misleading an enemy and reading his intentions are one of the major tasks for a commander on a battlefield or in planning a strategic move.
The scientific development of psychology has broadened the scope of Psychological Warfare in the 20th century. In an Information Warfare setting, Psychological Warfare Operations, or PsyOps, have a different meaning. FM 100-6 defines them as

> Operations to convey selected information and indicators to foreign audiences to influence their emotions, motives, objective reasoning, and, ultimately, the behavior of foreign governments, organizations, groups, and individuals. The purpose of PSYOP is to induce or reinforce foreign attitudes and behavior favorable to the originator's objectives.

PsyOps are explicitly meant to "magnify the image of US technological superiority" and provide content that is to be inserted into an enemy's communications infrastructure.

[34] David A. Fulghum: U.K. developing, testing directed energy weapon, *Aviation Week & Space Technology*, July 29, 2002, pp. 26–27.

[35] Sun Tsu: *Die Kunst des Krieges*, München 2001, pp. 53f.

[36] For a detailed analysis see S. Bergstein, this volume, pp. 183–215.

[37] Carl von Clausewitz: *Vom Kriege*, Hamburg 1963, pp. 93ff.

- *Integrated C³I-infrastructure*
Computerized command, control, and communications networks can be found in armies ranging from South Korea over Israel to the United States. The degree of integration, however, varies significantly. The German Bundeswehr, for example, still has problems with the interoperability of different generations in its army command and control system HEROS[38]. Efforts to modernize the Luftwaffe command and control system EIFEL were even abandoned[39]. The only military force with an operationally significant degree of overall integration of data communication today are the U.S. Forces that began interoperability efforts in the 1960s.

- *IT-based coupling of intelligence and warfighters*
The concepts for an integrated battlefield, developed for the AirLand-Battle doctrine in the 1980s, have been put into practice in several NATO forces. Lacking infrastructural elements for intelligence gathering such as space-based assets, the intelligence capabilities of most European forces are limited. The only force with a truly global collection and dissemination capability for various kinds of intelligence are the U.S. forces that daily collect several terabytes of data. The U.S. Army alone projects as the rate of data collected and disseminated daily by the year 2010, when a force of the size of the Gulf War expeditionary forces may be in combat, to an amount of 268 terabit; when including the data of all military units the estimates sum up to a daily rate of 570 terabits[40]. These data are to be made available to an increasing extent to troops on the battlefield, according to Information Warfare doctrine that demands better "situational awareness" through superior data on the battlefield for the soldier.

- *IT as a warfighting capability*
Computer viruses, hacking, and other remote forms of disruption of enemy command and control networks is the most speculative area of Information Warfare today - speculative because of the specific combination of real background and completely implausible assertions.
On the side of reality, intrusion into enemy computer systems dates back into the 1970s, when U.S. intelligence agencies and special forces admittedly were successful in gaining physical access[41]. Implausible is the idea that professionals responsible for communications security would take the extremely negligent step to connect a sensitive computer system to an insecure network. This risk is only

[38] See Stephan Söffing: DV-Unterstützung für die Führung des Heeres, *Wehrtechnik*, 8/94, pp. 33–37, p. 33.

[39] Hans-Josef Salm: Was lange währt..., *Wehrtechnik*, 6/92, pp. 74–76, p. 75.

[40] W. E. Howard, D.K. Evans: Growth in data speed creates opportunities and bottlenecks, *Signal*, Sept. 1994, pp. 67–68.

[41] One of the rare public statements can be found in: Jay Peterzell: Spying and sabotage by computer, *Time*, March 20, 1989, p. 41. From other sources reporting on their work on computers of western origin in the former Soviet bloc, it can be deduced that sensitive computer systems of the former Soviet Union military command were accessible to western personnel.

taken when military communications links are destroyed in war and the civilian infrastructure is the only available way to transmit the data. When one adds to this the manifold of command and control computer systems of most countries that only interconnect partially, one must cautiously differentiate between possible ways and means of attack, such as hacking into Internet computers in some other country, and fiction, when it comes to hacking the computers of a military command network.

- *Fighting by Information Warfare principles*
  Many singular elements of Information Warfare doctrine are already in use; some experimental U.S. units such as the Force XXI Army unit or the U.S. Air Force's 609th Squadron already wargame by Information Warfare doctrine. Waging war completely by Information Warfare principles, however, is not in sight.

This overview on capabilities making up Information Warfare and Operations should lead to a differentiated look on Information Warfare. Many military and paramilitary organizations have capabilities and opportunities to use electronic and psychological warfare or even to hack computers. If we interpret this as Information Warfare, it would already be a fact. But: Information Warfare is not meant and defined to be just that simple. We therefore conclude that Information Warfare rests on military capabilities already well developed, and is being developed further to acquire a greater role for military forces. What we can observe is a process of gradual development of a new form of military operations into maturity.

Now that the use of computers by military organizations and their capabilities concerning different elements of Information Warfare has been elaborated, the question remains, to what extent the manipulation of computers by digital means – instead of explosives – plays a role in actual Information Warfare.

3 War by Remote Control – on Computer Network Attacks and IT Security

The media have declared Information War several times in the last years. Following the conflict between the U.S. and the People's Republic of China over a U.S. spy plane in April 2001, Chinese groups vowed revenge and began to hack into web sites hosted in the U.S[42]. The media reported on some activities and raised fears about the advent of a so-called Information World War I. Disenchantment set in when, after some days, it became clear that nothing much had happened beyond the defacement of web sites. This kind of activity is nothing new and has

[42] On the incident: http://www.cnn.com/2001/US/05/01/china.us.plane.04/. On the reactions on the internet, cf. http://www.wired.com/news/politics/0,1283,42982,00.html; http://www.heise.de/tp/deutsch/special/info/7382/1.html.

even reached such proportions that a mirror site dedicated to tracking this kind of incident declared it could no longer keep track of the numerous hacks[43].

Similar stories are nothing new. In the Gulf War, faked news was spread about computer viruses in Iraqi military computer networks inserted through a load of computer printers imported from France. In 1995, the revolt of Mexican Zapatist Groups was accompanied by e-mail battles for the mailboxes of the media between Zapatistas and the Mexican government, just like the mail battle between Peru and Ecuador accompanying their border dispute in the same year. Similar kinds of online battles have happened regularly since as a part of the conflicts in Northern Ireland, with Basque separatists and Tamil Tigers in Sri Lanka. Between Israel and the Palestinians, the so-called "inter-fadah", the Information War begun in October 2000, rages on as well as the online quarrels between the People's Republic of China and Taiwan.

These visible, but rather unsophisticated quarrels are cited as the beginning of a more dangerous kind of computer intrusion by state actors to disrupt the civilian and military infrastructure and to spread insecurity. These fears are fueled by the term "computer network attacks", defined by the U.S. Army as:

> Computer network attack consists of operations that disrupt, deny, degrade, or destroy information resident in computers and computer networks. It may also target computers and networks themselves.[44]

This definition neither distinguishes between civilian or military computer systems nor between the kind of information it aims to destroy or degrade. By this definition, any information on any computer system might be a target for a computer network attack.

Since Cyber Warfare today more often has elements of a publicity activity than as a serious security threat, the most important effort is to differentiate between the media hype, the potentials of computer network attacks and the IT security problems that worry security experts. This part will take a closer look at the players, their weapons, and the fragile IT security they exploit. The aim is to differentiate between the dangers cited, the intentions declared by cyber warfare proponents and the realities behind these assertions.

3.1 IT security still fragile

The potential for the exploitation of IT security holes and the belief of the general public in the threat posed by it is founded on the fact that IT security is rather underdeveloped. Regularly, computer viruses and Internet worms demonstrate to a broad audience the vulnerability of networked computers. It thus need not be elaborated further that severe security problems exist in IT systems opening up leaks in the security even of sensitive systems.

[43] Press statement from attrition.org at: http://www.attrition.org/news/content/01-05-21.001.html.

[44] U.S. Department of the Army: *Field Manual 3-0, Operations,* Washington, June 2001, pp. 11–19.

For two reasons, many of the real problems are only rarely discussed outside expert circles. The first reason obviously is the lack of interest of the organizations affected in discussing their IT security problem in the open. Organizations with IT security problems quietly try to fix them. Hacker attacks on companies have a high estimated number of unknown cases because police authorities are rarely contacted. The second reason is less obvious. IT security problems in computer systems usually are quite complex. The understanding of the risks and their implications often requires a deep insight into information technology. These conditions are a severe hindrance for a discussion in public media.

Real problems with IT security and manipulated computer systems thus are reported mostly in expert circles. The rare instances discussion problems in public can be found on the internet, most notably the online information service – relayed by means of a newsgroup – comp.risks. Computer emergency response teams (CERTs) and government agencies such as the German "Bundesamt für Sicherheit in der Informationstechnik" (BSI), responsible for IT security in Germany's public sector, share their knowledge with interested parties[45]. But few of even the most alarming findings are ever reported in the public media.

Surveys over the last 15 years have shown however, that most of the security problems do not result from malicious hackers but from badly designed or dysfunctional software or the incorrect handling of IT systems. Connecting badly designed or administered systems to the internet inevitably results in security leaks giving hackers access to sensitive data such as files of credit card numbers[46]. Of course, credit card numbers are not the only valuable data on the Internet, but unlike any other kind of threat assessments, there remains no doubt about the potential damage.

This should leave no doubt that IT security still has to cope with huge deficits in knowledge, consciousness, and technology. Even without malicious activities, the security situation is fragile and far from standards established in mature technologies.

3.2 Mismatch between media coverage and threats

The relation between IT security problems and media coverage always has been a quite irrational one. Periodically, reports on hackers altering web sites or breaking into Pentagon computers hit the news. A closer inspection of the cases often reveals that no real damage has been inflicted beyond unauthorized access to a system or the defacement of some web pages.

What often is seen as a serious damage will seem less so when viewed from a security perspective. From this viewpoint, a web server is a computer connected to

[45] See www.bsi.de, www.cert.dfn.de in Germany or www.cert.org for international problems.

[46] Florian Rötzer: Bisher größter Kreditkartenklau im Internet; cf. http://www.heise.de/tp/deutsch/inhalt/te/5665/1.html. Sheer neglect was the reason for accessible credit card numbers at a Berlin software company: Internet-Shop offenbart Kreditkartennummern, cf. http://www.heise.de/newsticker/data/jo-22.10.99-000/default.shtml.

the Internet for the sole purpose of providing data to web users. It should not contain any data that better should not be published. Under security considerations, web servers are considered as external resources to a company network, guarded by a firewall against outsider attacks. Because this alone cannot be considered as completely safe, the internal network of a company is additionally shielded from the company's web server by a second firewall. Attacks on web servers thus should not interfere with any sensitive activities and especially should not reveal any sensitive data.

Web servers are semi-public places that should rather be compared to the physical equivalent of an office building lobby or a building's entrance area: the guarded figurehead of a company with a high number of visitors, where no one would store sensitive data or hold sensitive meetings. In this picture, the defacement of a web page compares to a graffiti on a wall or in the lobby. It is a mark of the attacker, a nuisance for the site owner, and of course an annoyance for the company's security unit – but neither a real safety problem nor something that has to be prevented at any price. Semi-public places like company lobbies or web sites must be open to visitors and therefore cannot be completely secured. Their high symbolic value makes them an attractive goal for activities to achieve public awareness. The defacement of web sites as the electronic equivalent of graffiti is often optimized for public impact with either intelligent or outrageous content that still achieves maximum media coverage.

On the other hand, real problems with IT security have a hard time getting public attention at all. It is hardly understood by the public that IT security aims to ensure the availability, *confidentiality* and *integrity* of IT systems.

- The best-known example of severe deficiencies in the *availability* of IT systems to date has been the Y2K-bug, the inability of software to cope with the year 2000. Reprogramming old software prevented serious damage. Unchecked, this bug would have reduced the availability of computer systems to a fairly low level.
- The *confidentiality* of IT systems most often is endangered by sloppy administration or sheer neglect, as in one of the credit card examples given above where credit card data stood openly accessible in a file on the WWW-server of an e-commerce shop.
- Endangering *integrity* and *confidentiality* of IT systems is the "back orifice" software for the exploitation of security deficits in the Microsoft NT operating system. "Back orifice" is seen by critics as a trojan software and can be downloaded from the internet as freeware. It can give its users – just at the click of a mouse – total control over poorly managed NT systems online including the erasure of all data. "Back orifice" is seen as one of the most dangerous trojans freely available, but has been a lesser topic in the public media than the problems of software bugs in general. "Back Orifice" only works because of the mechanisms of software distribution on the internet. Administrators in many companies use freeware or shareware downloaded from the internet that might contain erroneously programmed or even malicious functions that may irreversibly damage the IT system of a company. But even completely refraining from using anything that is not from a big software company leads to problems when – from time to time

– some updates and patches put on the net by large software companies contain software bugs that severely endanger system integrity. The certificates used for software accuracy and authenticity do not guarantee error-free software. This at the extreme end sometimes leaves only the choice to download patches for a software bug with unknown consequences or not to download them, thereby leaving severe security holes unpatched.

Problems like these are the bread-and-butter job for IT security experts and the main risks for computer users, potentially causing the loss of data confidentiality and integrity and maybe even total system loss. The problems are real and threatening, but lack vital elements to arouse media interest. So, the most serious problems largely go unnoticed while minor troubles are willingly reported. The consequence for the public opinion is not surprising. IT security problems are seen mostly as a threat to personal data or as the spread of viruses. More complex issues are beyond the attention of the public.

Anyone trying to explain the risks and implications of computer network attacks as a tool in conflict will have to take into account the multitude of misperceptions in the general public on the relevance and the dangers associated with activities aimed at exploiting security weaknesses.

But the public debate is just one side of the coin. Many papers from security advisers and the military alike are not very helpful in bringing more insight. Dangers tend either to be somewhat exaggerated with the obvious intent to promote certain measures or remain vague in order to prevent outsiders from exploiting known bugs and security leaks. After more than a decade of papers on the feasibility of cyber warfare and little at hand to demonstrate its effects, it has become even more difficult to give a down-to-earth assessment on computer network attacks and the many other forms of cyber warfare currently being discussed.

3.3 Computers as a "weapon"

Computer networks attacks are seen as one of the dangers to IT infrastructures from a defensive point of view and as a potential means of attack from an offensive one. Beyond the level of papers and doctrines and after trying to sort out fact from fiction, the involvement in manipulations of computer networks remains elusive.

Computer systems of adversaries have been a target for intelligence and military services in the last 30 years at least. The activities concentrated on the espionage of data and the protection against it. The development of TEMPEST equipment in the late 1960s is a vivid proof. TEMPEST stands for *Temp*orary *E*manation and *S*purious *T*ransmission and aims to prevent the transmission of data that electronic systems usually emanate, for example, when a computer terminal displays data or a printer receives data from a computer. The picture displayed on unshielded terminals can easily be received and reproduced on a TV set from a building across a street. Computer communication sent through a cable can also be inductively intercepted when the cable is physically accessible.

Emanations from unshielded electronic equipment were exploited as early as the 1950s, when British and American intelligence tapped the power cable to the cipher room of the French embassy in London. The faint signals of the electronic typewriters of the cipher machines made the original text accessible, which in turn – compared against the cipher code from these texts – compromised the crypto keys of the French system[47]. In many countries, the use of TEMPEST and the shielding of equipment became a norm in the 1960s for electronic equipment handling sensitive data. The predecessor of the German BSI – which developed from a department of the Federal Intelligence Service Bundesnachrichtendienst (BND) – TEMPEST-proved systems officially since the early 1970s.

Before the widespread use of the Internet, physical access or at least proximity to IT systems was the precondition for data gathering and system manipulation. Reports of U.S. activities date back into the 1980s[48]. Targeting computers as tools for the coordination of military activities and their communication infrastructure is the reason the Internet was built upon. Command and control of military forces has been the domain of networked computers since the early days of their development. Whirlwind – one of the first large computer projects in the early 1950s – was built to control and coordinate the status of the U.S. nuclear forces. The mainframe computers of that time were buried in hardened bunkers or mountain sites to prevent their destruction in conflict. In the 1960s, nuclear detonations were carried out in the upper atmosphere to test the effects of the electro-magnetic pulse (EMP) on electronic devices, making even sheltered systems a target as long as they were connected to a medium working like an antenna.

Since networks of computers cannot likewise be sheltered, the Internet was especially built to compensate attacks on its communication infrastructure by dispersing and rerouting data traffic around destructed network nodes. Command and control networks and their computing power have only grown in importance since. Their destruction is the prerequisite for most other military operations, making telecommunications installations, command sites and all of their respective electronic equipment and computers primary targets in any conflict.

On this level of conflict, the manipulation of computers and the use of force against installations is not at all a novel idea, but has been a military doctrine for years. Many of the activities described in modern information warfare doctrines do not go beyond the levels of data gathering on the one end of the spectrum and forcible destruction of computers and their network infrastructure on the other. If seen from this perspective, computer network attacks have been practiced for decades as a fairly lethal type of warfare. The non-lethal appearance of Information Warfare today results from the large area in between two historical poles of Information Warfare mentioned.

47 Peter Wright, Paul Greengrass: *Spy Catcher. Enthüllungen aus dem Secret Service*, Frankfurt 1989, pp. 115ff.

48 Jay Peterzell: Spying and sabotage by computer, *Time*, March 20, 1989, p. 41.

3.4 Targets for computer network attacks

Information warfare and computer network attacks as a novel approach set in on a different level. While Information Warfare concepts always stressed the need for a defense of highly IT-dependent countries against attacks on their communications and IT infrastructure, the computer network attack concept most interestingly stresses the offensive use of computer manipulations for military purposes.

Although many of the potential activities for computer network attacks hinted at in official papers will ultimately result in the same use of force as before, the idea of computer manipulations as a tool in war is not as straightforward as it might seem.

First of all, it should be remembered that security critical computers should not be networked and networked computers should not be the backbone of a critical infrastructure. Military IT systems, consequently are either connected to dedicated networks or not networked at all. How then should a U.S. military hacker enter enemy systems via the Internet when they are not connected?

The second point is the diversity of military systems. Even in NATO countries, the programming of today's military IT systems often started decades ago. Many systems are not interoperable because they work only with legacy software especially developed for solely this system. When interoperability of military IT systems is impossible between an artillery fire control system and other systems, or, for example, even between different layers of the German army's tactical Command and Control System HEROS[49], how high can the likelihood be of disrupting military IT systems with unknown source code and their interconnectivity reduced or non-existent?

The easiest way to disrupt an adversary's infrastructure is to manipulate civilian infrastructure systems, those infrastructures of the networks of communications and public utility companies, the banking system, transportation and the like. The common element of these infrastructures is their high dependency on computers and the dependency of an industrialized society on them. The proper function of these infrastructures is critical to society, which is why they are called critical infrastructures. Attacking and disrupting these computers is possible because they are interconnected to a rather high degree and are not as well guarded against attack as military systems. Critical infrastructure protection thus is a high priority in different countries.

In the U.S., the protection of critical infrastructures was assessed by a presidential commission and is a policy goal according to the Presidential Decision Directive 63 dating from May 1998. A Critical Infrastructure Protection Office (CIAO) was established[50]. In Germany, critical infrastructure protection was assessed by the "AG KRITIS" Group of BSI after a parliamentary enquiry[51] – although no

[49] As is the case for HEROS 1/2 for mobile command posts and HEROS 3 for the brigade or above; Stephan Söffing: DV-Unterstützung für die Führung des Heeres, *Wehrtechnik*, 8/94, pp. 33–37, p. 33.

[50] http://www.ciao.gov.

[51] Answer of the Federal Government to a parliamentary enquiry of rep. Dr. Manuel Kiper: *Lage der IT-Sicherheit in Deutschland*, Bt.-Drs. 13/7753, Frage 38.

results were published. Germany, however, has a more effective instrument to prevent IT security deficiencies than the U.S. since German privacy laws require organizations to follow certain security rules diminishing risks and allowing privacy institutions to check the level of safety technically implemented. The only effective measure against attacks on the critical infrastructures has basically been the establishment of new CERT groups in different countries.

The concept of computer network attacks builds on the belief that the disruption of computer systems can severely hinder any modern organization and unsettle industrialized societies. The comparison of the security precautions on civilian and military computer systems has led to the idea that it is far more effective to disrupt non-military systems, which in turn led to the establishment of organizations in several countries that – to a varying degree – try to prepare against attacks on the civilian IT infrastructure.

Since – at least until now – attacks on civilian computer networks were the acts of youthful computer enthusiasts rather than military actors, it cannot be distinguished if a computer network attack is meant as a prank of computer hacking youths or a military attack. If computer hacking in the future will be seen solely under military categories, a consequence is the extension of military thinking and military activities well into the civilian domain.

4 Political and Military Implications of Information Warfare

As we have seen until this point, Information Warfare is transforming military thinking and organizations. In most countries, this transformation process is not very structured and thorough. The activities in the U.S. military however, are far broader in scope reaching from the battlefield up to the geostrategical level.

On the *operative* level, the military doctrines for Information Warfare are refined down to the level of operative planning in the Field Manuals already referenced. The experimentation with new tools and tactics in units like Force XXI is surpassed today by the deployment of computerized special forces units on the battlefields of Afghanistan. In this conflict, and in Bosnia before, the integrated use of computers and networks for intelligence sharing and the exertion of command and control has reached a level unseen before and rather close to the concepts of Information Warfare. There still is no single integrated command network in sight, easily spanning all separate forces and the single warfighter in the whole theater of war, but the relevant data are transmitted with increasing ease between different units.

On an *organizational* level, structures for Information Warfare were already shaped in the 1990s. The Joint Chiefs of Staff installed their Joint Information Warfare Center, Air Force and Army their respective units. After first experiences with the newly set up command centers and Information Warfare units, some restructuring has begun. Information Warfare is now the task of so many units in the U.S. forces that an overview has become confusing[52].

Even on the *strategic* level Information Warfare has become an option for a time without a bipolar world dominated by nuclear deterrence. Although the political activities to end the anti-ballistic missile treaty, ongoing since the Reagan Administration, show the unparalleled significance of nuclear weapons in political thinking, most military activities cannot rely on nuclear force, but are fought with conventional means. Since the Reagan Administration spent large sums on the development of IT-based weapons and communications systems, the advantage achieved is also used on a strategic level. The basis for the ability of the U.S. forces for a power projection rests only in part on the warplanes that can fly half around the globe or fly from bases world wide, but rather on the intelligence satellites, communications networks and precision weapons powered by computers. The ability to react with military force to achieve political goals in a short period of time has led to the notion of a "strategic army"[53] standing at the ready and being controlled from central command centers. Information Warfare in the eyes of high ranking military and political planners has become a strategic doctrine shaping alliances and operations[54].

This development is far advanced in the U.S., but has also begun in other countries as well:

- Russia concentrates on psychological and electronic warfare instead of computers[55].
- The People's Republic of China claims the concept of Information Warfare as its own idea[56] and follows similar ideas as the U.S. with an emphasis on "People's Information Warfare"[57].
- Taiwan uses the strength of its electronics industry and explores computer viruses and other means of manipulation[58].
- India is adapting ideas from the U.S. and develops similar plans for regional conflicts[59].
- In Germany, the Bundeswehr has worked on protection against Information Warfare attacks and develops Information Operations concepts for an operations planning level[60].

[52] A list dedicated to activities and actors in the U.S. forces spanning some pages can be found at: http://www.terrorism.com/infowar/j6kdefense.html.

[53] US Department of the Army: Tradoc 525-5, p. 12; http://www-tradoc.army.mil/tpubs/pams/p525-5toc.htm

[54] J. S. Nye, Jr., W. A. Owens: America's information edge, *Foreign Affairs*, March/April 1996, pp. 20–36.

[55] Igor Panarin: InfoWar und Autorität; in: G. Stocker, C. Schöpf (eds.): *Information. Macht. Krieg*, Wien 1998, pp. 105–110.

[56] Shen Weiguang: Der Informationskrieg – eine neue Herausforderung; in: G. Stocker, C. Schöpf (eds.), *Information. Macht. Krieg*, Wien 1998, pp. 67–91.

[57] Wei Jincheng: Der Volksinformationskrieg; in: G. Stocker, C. Schöpf (Hg.): *Information. Macht. Krieg*, Wien 1998, pp. 92–104.

The political and military implications of Information Warfare concepts and ideas can only be assessed when the multitude of change is taken into account. Information Warfare is by no means a singular development, but can be translated into various forms of military activities. Just as new weapons systems like tanks and airplanes or modern weapons have led to mechanized warfighting as well as guerilla warfare concepts, Information Warfare ideas will be adapted differently according to specific military strengths and weaknesses. Especially, highly asymmetrical conflicts between powerful military forces on one side and militarily very weak groups will be affected by Information Warfare.

The broad use of Information Warfare concepts and the strong effects of Information Operations against civilian targets together will lead to a new challenge in security politics.

At the basis of the problem is the temptation of attacks against civilian IT infrastructures by weak adversaries and the idea of military actors to interpret attacks on IT systems according to military terms. Although almost 700 unauthorized attempts to access Pentagon computers are recorded per day, leading to the number of 250.000 attacks published by the Pentagon in 2001, a conflict constellation on a military level is extremely rare today. A change in the nature of the mostly youthful attackers is not plausible any time soon. The interpretation of intrusions into computer systems as military attacks although would quickly and significantly change the whole picture. Hacking, which is seen under German law as a crime only when guarded systems are attacked or data are altered (a restriction which aims at keeping a naive youngster from involuntarily starting a criminal career), could immediately mutate into an act of war. Although this interpretation is proposed by no one in Germany, and even in the USA the disproportion between the deed and a military reaction is still seen clearly, the discussion in the last years has slowly developed in this direction. A clear distinction between the civilian realm of IT security and its military side of Information Warfare is clearly missing.

More important for securing IT systems is the interest of military proponents of Information Warfare in keeping weak spots of IT systems open for manipulations. The most prominent example for the diverging interests of the civilian and the military side is the use of crypto systems. Since the 1970s, U.S. intelligence agencies and the military first worked against the development of strong encryption systems[61] and, after losing the battle for control over cryptological research in the USA, the export of crypto systems and know-how.

[58] Florian Rötzer: Taiwans Militär probt Angriffe mit Computerviren, *telepolis*, 8.8.2000, http://www.heise.de/tp/deutsch/special/info/6955/1.html.

[59] C. Uday Bhaskar: Trends in warfare: a conceptual overview, *Strategic Analysis*, Dec. 2000, pp. 1577–1589. See also Ajai K. Rai: Media at war: issues and limitations, *Strategic Analysis*, Dec. 2000, pp. 1681–1694; and Vinod Anand: An integrated and joint approach towards defence intelligence, *Strategic Analysis*, Nov. 2000, pp. 397–410.

[60] Ralf Bendrath: Informationstechnologie in der Bundeswehr, *telepolis*, 25.7.2000, http://www.heise.de/tp/deutsch/special/info/6933/1.html.

[61] Explicit against the development of strong encryption was the former NSA director Admiral Bobby Inman: The NSA perspective on telecommunications protection in the nongovernmental sector, *Cryptologia*, July 1979, pp.129–135, p. 129.

Transferring this view to IT security means that the interest in Information Warfare opportunities obviously leads to an interest in IT security weaknesses, since only insecure systems are vulnerable to attack in Information Operations. Instead of securing IT systems against manipulation this policy would lead to insecure systems that most likely will be hacked by youths with plenty of time and know-how. Even harsh punishment will not keep IT systems safe and the kids from experimenting. This line of thought fueled by military interests only leads to an increasing cycle of IT insecurity where IT security weaknesses open up opportunities to so-called cyber terrorists that in turn demand decisive military activities leading to an escalation of activities with no improvement in security.

IT insecurities that lead to conflict escalation instead of more stability are the least attractive point when Information Warfare implications are concerned. Even military planners concede that it would be dangerous to make the reactions in a political crisis dependent on the development of attacks on computer systems. The different clashes on the internet between opposing parties have only shown that in a case of tension a number of individuals seem determined to show their opinion through the manipulation of other people's computers. When a defacement of web sites or even the unleashing of a computer virus can be an act of a private person in a time of tension, and since it cannot be distinguished from any organized activities, the whole concept of Information Warfare beyond its use in open military action is at best useless and at worst dangerous.

The only way to prevent such a development is a strictly non-military view on IT security, which at least seems the preferred way within Europe. When it comes to catching the author of a computer virus, a cooperation of police forces of different countries clearly is a better strategy compared to a conflict between military forces. This is all the more sensible when we look at the statistics that show that military actors have played no role so far in IT security problems.

The best ways to improve IT security would be to fund research for secure systems and to find some new mechanisms for some of the internet communications protocol features, which today are necessary elements for internet message transfer but are also used as tools in distributed denial of service attacks. If the solution would be a division into a secure and an open internet or a new set of protocol routines is open to discussion. It would also be sensible to include critical IT infrastructures in arms control agreements or other means of international cooperation that lead to international stabilization[62].

Political and technical means for stabilization are at hand also in the age of Information Warfare. The only precondition is the political will to realize the dangers of inactivity and the opportunities for the solution of the problems. It should be clear that there are possible solutions to the problem instead of the mere solutions to the symptoms as the proponents of Information Warfare favor.

[62] On the minimal steps necessary, cf. Ingo Ruhmann: Back to the Roots; in: Jörg Tauss; Johannes Kollbeck; Jan Mönikes et al. (eds.): *Deutschlands Weg in die Informationsgesellschaft*, Baden-Baden 1996, pp. 403–412, pp. 411f.

More or Less Exposed Non-Combatants and Civilian Objects under the Conditions of “Modern Warfare“

Systematic violation of the rules of international humanitarian law as seen from the example of the NATO war against Yugoslavia

ELMAR SCHMÄHLING*

This report shows how the armed NATO forces, with their attacks on the territory of the BR Yugoslavia, violated the rules of the humanitarian international laws of nations (ius in bello).

Basis are the
- Nuremberg principles (crime against peace, war crime and crime against humanity),
- the Geneva convention of August 12th, 1949 on the protection of victims of international armed conflicts including the supplementary sections.

This article will not take into account the violations of international laws of nations resulting from planning warfare, distribution of warfare, as well as the indirect infringements of law that were not caused by the use of arms. The systematical violations of law by armed NATO forces against the international humanitarian laws of nations will be shown on the basis of six examples of concrete attacks on civilian objects or respectively on civilian population.

Example 1: The attack on the building of the Serbian state TV (RTS), Belgrad

The attack was effected by missiles (cruise missiles). The missiles (number unknown) hit the transmitter mast and the building in the entrance area. The upper floors collapsed. A fire developed which burned until morning. At the time of the attack about 100 civilian people were in the studios and editorial department; 16 people died, three were seriously injured and 13 people were injured slightly. The connecting building, which forms the entrance area, was almost completely destroyed as well as the transmitter mast.

* Former Rear Admiral, Berlin. Email: elmar_schmaehling@t-online.de
Translated from the German by Nicola Hellmich.

Figure 1.
Broad view of demolished RTS office building at 1 Aberdareva Street, 23 April 1999. [Source of all photographs in this article: NATO Crimes in Yugoslavia. Ministry of Foreign Affairs, Belgrade, 1999]

The arguments of NATO: In Yugoslavia technical devices of civilian radio- and TV stations are generally used for military communication. Moreover this station transmitted propaganda for the Milosevic regime. There can be no doubt that the studio was attacked directly and on purpose due to the fact that the attack followed after threats spread by NATO a few days earlier, regarding the transmission of the TV station mentioning off and on that an attack on the studio would follow, if the RTS TV station did not broadcast western reports and news for six hours per day.

The NATO speaker, colonel Konrad Frytag, GEAF, SHAPE, said a few hours after the attack and in direct connection to the bombing of the RTS studios:

> Attacks on TV stations and radio equipment are part of our campaign to dismantle the Yugoslavian propaganda machine, that plays a vital role in the control mechanisms of President Milosevic. (NATO Press conference, Washington D.C. , 23rd of April 1999).

Tony Blair, prime minister of the United Kingdom, explained: "It is this machinery that keeps him (President Milosevic) in power, and we as NATO allies have a legitimate right to damage and attack these targets" (quoted by *The Times*, London, 24th of April 1999). On another occasion Tony Blair said:

> We have to directly attack his military machinery and the whole apparatus of dictatorship. The media controlled by the state is a part of this, and I believe it is a correct and justified target for us. We knew for sure that these things (e.g. TV institutions) were legitimate targets, and they were absolutely legitimate targets. (Recording of an interview "The News Hour With Jim Lehrer" (PBS), Washington, D.C., Friday, April 23rd,1999).

George Robertson, British minister of defence replied to the question regarding the RTS studio bombing:

> It's a fact that many of these targets really are the brain behind this brutality that takes place in the Kosovo today, a firm component of the apparatus that

> carries out ethic genocide that takes place in this part of Yugoslavia, and as long as this continues, it is clear that we have to attack these targets.
> (Briefing of the defence minister , 30th of April 1999).

Article 51 of the First Geneva Protocol forbids attacks on civilian populations. In article 50 of the first protocol a civilian person is defined as a person who does not fall into the category of persons under articles 4A (1), (2),(3) and (6) of the Third Geneva Convention of 1949 and under article 43 of the protocol. Employees of the media do not fit into any of these categories; Moreover, war correspondents are mentioned in article 4A(4); and this group is specifically excepted from the list of military personnel in the First Geneva Protocol article 50. Additionally, in article 79 of the first protocol it is stated that journalists in the area of armed conflicts have to be regarded as civilian population. The attack therefore was a violation of the humanitarian international law of nations as far as it was directed at this group of people.

It can also be argued that the NATO attack was focussed on the building in which the people were working, meaning the TV studios. The First Geneva Protocol, article 51(5) forbids random attacks, "which could lead to the loss of human life, injuries of civilian personnel, damage to civilian objects or a combination of these, that in proportion to the expected immediate military advantage would be over dimensioned."

The concrete and direct military advantage was minimal, especially since RTS was able to resume transmission only six hours later: It is highly improbable that the NATO leaders did not know about this capability. The loss of human lives makes this attack a random attack. Moreover, the RTS studio could be regarded as a civilian object and was therefore not a legal target.

If the military target had been to prevent the transmission of "propaganda" then NATO could have attacked the transmission masts located separately in order to prevent the loss of lives; according to the First Geneva Protocol, article 5(3) a target like this has to be preferred.

For authorities on military structures there is no doubt that armed forces with regard to their need for military communication at any time will never tie themselves to local stations. Therefore the claim that the RTS stations were important for military communication is not plausible.

Furthermore, there can be no doubt that the NATO personnel knew that people worked in the RTS building during the night. At the time of bombing, RTS was transmitting and it was known that they did a 24 hours shift. It has been reported in detail that NATO had sent messages to foreign journalists, who normally used to send their articles from the RTS building and who filmed there, not to be present during that specific night, which proves that the NATO responsibles knew that work in this building continued through the night. Even if the killing of the RTS employees was not intended, the death and injury were at least taken into account in an approving way.

NATO's attempt of justification that the RTS building was a "legitimate target" because it supported the propaganda emission of the "Milosevic-regime" proves two things: the NATO responsibles knew that the attack was unlawful, the humanitarian international law does not know the term "legitimate". Either an object can be attacked in war because it is a military target, then the attack is legal, or it can not be attacked because it is a civilian object then the attack is illegal. Moreover, it is an unlawful intrusion upon the right of free information.

Example 2: The attack on the refugee convoy in the road Djakovica - Prizren on 14th April 1999

A convoy of Kosova-Albanian refugees was attacked over a distance of 12 miles at four different localities on the way from Prizren to Djakovica close to the villages Madanaj and Meja. In the convoy there were mainly women, children and old people who wanted to go back to their home villages in cars, tractors and trailers. The first attack on the marching column of more than 1000 people took place when the convoy moved through the village of Meja. The people of the convoy scattered and tried to find shelter in the houses close by.

But NATO planes also fired missiles onto these houses. The attack continued along the street between the villages Meja and Bistrazin. A tractor with trailer was completely destroyed. During the ongoing attacks on refugee vehicles another person was killed.

Twelve people were killed during the first attack. During the second attack seven people were killed. A further attack killed 20 more people including five children who had sat on a tractor. The total number of people killed during the attack of the refugee convoy was 74. The number of people wounded was 36.

Reasoning of NATO: Misconception of the pilot. Foreign journalists noticed a great amount of NATO weapon material right after the incident. NATO had denied their participation at first. On April 15th two NATO speakers , the brigadier general Giuseppe Marani and Dr. James Shea confirmed that the altitude of the attacking planes was 15.000 feet or about 5.000 meters.

Figure 2.
A view of the bombed refugee column in Meja village, Djakovica Municipality, 14 April 1999.

On April 19th NATO brigadier general Daniel P. Leaf confirmed that no plane descended to a lower altitude during the attacks. Due to the fact that the Yugoslavian anti-aircraft defence is not able to reach targets above an altitude of 13.000 feet it is probable that NATO's choice of altitude was calculated to take this into account to avoid the threat by anti- aircraft defence.

The brigadier general described the event of the attack as follows:

> As we are watching these videos (e.g., the Cockpit videos) on the big display in the comfort of this room it seems possible that the vehicles are tractors. When I watched the videos together with the pilots they agreed.

But, as they emphasized they seemed to be military vehicles for the naked eye seen from the altitude during the attack, the physical characteristics of the vehicles only being a factor for the forward air controllers target identification matrix. The convoys' guiding elements showed, seen from the air, various characteristics of military movement like similar height, form and colour, moreover consistent distances between the vehicles and a relatively high speed even shortly before the attacks.

> Moreover there had been reports on Serbian military using civilian vehicles for military and paramilitary operations. (Press conference, NATO-Main quarter, 19th April 1999).

Brigadier general Leaf ended with the statement:

> It is our estimation that the NATO military forces unintentionally hit vehicles of civilian type and maybe civilian persons during the attacks on the long convoy. (Press conference 19th April 1999).

The Yugoslavian army has recorded the following radio conversation on April 14th:

Pilot: I am now leaving the clouds. I still don't see anything.
Base: Continue your flight. Direction North 4280.
Pilot: I am lower than 3000 feet. Below me a column of vehicles. Some kind of tractors. What is this? I require instructions.
Base: Where are the tanks?
Pilot: I see tractors. I don't suppose that the reds have disguised the tanks as tractors.
Base: What kind of strange stories are these? What a mess! There must be the Serbs behind it. Destroy the target!
Pilot: What am I supposed to destroy? Tractors? Common tractors? I repeat: I don't see any tanks. I ask for further information.
Base: It is a military target. Destroy the target! I repeat: Destroy the target!
(Source: Video of the Belgian journalist "15 Belgians in Yugoslavia. Under the bombs of NATO", translation from the French.)

Regarding the NATO version of the events it is to be established: The pilots were not able to see clearly what kind of vehicles there were in the convoy and had to judge the situation seemingly from the weakest form of clues, like from the velocity of the movement and the similar distances between the vehicles, as brigadier general Leaf gave to understand in his briefing.

The altitude during the attacks was the other critical point. In the words of brigadier general Leaf "they seemed to be military vehicles as seen with the naked eye from the altitude of the attack". If the words of the brigadier general can be taken literally, then he has admitted that the NATO pilots judged their target by seeing it with the naked eye only, from a target distance of 5 km. This negligence over the choice of a target is, as this attack has shown, apt to result in a high number of casualties, due to the fact that even the sharpest eye is not able to distinguish between a tank and a tractor from a distance of 5000 meters.

The contradiction between the assertion that the planes did not go below an altitude of about 5.000 m and the statement of the pilot that he went down to 3000 feet – given during the NATO press conference – remains unsolved. In any case, it can be assumed that the NATO planning had failed to distinguish between civil and military targets as it is bindingly stipulated by the humanitarian international laws of nations.

Therefore, the described actions of attack by NATO unmistakably violated the regulations of the humanitarian international law of nations.

Example 3: The attack on the automobile plant "Zastava", Kragujevac

On April 9th, 1999, the terrain of the plant Crvena Zastava Ltd., a producer of vehicles (about 30.000 employees) was attacked. Two missiles hit and completely destroyed the production unit for the automobile Jugo. Other parts of the factory plant were badly damaged. The body paint plant, the forge plant and the "Zastava" power plant were each hit by a cruise missile. Part of the factory and the complete automobile production unit were completely destroyed. From the detonation effects 64 residential and office buildings, private and public, in the center of the city and right next to the factory terrain were damaged.

The factory was said to have a military function. NATO's claim that in the factory that was attacked and destroyed deliberately, military products were pro-

Figure 3.
Inside view of demolished *Zastava Trucks* plant, 9 April 1999.

duced, has been proven to be wrong. Due to the fact that the factory gave access to visitors before the war it was a known fact and must have been known to the US secret service, that there were no military objects stored or produced in the attacked halls. A journalist had stated that during his recent visit in the production plant he did not notice any military use of the factory.

Example 4: The attack on a passenger train on the bridge across the Grdelicka ravine on 12th April 1999

On 12th April 1999 the passenger train No. 393 was on its way from Beograd to Ristovac. The attack on the train occurred during its passage across the bridge.

Two planes approached the bridge from the west. The second fired two guided missiles type AGM –130 on a passenger train on the railway bridge, the first hitting the centre part of the bridge and the second hitting the second wagon of the moving train. According to the eye witness report of Zivojin Stanojevic, shortly afterwards another plane approached and also fired two missiles, this time directed at the so-called Sarajevo bridge, a road bridge. This description of the attack differs from the NATO report. The passenger train with a locomotive and four passenger wagons was completely destroyed. There were 21 dead and 16 injured people.

Reasoning of NATO: The essential parts of the explanation given by NATO commander in chief for Europe, general Wesley Clark, with regards to this incident is here completely printed:

> This was a case were a pilot had the task to hit a railway bridge that formed part of an integrated net of traffic and supply in Serbia. He released his missiles from a plane that was many miles away not being able to set his eyes onto the bridge, it was a remotely directed attack. And as he tried to make out the aspired target on the bridge, – and I talked to the team in Aviano, who were directly involved in this operation – well, as the pilot was trying to make out the aspired target on the bridge and was trying to work it out and worked it out and worked it out, and suddenly in the last moment, with less than a second to go before the launching he caught a flash of movement appearing on the monitor and that was the train that came in. Unfortunately, he couldn't dump the bomb at that point, it was locked, it was going into the target and that was a very unfortunate incident that he and his team and all of us are very sorry about. We surely don't want to cause a collateral damage. The order was to eliminate the bridge. He realised, after it had happened, that he had not hit the bridge, but that what he had hit had been the train. He had another target point on the bridge – it was quite a long bridge – and the pilot believed, that he still had to accomplish his mission, and so he circled back around. He put his aim point on the other end of the bridge from where the train had come. By the time the bomb got close the bridge was covered in smoke and clouds, and again at the last minute in an uncanny accident, the train had slid forward from the original impact and parts of the train had moved across the bridge and thus by hitting the other end of the bridge he indeed caused further damage on the train.

Figure 4.
Left: Pilot sighting on passenger train no. 393, 12 April 1999. *Right*: View and damage of the interior of the second carriage of passenger train no. 393 on railway bridge near Grdelica attacked by NATO bombers, 12 April 1999.

Clark then showed the cockpit video of the plane that fired on the bridge:

> The pilot inside the aircraft watches an about 5 inch sized monitor, – he can see about this much: here you can see: this is the railway bridge, and this is a much better view than he really had. You can see the rail tracks that run this way. Look hard on the target point, concentrate exactly on this, and you can see how, if you are focussed on the job like the pilot was, this train suddenly appears. It really was unfortunate. Here he came back to try to hit another part of the bridge, because he was trying to do a job, to eliminate the bridge. Look at this target point – you can see smoke and other obscurations - he could not make out clearly what that was exactly. Focus hard to the right on the centre of the cross. He is about to bring those two crosses together and suddenly, at the last moment, he realizes that the train that had been hit had continued to move across the bridge and this way seemingly, the towing engine was hit by the second bomb. (Press conference, NATO-Main quarter, Brussels, 13th of April).

There is a detailed expert's report that proves that the video that showed the attack was manipulated und due to this the description of the action did not correspond to the truth.

It can be assumed though, according to the video, that the gunner at first really aimed at the bridge and wanted to hit it. But when he saw the approaching train and realised that he would hit the train he didn't take any action to break off the attack. The company who produced the guided missiles have contradicted the description of the NATO commander in chief that the weapons system officer only saw the small inner section of the monitor. According to this he should have noticed the train much earlier on the monitor than it was claimed and could have taken the guided missile out of "lock on" in time and steered it into another direction. A few days before NATO had shown to journalists how, with the same weapon, it had been possible to stop an attack in the last minute to prevent a collateral damage.

General Clarks description is grossly misleading:

> The videos that were recorded automatically during the attack were running at triple speed which Phillips (a NATO speaker, in the beginning of January 2000) tried to excuse as an unplanned technical problem during the transmission from one system to another. The day after the attack the two videos had been shown in Brussels, giving the impression that the train had appeared on the bridge surprisingly fast so that the pilot would not have been able to stop the attack. The American NATO commander in chief Clark moreover had given the impression that the pilot himself was responsible for the steering of the guided missiles and therefore had had his hands full. But actually a weapons officer, who was in the plane, was responsible for the launching. These facts, said Philips, could maybe not have been described correctly by Clark. The accusation of manipulation though he rejected strongly. (Ho., Frankfurter Allgemeine Zeitung, 8th January 2000).

On a heavily frequented railway section the Leskovac bridge was crossed frequently; but as General Clark obviously admits the pilot did not look at the bridge at all, because he was flying a remotely directed attack.

Secondly, general Clark admits that the pilot knew that he had hit the train with his first attack. In spite of this, the pilot continued to effect a second attack. He still tried to eliminate the bridge, although at that moment there was a train that had been bombarded. Obviously the pilot tried, according to general Clark's own confession, to destroy the bridge, although he knew that it carried a train at that moment. It is hard to imagine how this could have been achieved without the loss of more lives.

Moreover, the second attack was no instant reaction, excusable with nervousness or error; indeed: "to accomplish his mission, the pilot circled back around".

Thirdly, general Clark explains that the pilot could not have seen at all what he attacked. During the presentation of the video general Clark said: "You can see smoke and other obscuration there – he couldn't tell what this was exactly." This means that the pilot fired onto an object that he could not even identify due to "smoke and clouds".

These three factors indicate that the pilot has acted unlawfully. Firstly, with the pilot not checking what approached the bridge at that time, he violated article 57(1) of the First Geneva Supplementary Protocol, which says that "at any time it has to be respected that civilian population, civilians and civilian objects have to be spared", see also article 57(4). Secondly, with the pilot attacking a bridge for the second time knowing that it carried a passenger train, he violated article 57 (2) (b), which states that "an attack [...] has to be ended finally or temporarily, if it is proven that the target is not military, since it is under special protection [...]", according to the principle of proportionality.

Summary

Regarding this investigation it is important that the time honoured principles of warfare that are aimed at the fast, cost-effective victory over the enemy, meaning that to a certain extent the "military necessities" for military defeat and political conquest of the enemy, have continued validity before the rules of international martial laws. Therefore, with every one of the investigated cases it has to be evaluated with what kind of military measurements and if applicable political reasoning the target or steps had been chosen, and what contribution an attack or the choice of weapons achieved or was supposed to achieve to reach the goal of the war.

The war aim

NATO commander in chief Wesley Clark has explained the goals of the NATO-aerial warfare against Yugoslavia as follows: Attack, interrupt, wear out, deter further Serbian actions and neutralize the Serbian military potential. These goals were supposed to be pursued in two lines of operation:

1. Operations against the Serbian armed forces and the security forces in Kosovo and in the adjoining areas in order to destroy them, to isolate and to prevent them from continuing or intensifying their campaign.
2. Operations against a selection of strategic target categories. These included the supply bases, the integrated defence from the air, that protects the most important targets in the country, the upper levels of decision and leading, plants of production and stocks of fuels (POL = Petroleum, Oil, Lubricants), all institutions that support the "military- and security moloch".

The air raids would, according to general Clark, be carried out "systematically, methodically and with increasing intensity". This is another way of describing the clear assessment of the kind and importance of each and every target, e.g., the intent to attack definitely civilian objects. Doubt is not accepted. The leaders of the armed forces of Germany tried to get out of the responsibility, to be involved in the destruction of forbidden civilian objects, with the claim that legal advisers had evaluated every single target before the attack, if it really was a permitted military target. This is a bizarre assumption that was not even used by NATO speakers. This excuse though makes clear that the awareness of the problem was there and yet civilian objects were attacked, destroyed and due to this civilian people were killed. The behaviour of the authorities shows a considerable amount of "criminal energy".

Streets, railways and bridges

Naturally, streets, railways and bridges are also used for transport of military units and supply.

But to declare the infrastructure of a whole country as military target, goes far beyond the precept of relativity. This is true even more for the streets and bridges in the north of Serbia where avowedly the Serbian military and security forces should have been obstructed in their freedom operation in Kosovo.

Assessment:

"Military necessity" and legal protection are here blatantly contradictory. Especially the destruction of the bridges of the important international water route Donau with far reaching economic consequences for some neighbouring countries that were not involved in the conflict, was far out of the proportion and therefore was illegal. (Violations of articel 3(1), articel 48, articel 52, articel.54, articel 57 of the Supplementary Protocol I.)

Attacks on non-combatants

The affirmations of the NATO representatives run like a leitmotif through their press conferences that the air strikes were not directed against the Serbian people.

It is true that this assertion seems believable, if you look at it superficially, because it was feared that direct and very obvious unlawful attacks against innocent people might have negative effects on political support in the home countries. And yet this assertion is rather hypocritical, because on the one hand the indirect effects on the life and health of the civilian population by destructions of e.g. Pancevo (localitywise and timewise transferred) were simply ignored. On the other hand, Wesley Clark himself has admitted that it was important that the civilian victims of attacks would direct their anger against their president and would know whom they had to thank for their expenses.

This statement shows that part of the Serbian population really was used as hostages against the regime. Regarding the night time attack on the RTS broadcasting building the responsibles accepted knowingly that civilians could become victims, since it was known that the editorial section, the studios and the technical section would be busy around the clock .This way, with partial intent, 16 of the 150 civilian employees, who were in the building, were killed.

Violations of article 35, article 48, article 50, article 51 and article 52 of the Supplementary Protocol.

Collateral damages

A collateral damage is, according to the humanitarian international law of nations, an unintended minor damage that in order to achieve a legal attack on a military target is unavoidable, like broken windows of a civilian building close to the target. The collateral damage has to be kept limited. If the attack on a military target would result in an immensely high collateral damage, then the attack should not be carried out. Here applies the necessary weighing between military necessity and legally prescribed protection of civilians and civil objects. The intended attack on civil objects doesn't cause collateral damages but main damages. In this war NATO has tried to avoid real "collateral damages" by e.g. using precision weapons.

Violations of article 57 of the Supplementary Protocol I.

"High value targets" and "strategic targets"
With the introduction of the target category "high value targets" and "strategic targets", unknown to the humanitarian international law of nations, NATO has circumvented systematically the ban on warfare against civilian infrastructure of the enemy. By adding the adjective "legitimate" in explanations of questions from journalists, in order to defend their attacks on apparently civil objects, NATO practically declared the whole country as outlawed.

The reasoning lies in political-psychological aspects: In a dictatorship it would be hard to distinguish between civilian and military infrastructure. This was also the argumentation of German defence minister Rudolf Scharping in a ZDF broadcasting. Since the dictator would control all resources of the country, including human resources, all institutions of the state would serve to execute and maintain power. If the goal was to hurt the regime one would have to eliminate all components of the state. This was the reasoning behind the destruction of the USCE office building where the head offices of the ruling party and of the socialist party were supposed to have been.

NATO also classified the automobile factory Zastava in Kragujevac and many other industries with purely civil goods production as a category of so-called strategic targets. In so far as the NATO responsibles were asked about the doubtfulness of the destruction of such institutions, as for example a cigarette or a food factory, the speakers claimed that these objects had been military targets. The NATO intelligence services had ascertained their military properties and function. In not one of the cases though could the speakers determine, in spite of repeated enquiries, what type of military products allegedly had been produced there.

Assessment:
The humanitarian international law does not allow this kind of change in the interpretation of "civilian objects". Moreover article 52 of the Supplementary Protocol applies here, which states that an institution has to be of direct use for warfare. The NATO speaker upon being asked if he believed that after a month of war, battles were still being waged from these party head offices in the USCE highrise, denied this possibility. He attributed the destruction of this civilian object that took place, nevertheless, to the psychological effect on the state leadership and the population.

Violations of article 48, article 52, article 54 and article 57 of the Supplementary Protocol I.

Figure 5.
Broken railway tracks on the Belgrade–Skoplje railway line in the vicinity of the demolished flyover near Trupalske forest, Nis, 14 May 1999.

"Zero-victim-strategy"
Military deployments in foreign countries which do not fall under the heading of "life interest" according to American understanding, need on the one hand a high amount of ideological preparation and emotional accompaniment; on the other hand the expenses for the society and the sacrifices (e.g., killed soldiers of one's own ranks) should not be high nor visible.

The USA had to end the Vietnam war without victory, due to the fact that high personnel losses put an end to support by society in the USA. To avoid its own war victims, its own planes, during finding the target and attacking, stayed – if possible – out of reach of the enemy's anti-aircraft defence. This resulted in the fact that the pilot and the combat observer were highly limited in finding and recognizing the target. As well for the attack on the civilian passenger train as for the attack on the refugee convoy, the later-on used excuse for the public was that such mistakes can happen when it is necessary to fly this high.

Assessment:
The humanitarian international laws stipulate that an attack must not be carried out if the attacked target can not be clearly identified as a military target. Weighing the "military necessity" against the legal protection for civilian population, the attack in these cases would have had to be stopped, because the military success did not depend on these military targets. According to the classical military code of honour this conduct of attack was the action of a coward.

Violation of article 48, article 51 and article 52 of the Supplementary Protocol I.

Elimination of supply with electrical energy and long-distance heating
During the air raids transformer stations were repeatedly attacked with new graphite bombs, as well as power stations and heating power plants. They were temporarily cut off or destroyed respectively. Many vital or life ensuring technical systems of the civil infrastructure depend on the availability of electric energy. There were reports that in hospitals people died, because life ensuring machines stopped working due to loss of energy. Emergency generator equipment did not

exist in every case or fuel was missing because of it having been destroyed as well. In the case of the heating plant in Novi Beograd "Beogradske elektrane", which was also destroyed, a connection to the direct support of warfare could not be constructed. In spring and summer heating energy is not needed, neither by the military nor by the civilian population.

Figure 6. *Left*: Transformer and distributor installation of Nikola Tesla thermal electric power plant upon which cluster bombs with carbon fibers were thrown, causing short circuits and breakdown of the system, 2 May 1999. *Right*: Carbon fibre insert from a cluster bomb.

Assessment:
There are two plausible explanations for the NATO attacks on this civilian infrastructure: Firstly, terror under the same (as usual wrong) assumption, that the civilian population of the enemy, frightened and terrorized, would stop following their political leaders. The other explanation, also supported by the destruction of a food production plant and a cigarette factory is this: NATO wanted and wants to destroy Yugoslavia's economy by destroying important parts of the civilian infrastructure, supported by maintaining, and in the meantime now planned tightening of the embargo.

This assumption receives further intensification with the massive attack on the oil refinery in Novi Sad with 108 bombs in the night from the 09th to 10th June 1999, at a time, after the conditions for the stop of air raids had been signed and had been kept by the Yugoslavians. Also in this case "military necessity" could not be claimed.

Violations of article 54 and article 57 of the Supplementary Protocol I.

Environmental damages
With their attacks on facilities where especially dangerous substances were stored, NATO states, at the least, approvingly accepted serious damage to the natural environment and as a consequence thereof health and hereditary defects in people. Indeed, a considerable local contamination of ground, water and air has been diagnosed as consequences of war.

Assessment:
As with the turning off of the electricity also the destruction of power plants and supply of primary energies is aiming at the basic vital supplies of the humans considering that the military do not depend on local power plants, neither in short nor long term. Also in connection with attacks on those institutions which were unlawful under the international humanitarian law, the NATO representatives simply claim that those were military "high value" or strategic targets.

Violations of article 35, article 54 and article 55 of the Supplementary Protocol I.

Use of poisonous and extremely cruel weapons
The NATO armed forces have again used in Kosovo, as before in the second Gulf war, DU shells and scatter bombs which are outlawed by the UN. After many diseases and hereditary defects of people, especially children, were recorded and undoubtedly connected to the physical contact with Uranium 238, the renewed use of this ammunition was highly cynical and inhumane in a special way.

The same is valid for the scatter bombs, whose anti-person ammunition works like the illegal shrapnel shells having an officially admitted failure quote between 5 and 15%. The submunition that did not detonate contaminates the area of action for a long time. Differently to the landmine fields, whose locations often are known, the bomblettes of the scatter bombs lie in unknown places in the countryside, in towns, on the paths, etc. Many people in Kosovo have already fallen victim to detonating bomblettes. Many will still follow.

Assessment:
The use of poisonous uranium and of the modern "DU shells" is clearly a violation against the international martial laws and a crime against humanity

Violations of the Nuremberg principles and of article 35, article 48, article 51, article 52, article 55 and article 57 of the Supplementary Protocol I.

III
Ethical Issues

Niels Bohr's Political Crusade during World War II[1]

FINN AASERUD*

Niels Bohr's "Open Letter to the United Nations", published in 1950 and pleading for an "open world" between nations, is well known. It is also well known that Bohr took part in the Manhattan Project during World War II. This article describes in some detail how Bohr's idea of an open world not only matured during his war-time exile, but even constituted the basis for a veritable crusade on Bohr's part in the course of the war to convince statesmen of the necessity to think differently in the post nuclear-bomb era. How and why did Bohr come to these ideas? Why were they so important to him? How did his crusade relate to his simultaneous participation in the Manhattan project? What means did he use during wartime to convince the statesmen? How successful were his efforts? A treatment of these specific questions is intended to serve as background for a discussion of what general lessons, if any, can be drawn on the basis of Bohr's crusade with regard to the political role of the scientist, particularly in a war situation.

Unlike most historians contributing to this volume, my specialization is not in the history of mathematics. Likewise, Niels Bohr, the leading person in the present article, while known for many things, was not a mathematician (although his younger brother, Harald, was). Indeed, one contributor to this summer's newspaper debates in Denmark has even claimed that Bohr's influence inside and outside science has been greatly overrated and that, in particular, he was extremely bad at mathematics.[2] These claims have led another newspaper jokingly to suggest that the Danish 500-kroner bills, which carry a photo of Niels Bohr, must be withdrawn.[3] Although this debate is not to be taken all too seriously, it does underscore that physics and mathematics constitute separate activities and that Bohr was first and foremost a physicist. I am happy to observe that the editors nevertheless find my contribution relevant to the questions addressed in this volume.

* Director of the Niels Bohr Archive, Copenhagen. Email: aaserud@nbi.dk

[1] The article is based on Finn Aaserud, "The scientist and the statesmen: Niels Bohr's political crusade during World War II," *Historical Studies in the Physical and Biological Sciences 30* (1, 1999), 1–47, as well as Aaserud's editorship of Volume 11 of the *Niels Bohr Collected Works*, tentatively titled *Political Activities and Occasional Writings* and scheduled for publication by Elsevier in 2004.

[2] Ib Ravn, "Opgøret med Bohr," *Politiken* 18 August 2002.

[3] *Ekstra Bladet* 21 August 2002.

My article follows upon that of Andrew Hodges, whose book on Alan Turing I consider truly admirable.[4] Alan Turing (who *was* a mathematician) and Niels Bohr have recently shared the fate of being the main characters in two very well-received plays for the stage – Turing in Hugh Whitemore's *Breaking the Code* (based directly on Hodges's Turing biography)[5] and Bohr in Michael Frayn's *Copenhagen.*[6] Indeed, there were lectures by both Whitemore and Frayn at a symposium organized by the Niels Bohr Archive in September 2001 on the relationship between drama and history of science.[7] This is a topic outside the scope of this volume, but I'm happy to be able to present an article in the company of the person who was the true originator of Whitemore's excellent drama.

As a result of the great success of Michael Frayn's play, the focus with regard to Bohr's war experience has lately been on his meeting with Werner Heisenberg in Copenhagen in German-occupied Denmark in September 1941 (Fig. 1). Partly as a result of the great public interest in the play, my institution, the Niels Bohr Archive, has recently been able to place previously unpublished documentation on this meeting on the web.[8] Notwithstanding the indubitable qualities of the play, the result has been that the historical significance of the Bohr–Heisenberg meeting has become greatly overrated, whereas Bohr's other political activities in the later stages of the war, which were substantially more important, have largely been forgotten or at least ignored. It was these activities that led up to Bohr's most well-known publication on political matters, namely his "Open Letter to the United Nations" from 1950.[9] This publication and the views it represents, while not leading to concrete results, has been claimed to be prophetic with regard to national relationships in the modern world.[10] Bohr's article cannot be understood without reference to Bohr's wartime activities, which is the topic of this article, and which may well be argued to have had greater impact than his subsequent publication.

But in order to understand Bohr's political activities and viewpoints we need to go yet further back in time. Neither the "Open Letter" nor Bohr's main wartime activities can be understood without an appreciation of Bohr's earlier experiences. I will therefore start by summarizing the most relevant aspects of his earlier career.[11]

[4] Andrew Hodges, *Alan Turing: The Enigma* (London: Simon and Schuster, 1983; Vintage, 1992).

[5] Hugh Whitemore, *Breaking the Code* (London: Samuel French Ltd, 1987).

[6] Michael Frayn, *Copenhagen* (London: Methuen, 1998; Anchor Books/Doubleday 2000).

[7] The symposium, including most of the talks, is presented at the website of the Niels Bohr Archive, www.nba.nbi.dk.

[8] *Ibid.*

[9] Niels Bohr, "Open Letter to the United Nations," private print in Danish and English 1950. Posthumously published in *Niels Bohr: His life and work as seen by his friends and colleagues*, Stefan Rozental, ed. (Amsterdam: North-Holland, 1967), pp. 340–352.

[10] See, for example, Abraham Pais, *Niels Bohr's Times, in Physics, Philosophy and Polity* (Oxford: Clarendon, 1991), pp. 473.

[11] In the narrative that follows, references are given for direct quotations only. More detailed documentation can be found in Aaserud, "The scientist and the statesmen" (note 1).

Figure 1. Heisenberg (left) and Bohr at Bohr's institute in 1936, five years before their fateful encounter during the German occupation of Denmark. [Courtesy of Niels Bohr Archive]

These are:

1. Bohr's background in Denmark's social and cultural environment.
2. His experience and activities within the international physics community.
3. His establishment and running of Copenhagen University's Institute for Theoretical Physics.
4. Bohr's philosophical outlook.

As for the first point, Bohr was born and raised within a prominent intellectual family. His father was professor of physiology at the University of Copenhagen, and from an early age Bohr and his siblings were bred on general intellectual discussions in the home. Niels got the very best *Gymnasium* education, the majority in his class subsequently receiving entries in *Kraks Blå Bog*, the Danish Who's Who of influential persons. He decided early to study physics at the University of Copenhagen, the Danish newspapers hailing his doctoral dissertation and defence in 1911 as a major scientific achievement. Bohr's continued association with the Danish intellectual elite can be seen from his membership on the committee to rebuild the Danish National Museum as early as the mid-1920s and from his long-time presidency (from 1939) of the Royal Danish Academy of Sciences and Letters.

Bohr's introduction to the international physics community took place immediately after his doctoral defence, when he went to England to do post-doctoral work. His first main – and perhaps his very greatest – contribution as a physicist stemmed from his work with experimentalist Ernest Rutherford in Manchester. Having returned from Manchester in the middle of World War I to take up a professorship established for him at the University of Copenhagen, Bohr wrote a long letter at the end of the war, congratulating Rutherford with the happy outcome. In his letter he stated, first, the opinion, typical of the time, that the nations must now have learned from the terrible events never to make war again. He thus believed in the establishment of new "principles ... of international politics" and "a new era in history." Second, he considered Rutherford a role model for having played a comforting role for the younger guard of physicists throughout the war. Third, he expressed his awareness of and respect for Rutherford's application of science for the war's outcome.[12] In the years immediately following, however, Bohr did not show much interest in international politics, except by taking up earlier and closer relations than most with the physics community of defeated Germany.

In 1921 Bohr's new Institute for Theoretical Physics at the University of Copenhagen was inaugurated. Henceforth, it was this institution that would provide the setting for Bohr's interaction with the international physics community. In his inauguration speech Bohr underlined two main points: first, that experimental activities at the institute were essential in order to test the theoreticians' predictions, and, second, that the institute would be a meeting place for the younger generation of theoretical physicists from around the world. Both desires were fulfilled to the utmost. As for the latter, the younger physicists came to remember Bohr's institute as the site of the "Copenhagen spirit," where Bohr and his disciples discussed physics without regard for anything else. In fact, Bohr was psychologically dependent on having a "helper" - sometimes half in jest called a "slave" - to throw back his ideas on. Thus Bohr was able to play the role of the full-fledged physicist, keeping his younger visitors completely occupied by physics and completely unaware of Bohr's quite considerable fund-raising activities and direction of his institute.[13]

Having physicists visiting from Germany as well as the Soviet Union, Bohr became intimately familiar with the effects of the political developments in both countries. He wholly disapproved of Hitler's taking power in Germany, making his institute a haven for the young guard of Jewish refugee physicists. Being more ambivalent to the Soviet system, he eventually accepted that the Russian physicist Peter Kapitza - like Bohr a disciple of Rutherford - was not allowed to return to England and instead given a physics laboratory in Moscow. As for experimental work at the institute, Bohr was able in the 1930s to change from spectrography - the basis for atomic physics - to accelerators, thus redirecting his institute to the field of nuclear physics well in time for the discovery of fission. Indeed, Bohr was

[12] Draft of letter from Bohr to Rutherford 24 Nov 1918, Niels Bohr Scientific Correspondence (BSC), Archive for the History of Quantum Physics, microfilm collection retained in several repositories, microfilm 6, section 3. The original BSC is retained in the Niels Bohr Archive.

[13] The "Copenhagen Spirit" as well as Bohr's fund raising activities are described in some detail in Finn Aaserud, *Redirecting Science: Niels Bohr, Philanthropy and the Rise of Nuclear Physics* (Cambridge, etc.: Cambridge University Press, 1990).

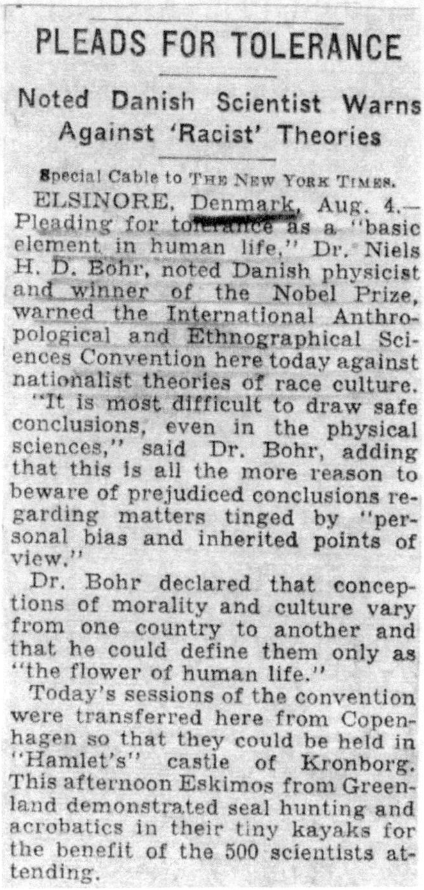

PLEADS FOR TOLERANCE

Noted Danish Scientist Warns Against 'Racist' Theories

Special Cable to THE NEW YORK TIMES.

ELSINORE, Denmark, Aug. 4.—Pleading for [redacted] as a "basic element in human life," Dr. Niels H. D. Bohr, noted Danish physicist and winner of the Nobel Prize, warned the International Anthropological and Ethnographical Sciences Convention here today against nationalist theories of race culture.

"It is most difficult to draw safe conclusions, even in the physical sciences," said Dr. Bohr, adding that this is all the more reason to beware of prejudiced conclusions regarding matters tinged by "personal bias and inherited points of view."

Dr. Bohr declared that conceptions of morality and culture vary from one country to another and that he could define them only as "the flower of human life."

Today's sessions of the convention were transferred here from Copenhagen so that they could be held in "Hamlet's" castle of Kronborg. This afternoon Eskimos from Greenland demonstrated seal hunting and acrobatics in their tiny kayaks for the benefit of the 500 scientists attending.

Figure 2. Article in the *New York Times* of 5 August 1938 about Bohr's lecture at the International Congress of Anthropological and Ethnological Sciences. [Courtesy of Niels Bohr Archive]

one of the first to foresee the implications of this discovery, contributing importantly to the understanding of the phenomenon during a visit to the United States in the spring of 1939.

As for Bohr's philosophical outlook, he returned to his early interest in philosophy by formulating the "complementarity argument" in 1927. Arising from Bohr's involvement in the interpretation of quantum mechanics, the argument was subsequently applied by Bohr to other fields, notably psychology and biology. In 1938 Bohr gave the opening lecture at the second meeting of the International Congress of Anthropological and Ethnological Sciences. Speaking at Kronborg Castle, Bohr argued that man is shaped by his culture and not by race.[14] Since cultures are complementary to one another, no culture is superior to any other culture. The talk was even reported by the *New York Times*, which took the talk to be a criticism of Nazi race theories flourishing at the time (Figure 2).

Bohr's close involvement in the theoretical work on fission had convinced him that although an atomic bomb was theoretically feasible, it would be technically impossible to develop one before the end of the war. Indeed, he made this point publicly in more than one talk shortly before the occupation of Denmark. Heisenberg's mention of bomb work during his 1941 visit to Copenhagen did not change Bohr's mind, and when James Chadwick secretly invited Bohr to England in early 1943 to help develop the bomb, Bohr refused on the grounds that he considered the task impossible (Figure 3). He escaped across the Sound to Sweden only when compelled to as a Jew at the end of September that year, whereupon he did accept the repeated British invitation. While in Stockholm he prepared a talk to a British audience

[14] Niels Bohr, "Natural philosophy and human cultures," reprinted in *Niels Bohr Collected Works*, Finn Aaserud, general editor, Volume 10: *Complementarity Beyond Physics*, David Favrholdt, ed. (Amsterdam: Elsevier, 1999), pp. 240–249.

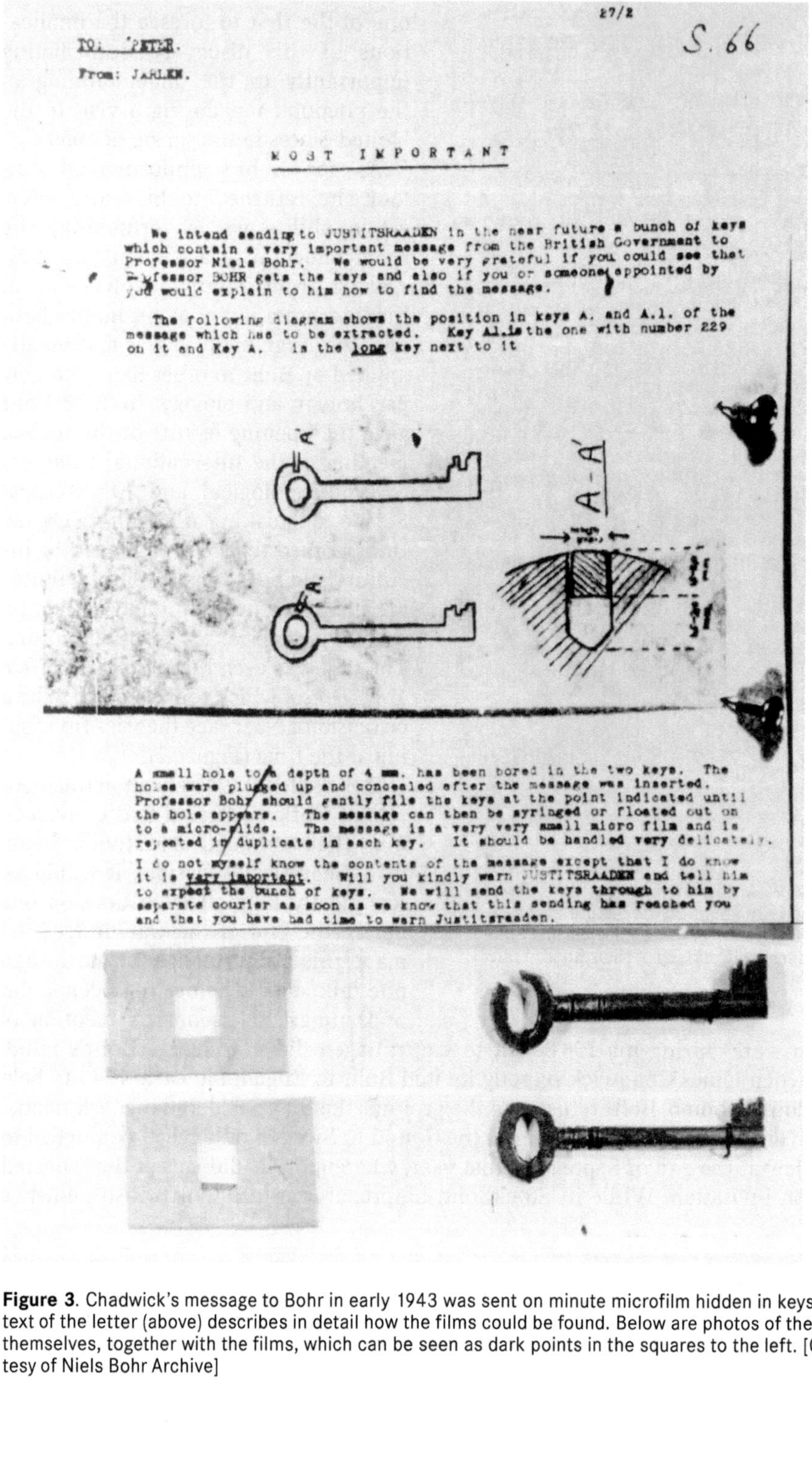

27/2 S. 66

TO: PETER.

From: JARLEN.

MOST IMPORTANT

We intend sending to JUSTITSRAADEN in the near future a bunch of keys which contain a very important message from the British Government to Professor Niels Bohr. We would be very grateful if you could see that Professor BOHR gets the keys and also if you or someone appointed by you would explain to him how to find the message.

The following diagram shows the position in keys A. and A.1. of the message which has to be extracted. Key A1. is the one with number 229 on it and Key A. is the long key next to it

A small hole to a depth of 4 mm. has been bored in the two keys. The holes were plugged up and concealed after the message was inserted. Professor Bohr should gently file the keys at the point indicated until the hole appears. The message can then be syringed or floated out on to a micro-slide. The message is a very very small micro film and is repeated in duplicate in each key. It should be handled very delicately.

I do not myself know the contents of the message except that I do know it is very important. Will you kindly warn JUSTITSRAADEN and tell him to expect the bunch of keys. We will send the keys through to him by separate courier as soon as we know that this sending has reached you and that you have had time to warn Justitsraaden.

Figure 3. Chadwick's message to Bohr in early 1943 was sent on minute microfilm hidden in keys. The text of the letter (above) describes in detail how the films could be found. Below are photos of the keys themselves, together with the films, which can be seen as dark points in the squares to the left. [Courtesy of Niels Bohr Archive]

in which he expressed himself in uncharacteristically bitter terms about the Nazi occupation of Denmark, yet expressed "the hope" – as opposed to the certainty in the letter to Rutherford after World War I – "that the present crisis through which mankind is passing will, how[ever] great the sacrifices have been and will still be, have to enforce the understanding of the unity of mankind."[15]

Bohr was taken to England after a week's time, being joined by his son Aage a few days later. He was briefed immediately about the development of the atomic bomb and seems to have been convinced immediately of the project's feasibility. He agreed to help in the project while at the same time developing his own thoughts about the post-war implications of the bomb and what could and should be done about it. He found sympathy for his thoughts in particular with John Anderson, the political leader of "Tube Alloys" (the British part of the bomb project) and Winston Churchill's Chancellor of the Exchequer. The first contemporary documentation of Bohr's views and Anderson's involvement in them is a long letter from Bohr to Anderson written from the United States in mid-February 1944. Here Bohr laid out in some detail the need for sharing the atomic secret with the Soviet Union before making use of it, arguing in his typically complementary fashion that "the impending realization of the [atomic bomb] project would not only seem to necessitate, but should also, due to the urgency of confidence, facilitate, a new approach to the problem of international relationship."[16] This was the core of Bohr's viewpoint, which he would repeat in his innumerable subsequent communications and memoranda.[17] Thus Bohr seems to have considered the existence of the atomic bomb a welcome instrument by which "to enforce the understanding of the unity of mankind," as he had expressed it when writing his speech in Stockholm before learning about the advanced state of the bomb project.

Anderson encouraged Bohr to contact the British Ambassador, Lord Halifax, about his political views. Halifax's reaction to his meeting with Bohr is telling: "These scientists find it very difficult to make their thoughts precise on political problems, and … I have had to do a lot of work with B[ohr] to get any clear idea of how his thoughts worked. But I think we succeeded in doing it fairly well in the end." Indeed, he was appreciative of Bohr's idea: "His thought is, I think, that, as long as no competition has begun (and the situation in this respect may change very rapidly after the surrender of Germany), Great Britain and the United States would seem to have in their hands a card which they could use in their negotiations with others of the United Nations for the improvement of the world situation with a view to the assurance of future peace. This idea of his about the political treatment of the project seems sensible, if it can be translated into a practical course of action."[18] This was a translation of Bohr's ideas into the language of the politician; yet Bohr approved of it.

[15] "1.4 Regarding the persecution of Jews," Niels Bohr Political Papers (BPP), Niels Bohr Archive.

[16] Letter from Bohr to Anderson 16 Feb 1944. Original in Records of the [British] Cabinet Office, series 126, folder 39 (henceforth CAB 126/39), Public Records Office (PRO), London. A carbon copy can be found in "2.2 Correspondence with John Anderson," BPP.

[17] Much of this material will be published in the *Niels Bohr Collected Works*, Vol. 11 (note 1).

[18] Letter from Halifax to Anderson 18 Feb 1944 (CAB 126/39).

Figure 4. The only photo of Niels Bohr from Los Alamos during the war. [Courtesy of Niels Bohr Archive]

Because of the strained relationship with the Manhattan Project – the U.S. side of the effort to develop the atomic bomb in Los Alamos, New Mexico – Bohr was expected not to tell the American side of his receiving advice in these matters from the British; the crusade should be perceived entirely as that of a concerned private citizen. In the same vein, Bohr took care only to discuss purely scientific issues with the physicists working on the bomb and not inform them about his political activities. This was a difficult task for Bohr, although, as we have seen, he had made a similar fruitful separation between scientific and administrative work when working with fellow physicists at his institute between the wars (Figure 4).

Halifax agreed that the best approach might be to seek contact as high in the political system as possible, that is, to arrange an interview with President Roosevelt. In the circumstances, such an arrangement could not be made through the British. Nevertheless, an opportunity arose when Bohr met Felix Frankfurter, a former acquaintance, who was both a member of the U.S. Supreme Court and a close friend and adviser of the President. When in mid-February Bohr found that Frankfurter knew about the bomb project in outline, he explained his political objective. Frankfurter was enthusiastic, and at the end of March told Bohr that the President had reacted approvingly to Bohr's views and even wanted him to bring his concern about the post-war implications of the bomb directly to Churchill. Frankfurter drew up a "message" to this effect, whereupon Bohr went back to the British asking for passage back to England.

Figure 5.
Peter Kapitza (left) and James Chadwick during happier times. Cambridge 1923.
[Courtesy of Niels Bohr Archive]

Anderson had already advised Churchill to tell the Soviets about the bomb project, getting a very negative reaction. He now agreed to let Bohr carry the message to England, whereupon Bohr and son arrived in London on 17 April. In London, a letter was waiting for Bohr at the Soviet embassy. It had been written several months before by Bohr's Russian colleague Peter Kapitza, and was in effect an invitation to Bohr to work on physics in the Soviet Union. Both Bohr and British intelligence suspected that the letter meant that the Russians were working on an atomic bomb and wanted Bohr to join in the effort. Bohr wrote a guarded response to Kapitza, which was cleared by British intelligence before it went off (Figure 5).

Bohr was finally successful in arranging a meeting with the Prime Minister, to a great extent through influential friends. The meeting took place on 16 May and was a dismal failure. Bohr hardly got in a word while Churchill and Lord Cherwell – the Prime Minister's personal science adviser – argued over the relationship with the Americans. Most likely, Churchill's mind was elsewhere, as he was in the midst of preparing D-Day at the time. When at the end of the meeting Bohr asked Churchill whether he could write him a letter setting out his views, Churchill responded that he would be grateful to receive a letter from Bohr, only not about politics. Having written a couple of more memoranda for Anderson, Bohr and son left Britain quite disillusioned on 9 June, some time after Leslie Groves, the military leader of the Manhattan Project, had expected Bohr to return to his duties at Los Alamos.

Back in the United States Frankfurter encouraged Bohr to write a "Memorandum" for Roosevelt, laying out his views. Bohr put in a great effort writing the document, but after giving it to Frankfurter expressed uncertainty as to whether it was formulated in a way that would be meaningful for the President. By that time, Frankfurter had already delivered the document to his boss, and he reassured Bohr that it was just right. This document was later regarded by Bohr as the most important of his writings during the war, and he reprinted a substantial part of it in the "Open Letter" of 1950.

Bohr was now waiting eagerly for the President's reaction, and a meeting was indeed arranged, quite unofficially and at short notice, on 26 August. In contrast to Churchill, Roosevelt was quite forthcoming, describing his role as a "mediator" between Churchill and Stalin, expressing great sympathy for Bohr's views, and promising to take up the matter with Churchill on their meeting in Quebec scheduled only two weeks later. Understandably, Bohr was in high spirits after the meeting with the American president.

However, the promise of success was again turned into failure. After the formal meeting in Quebec, Churchill and Roosevelt worked out an aide-mémoire at Roosevelt's summer residence in Hyde Park, New York, on 19 September (Figure 6). The document consisted of three numbered paragraphs, the last of which read: "Enquiries should be made regarding the activities of Professor Bohr and steps taken to ensure that he is responsible for no leakage of information particularly to the Russians."[19] In a letter to Cherwell written from his London residence soon after, Churchill thundered: "Bohr ought to be confined or at any rate made to see that he is very near the edge of mortal crimes." He went on to confide that, "I did not like the man when you showed him to me, with his hair all over his head, at Downing Street."[20] Until learning about these developments about a month later, Bohr continued to await a second call from the President in Washington. The explanation Bohr received was, first and most importantly, that the letter from Kapitza had made him appear like a Soviet spy and, second, that sharing the secret of the atomic bomb with Frankfurter, who was not formally cleared for this information, was considered a treacherous act. In Bohr annals, the incident is referred to with characteristic understatement as the "misunderstanding."

Had Roosevelt changed his mind, was he under the influence of Churchill, or had he been less than honest when speaking to Bohr? Only three days after the

[19] The aide-mémoire is reproduced in full in Margaret Gowing, *Britain and Atomic Energy 1939–1945* (London: Macmillan, 1964), p. 447. A copy of the original document is in CAB 127/201.

[20] Quoted in *ibid.*, p. 358. The original is in CAB 126/39.

Figure 6. Franklin D. Roosevelt and Winston Churchill. [Courtesy of Niels Bohr Archive]

Hyde Park meeting Vannevar Bush and James Conant – two of the main civil administrators of the American scientific war effort – were distressed by an utterance by Roosevelt consistent with his and Churchill's joint Hyde Park statement. Roosevelt thus stated that he and Churchill saw the atom bomb as a means for "policing" the world when the war was over. Bush and Conant were provoked to produce memoranda containing viewpoints similar to Bohr's. They too now wanted to see the President, but their papers never reached beyond Secretary of State Henry Stimson. Thus, unlike Bohr, they were never able to speak to President Roosevelt.

Eventually Bohr too was told that the President did not have time to see him and that he should rather see Bush about the matter. Cherwell, however, objected to such an approach, which meant that Bohr had lost contact not only with the American President but also, at least momentarily, with the American leadership of the bomb project.

By this time not only representatives of the bomb project such as Bush and Conant, but also scientists outside the project, expressed concern for an arms race after the end of the war. In this regard, Bohr reported to Anderson that he was striving "to discourage steps which might complicate matters for responsible statesmen … without disclosing information and by expressing general hope."[21] Bohr seems to have taken the comforting role that he had thanked Rutherford for in the letter celebrating the outcome of World War I. A particularly interesting case in point was Albert Einstein who on 12 December sent a letter to Bohr. The only solution to the arms race problem Einstein could envisage was that scientists in the U.S., England and Russia advise their respective governments to work for "an internationalization of military power." He suggested that with his international connections, Bohr would be instrumental in implementing this.[22] Bohr, who considered Einstein's suggestion to imply a severe breach of security, appeared personally in Princeton ten days later explaining his strong reservations and assuring Einstein "confidentially … that the responsible statesmen in America and England were fully aware of the scope of the technical development, and that their attention had been called to the dangers to world security as well as to the unique opportunity for furthering a harmonious relationship between nations."[23] Einstein proceeded to drop his proposal without a quibble.

On 5 March 1945 Bohr was back in England, after Groves had accepted that he made the trip for personal reasons. While there, he wrote another memorandum – in Bohr parlance known as the "Addendum" to the "Memorandum" previously written for Roosevelt – expressing his concerns. This is the second of Bohr's war writings that he subsequently regarded as important enough to include parts of it in the "Open Letter" of 1950. Again, however, the question came up of whom Bohr could show it to. Now it was Anderson who was sceptical of Bush, whereas Cherwell resisted showing the document to Frankfurter. After Halifax had recommended to Anderson that Bohr meet with Bush, there were three meetings between the two upon Bohr's return to the United States, the first taking place on 23 April. At this point, however, the so-called "Interim Committee," of which Bush was a member, had been formed to advise on political matters in relation to the atomic bomb. Even from Bohr's rather positive account of the meeting, it appears that Bush sought to keep him at arm's length. Bohr seems to have wanted to stay in the U.S. for a longer period, and after the capitulation of Germany, preparations were made to bring his wife Margrethe from Stockholm.

The unrelenting Frankfurter now sought to achieve a meeting between Bohr and Stimson. After Bohr had informed him on 23 June of Frankfurter's efforts, Roger Makins, the new Minister at the British embassy, reported back to England that the Americans were suspicious of Bohr's "always popping around" as well as, indeed, Bohr's continued relations with Frankfurter, who was considered suspect by many. Indeed, Makins considered that after the appointment of "the advisory group on the American side" there was no longer any need for Bohr's viewpoints.[24]

[21] "2.6 Notes regarding conversations with Sir John Anderson," BPP.

[22] Letter from Einstein to Bohr 12 Dec 1944, "10.1 Albert Einstein, 12 Dec 1944," BPP.

[23] "10.2 Note regarding letter from Einstein, 12 Dec 1944," BPP.

[24] Reports from Makins to Denis Rickett (War Cabinet Office, London) 25 May 1945, CAB 126/40 and 23 June 1945, folder 65, box 12, Chad4, James Chadwick Papers, Churchill Archives Centre, Cambridge, England.

Four days after the meeting with Makins, Bohr suddenly found himself back in England, to where his wife had now been diverted.

Indeed, although trying hard, evidently with Anderson's support, Bohr was not able to obtain passage back to the United States. He was duly informed, but only at a distance, of the bomb test in Alamogordo on 16 July as well as of the bombing of Hiroshima on 6 August. When he realized that a return to the U.S. could not be achieved, he called back his son, Aage, who had been waiting for him there. Bohr's only option now was to make his concerns public, and upon Anderson's advice, the article "Science and Civilization," written in collaboration with his son, appeared in *The Times* 11 August. Two weeks later Niels, Margrethe and Aage Bohr returned to Copenhagen.

The Military Use of Alan Turing

Andrew Hodges*

Alan Turing (1912-1954), British mathematician, was critical in the Anglo-American decipherment of German communications in the Second World War. This experience enabled him to formulate an original plan for the digital computer in 1945, based on his own 1936 concept of the universal machine. He went on to found the program of Artificial Intelligence research. This article discusses the relationship between these developments, and more general questions of mathematics and war illustrated by Alan Turing's life and work.

The British mathematician, Alan M. Turing (1912–1954) played a critical role in the Second World War, as the chief scientific figure in the Anglo-American decipherment of German military communications. Furthermore, his work was central to the emergence of the digital computer in its full modern sense in 1945. However, the secrecy surrounding his work was so intense that until the 1970s only hints of it were published. This secrecy was enhanced by the mystery of his sudden death in 1954 and the effective taboo, which prevailed for twenty years, on any public mention of his homosexuality. To those interested in the true history of the computer, Alan Turing's role remained as elusive as the myth of Atlantis. This secrecy has now almost completely been dispelled, partly through this author's work (Hodges 1983), but only to reveal a much deeper enigma of Alan Turing, who gave himself first to the purest and most timeless mathematics, but then applied himself to its most urgent and timely practice. What did Alan Turing think of his intellectual and moral involvement in the world crisis? And what is the true assessment of the impact of the war on his scientific work?

This article will review Alan Turing's mathematical work in the Second World War, discuss how this relates to the history and philosophy of computing, and then raise the wider question of his place in mathematics and war.

Turing's role in the Second World War was largely dominated by the particular form of the Enigma ciphering machine as elaborated for military and naval purposes by the German authorities. For a recent complete description of the Enigma see (Bauer 2000). Essentially, it was Turing who picked up the relay baton when the Polish mathematicians shared their brilliant cryptanalytic work with Britain and France. It then fell to him to pass on the baton, by sharing British achievements with the United States.

* Wadham College Oxford University OX1 3PN, UK. Email: andrew@synth.co.uk

Alan Turing's primary role stemmed partly from the fact that he was the first scientific figure to join the British cryptanalytic department, the so-called 'Government Code and Cypher School,' which until 1938 was staffed essentially by the language-based analysts of the First World War. (One reason for Turing's recruitment may have been that he had, through his Fellowship of King's College, Cambridge, personal connections with that older British generation; in particular J. M. Keynes may well have formed an important link.) Turing was brought into the work on a part-time basis at the Munich crisis period, and joined full-time immediately on declaration of war. Meanwhile an Oxford mathematics graduate, Peter Twinn, was recruited through open advertisement in 1938, and this belated acceptance of the modern world was shown also in the development of a modern communications infrastructure for the new headquarters at Bletchley Park, Buckinghamshire. Nevertheless, the Polish mathematicians were well in advance at the time of the now famous meeting in July 1939. They had used group-theoretic algebra to deduce the Enigma rotor wirings from information obtained by spying; they had noticed and used other group-theoretic methods and mechanical methods to exploit certain simple forms of indicator system that were then in use by the German forces. It is not clear to what extent Turing had discovered these independently in early 1939 – his report (Turing 1940) does not say – but in any case the Polish group had successfully made an all-important guess which had eluded the British: this was the order in which the keyboard letters were connected to the first rotor. They were in fact in the simple order ABCD.... This almost absurdly simple fact was the most critical piece of information imparted by the Poles.

In late 1939, Turing initiated the two most decisive new developments: he saw the 'simultaneous scanning' principle of what became the British 'Bombe', and he deduced the form of the more sophisticated indicator system that was being used for the German Naval communications.

Turing's 'Bombe' was an electromechanical machine of great logical and technical sophistication. Its property was this: given a stretch of ciphertext and the corresponding plaintext, it could search through all possible settings of the military Enigma and detect those which could possibly have been responsible for the encipherment. It is not difficult to see, from simple counting arguments, that a 'crib' of about 20 letters will generally serve to identify such a setting. (The reader may take it that, once some penetration into cipher traffic has been made, such a 'crib' is not impossible to find.) It is much harder to see that this theoretical possibility can be matched by any practical method. In particular, the Stecker or plugboard complication introduced in the military Enigma had so many possible settings that serial trial was impossible. In fact serial trial was indeed necessary for searching through the possible positions of the rotors. But Turing's great discovery was that the huge number of plugboard possibilities could effectively be tested in parallel, and virtually instantaneously. His idea was this: suppose we are testing, in the serial sequence, a particular rotor setting. A plugboard setting consists of a number of pairs like (AJ), (UY), representing the swapping of letters performed by the plugboard on entry to, and on exit from, the rotors. There are 150,738,274,937,250 possible settings consisting of ten such pairs, the choice normally made. However there is no need to work through such a number of possibilities. Instead, consider the smaller number of 325 basic pair-possibilities: (AA), (AB), (AC) ... (ZZ), where '(AA)' represents the letter A being left unchanged by the plugboard. Now, given any one basic pair-possibility, e.g., (AA), knowledge of the plaintext, ciphertext, and rotor setting will imply various other pair-possibilities, and these in turn yet more. Turing saw first, that finding a single 'contradiction' serves to eliminate a possibility: that is, if by following the implications (AA) can be shown to imply (AE), then (AA) must be false. He saw the less obvious point that by allowing the logical deductions to continue, (AA) would generally imply all of (AB) ... (AZ); hence all must be false; hence the rotor position being tested could not possibly be correct. Turing was proud of this counter-intuitive idea, of continuing to follow through the consequences of what must be false propositions. He said it was akin to the principle in mathematical logic that 'a false proposition implies any proposition.'

Turing also saw how to embody these 'deductions' simply in wired connections between rotors and terminals, so that the flow of implications would take place at the speed of electric current. (But it took another Cambridge mathematician, W. G. Welchman, to improve the circuitry with the 'diagonal board' which automatically identified (AB) with (BA), and so on; it is curious fact that Turing missed this simple idea.) Finally, it was essential that the machine could be equipped to detect the possibility of a correct rotor position, with logical switching capable of recognising an incomplete bank of pair-possibilities. With this achieved by the engineers of the British Tabulating Machinery company, Turing's Bombe yielded the central process on which Enigma decryption rested throughout the War. Its principle was highly non-trivial, and it apparently went unnoticed by G. Hasenjäger, the young German logician who was Turing's unseen opponent in this war of logic (Bauer, 2000).

As already noted above, in the business of searching through the 26^3 possible rotor positions, no improvement was possible on serial trial using the equivalent of moving Enigma rotors, so that improvements rested on having ever faster, more reliable machines produced in larger numbers. For the question of the choice of rotors and their order, however, which was particularly relevant for the Naval Enigma problem, Turing developed a method called 'Banburismus' which, particularly in 1941, much improved upon the simple trial of the possibilities. This method rested on the logical details of the turnover positions of the different rotors, but also on assessing the statistical identification of likely 'depths' – two different stretches of message, both sent on the same Enigma settings. For detecting depths objectively, giving a reliable probability measure to them, Turing developed a theory of Bayesian inference. The most striking feature of his theory was his measurement of the weight of evidence by the logarithm of conditional probabilities. This was essentially the same as Shannon's measure of information, developed at the same time. This theory was developed into sophisticated methodology (Good 1979, 2000). Whilst the logical principle of the Bombe, and its amazingly effective application, was perhaps Turing's most brilliant single idea, his statistical techniques were more general and far-reaching. In particular, they were also applied to the quite different Lorenz cipher system employed by Germany for high-level strategic messages. Thus it was Turing's theory of probability estimation that underlay the methods mechanised by the large electronic 'Colossus' built in 1943–45 to decipher this type of traffic.

The latter half of the war saw the relay baton pass on to the USA. Turing himself had to cross the Atlantic, at the height of the battle, in connection with naval Enigma crisis of 1942 and the building of American Bombes at Dayton, Ohio. Besides transferring British expertise in Enigma-breaking to the United States, he also had a new top-level role in inspecting and reporting on the American equipment for speech encipherment (to be used by Roosevelt and Churchill), and became fully acquainted with the use of information sampling theory as well as the electronic technology involved. It is fair to say that Turing most enjoyed a pioneering period of breaking into the unknown, and flourished best in such settings; he was not so happy at detailed follow-up or development. After 1943 he spent much of his time on a freshly self-imposed problem: the design of an advanced electronic speech scrambler of a much more compact form than the American equipment he had inspected (Hodges 1983, p. 273).

Sometimes it is blithely asserted that what one mathematician can do, another can undo. Not so: the possibilities of cryptanalysis are highly contingent on details; and even if a system is breakable in the long term, short term considerations may be of the essence. The German military adaptation of the Enigma might have made it unbreakable; the British version of the Enigma, which has attracted far less attention, had more rotors and a non-reciprocal plugboard, and was apparently invulnerable to German attack. The successful continuation of the Polish work may very well have depended critically on Turing as an individual. It was not that Turing merely played the expert part expected of him. Rather, it was at a time when a distinctly pessimistic attitude prevailed, that Turing took on the Naval Enigma problem precisely because no-one else thought it tractable, and so he could have it to himself. This individualistic approach was also necessary in developing from scratch a suitable statistical theory. Thus, it can well be maintained that the Anglo-American command of the Atlantic battle, crucial in the central strategy of the Western war, was owed to Alan Turing's work.

Turing did not create great new fundamental mathematics in this work, but he brought to bear the insights of a deep thinker. The Bletchley Park analysts compared their work with chess problems, and G. H. Hardy had famously called chess problems the hymn tunes of mathematics, as distinct from serious, interesting problems (Hardy 1940). Yet unusually these military hymns had some beauty. In a remarkable comment on an episode in mathematics and war, the contemporary chronicler of the Naval Enigma section ended by saying (Mahon 1945). In finishing this account of Hut 8's activities I think that it should be said that while we broke German Naval Cyphers because it was our job to do so and because we believed it to be worthwhile, we also broke them because the problem was an interesting and amusing one. The work of Hut 8 combined to a remarkable extent a sense of urgency and importance with the pleasure of playing an intellectual game.

It can, however, be argued that more important than the direct application of new mathematics was the influence of his wartime experience in leading Turing from being a pure theorist of computation to being the leading expositor of an actual electronic design for a modern computer. At the age of twenty-four, Turing had published what is now his most celebrated work, defining computability (Turing 1936–1937). Its motivation lay in pure mathematical logic, clarifying the nature of an 'effective method' with the Turing machine concept. It did not set out to assist practical computation in any way. Yet it did produce the constructive and highly suggestive idea of the universal Turing machine, and in fact Alan Turing was never entirely 'pure' in his approach: bringing 'paper tape' into the foundations of mathematics was itself a striking breach of the conventional culture. As we shall see, he took an interest in the practical application of his ideas even in 1936. But Turing's harmonious collaboration with the engineers of the Bombe took him very much further towards practicality, indeed into the most advanced practical engineering of 1940. Then his acquaintance with the world-leading technology of the Colossus showed him the viability of an electronic digital machine capable of embodying the idea of the universal machine – in modern terms, a computer with modifiable stored program. Turing was able to write a detailed proposal for such a computer, the ACE, in 1945–1946, and was able to survey from a position of considerable strength all the possible forms of data storage available at that point (Turing 1946).

In discussions of the origin of the computer there is an inevitable rivalry between the claims of mathematics and engineering. Engineers, often overlooked and accorded a low social status, do not take kindly to being treated as mere technical assistants when in fact they have contributed immensely skilled and creative solutions. Nevertheless, whether we speak of the computer, or of machines such as the Bombe or Colossus, no engineer originally conceived the nature and purpose of the machine: the conception was that of mathematicians. The point of the Bombe was to implement Turing's brilliant logic; the point of the Colossus was to implement ingenious developments of his statistical theory. The point of the modern computer is that it implements Turing's universal machine, on which instructions can be stored and manipulated in exactly the same way as data. It is usually said that von Neumann, in formulating the EDVAC plan in 1945, played the same mathematical role independently, but Martin Davis has recently argued that von Neumann could not have played this part without learning of Turing's work before the war (Davis 2000).

However, Turing was unusual in wishing to dominate every aspect of the computer design: not only its central principle and purpose, but the applications for which it would be used, and the details of its design. His report (Turing 1946) for the National Physical Laboratory, London, covered all of these. In every way it reflected the influence of his war work, but there was an irony. Official secrecy about the wartime success was total and prevented Turing from arguing from his extensive experience to establish some influence over the engineering side of the problem. So although his report alluded to important military applications in the solution of combinatorial problems, showed other influences of non-numerical Bletchley Park work, and claimed Foreign Office support, the NPL management continued to regard mathematics and engineering as belonging to two distinct planets and blocked Turing's efforts to argue otherwise.

Turing was particularly frustrated over the inability to command the necessary technology, because although vital for the trying out of a practical form of a universal machine, the actual form of the implementation was essentially a secondary matter. Turing correctly saw his ACE plan as necessarily obsolescent, leading the way to better machines, built using faster control circuits and larger storage systems. Of far greater significance, in Turing's plans, was the scope of what a practical universal machine would be doing: namely, implementing an open-ended array of software or as he called it, instruction tables.

In that first report he wrote a number of key remarks about the potential of the universal machine:

> It is intended that the setting up of the machine for new problems shall be virtually only a matter of paper work... There will be positively no internal alterations to be made even if we wish suddenly from calculating the energy levels of the neon atom to the enumeration of groups of order 720. Instruction tables will have to be made up by mathematicians with computing experience and perhaps a certain puzzle-solving ability. There will probably be a great deal of work of this kind to be done, for every known process has got to be translated into instruction table form at some stage. The process of constructing instruction tables should be every fascinating. There need be no real danger of it ever becoming a drudge, for any processes that are quite mechanical may be turned over to the machine itself.

In (Turing 1947) he elaborated on the potential for computer languages opened up by this last remark:

> Actually one could communicate with these machines in any language provided it was an exact language, i.e. in principle one should be able to communicate in any symbolic logic, provided that the machine were given instruction tables which would allow it to interpret that logical system...

These and other observations mapped out the scope of the computing industry of the future (although Turing did remarkably little to implement his prophetic ideas for computer languages). But, at a more profound level still, even this software manifesto was also secondary. He was, he said, building a brain: embarking on an experimental programme for what he called 'intelligent machinery', or what would now be called 'artificial intelligence.'

This deeper scientific programme was also influenced by his war experience. Before the war Turing had, in his work on ordinal logics, apparently accepted the standpoint of Gödel in which the mathematician does something beyond the scope of rule-following when seeing the truth of a Gödel sentence. It has been argued by the author (Hodges 1997, 2002), that Turing must have had a change of mind, or at least a definite clarification of mind, in about 1941. By that time the spectacle of the Bombes and of the mechanisation of human guessing and judgment by Bayesian statistical methods had supplied a vivid argument to the effect that the concept of being 'merely' mechanical was misleading. To capture the spirit of the time, it is worth noting from a recently declassified document (Good et al. 1945) the vivid impression that the beginning of the electronic information technology age made on its beholders:

> It is regretted that it is not possible to give an adequate idea of the fascination of a Colossus at work [...] the fantastic speed of thin paper tape around the glittering pulleys [...] the wizardry of purely mechanical decoding letter by letter (one novice thought she was being hoaxed); the uncanny action of the typewriter in printing the correct scores without and beyond human aid [...].

Turing, even though he perfectly understood what was going on, was doubtless also influenced by this magical power of the purely mechanical. Possibly, Turing was also encouraged by exposure to intellectual currents which suggested thinking of the nervous system in behaviourist terms.

In any case, it is clear that by 1945 Turing had decided that the brain did not actually perform any uncomputable operations of the kind suggested by Gödel's theorem. He had formulated his 'mistakes' argument: that in assessing mathematical proofs the brain will sometimes make mistakes, and that accordingly Gödel's theorem does not have any force. His guiding thought was that the brain must be effectively a finite machine and therefore performs operations which can, at least in principle, be run on a universal machine, the computer. A further report (Turing 1948) explored the question of how a machine could be led to perform actions which appear not to be of a 'mechanical' nature, as understood in common parlance. Turing discussed program self-modification, networks of logical elements as models of neuronal systems, the idea of 'genetical search.' His ideas for emulating human 'education' combined what would now be called top-down and bottom-up approaches to Artificial Intelligence. This report was the basis of his more famous philosophical paper (Turing 1950).

Turing spent a year at Cambridge to do this theoretical work in 1948, because the National Physical Laboratory was making such desultory progress with implementing his computer plan. While he was there, another opportunity appeared to open. The key figure was the topologist M. H. A. Newman, the Cambridge lecturer who had introduced Turing to the frontier of mathematical logic in 1935, the first reader of Turing's consequent logical discoveries, and indeed to some extent Turing's collaborator in logic. Newman took the idea of the universal machine with him when he went to Manchester University as professor of pure mathematics in 1945. He also took an acquaintance with electronics, since he had directed the development of the Colossus as applied to the Lorenz cipher problem at Bletchley. In fact, he was equipped with all the force of Turing's ideas except that he had none of Turing's interest in mastering electronics for himself.

Newman's fervent desire was that the concentration and investment of scientific resources that had been shown possible in the emergency of the war, should now be dedicated to pure science. He rapidly made a proposal to the Royal Society, and was successful in getting a large grant for a computer on which to do research in mathematics (including for instance the Four-Colour Theorem), on an ambitious and long-term scale. Newman was able to recruit for his project the leading electronic engineers who came to Manchester from wartime radar development work, and he imparted to them the essential stored-program ideas that had to be implemented.

Newman offered Turing the role of running this emergent Computing Laboratory. Turing accepted. But even before he arrived, the success in June 1948 of the engineers' prototype meant that it was demanded for the British atomic bomb development. The government contract went to the engineers, not to Newman; the Royal Society connection was abandoned. Turing was sidelined along with Newman himself, and had to relaunch himself with quite new work: his mathematical theory of morphogenesis. By 1950 it was the end of the road for the synthesis and collaboration achieved in the Second World War. What is now the engineering discipline of 'computer science' took on a life of its own and the connection with mathematics was largely lost, Turing's legacy being abandoned with it.

There is, however, an important caveat to be made here: we do not know what Turing did as secret work for GCHQ after 1948. This Cold War question is still as secret now as the Second World War operations were in 1970. And it opens an aspect of the subject of Mathematics and War where all is speculation, an unresolved enigma. Perhaps the most extraordinary aspect of Alan Turing's story is his role as a great innocent figure of pure science, yet at the heart of the most urgent world crisis. It is a magical picture, but this sequel to it is still incomplete and obscure.

Was he involved in the Venona problem, which involved tracking KGB messages to and from its western agents? If so, was he eager for participation in the work against Soviet influence? Or was he only responding to a strong personal plea from Hugh Alexander, who from being Turing's deputy in 1941 had risen to become director of GCHQ in 1952? Or was it more the challenge of another 'interesting and amusing' mathematical problem? What were the inner springs? We do not know. I turn now to those more general questions about the setting of Turing's mind in the world of war.

There is a common joke about Military Intelligence being a contradiction in terms, and the concept of the 'military use of mathematics' has a similar problem, for mathematics is not just used: it has to be created by awkward, non-standard, individual mathematicians. A subordinate cannot simply be ordered to do something extraordinary, unexpected and innovative, such as breaking a cipher system that is supposed to be unbreakable. Anyone approaching such work in the spirit of carrying out an order, is unlikely to be inspired. Turing exercised a wilful initiative and went against prevailing wisdom to attack the naval Enigma systems. Such an unorthodox success, however welcome in the desperate situation of 1940, clearly signified a certain danger for government control: if people act from self-willed initiative, rather than following orders and duty, they may progress to some other self-willed initiative less to the taste of the State. As later became well known, some of those who decided for themselves to fight Germany, also chose with equal force to assist the USSR. In the postwar period the emergent idea of 'security' had to take this into account, and after 1948 Turing was bound to find that the wartime personal networks, class-based trust and easy acceptance of his eccentricity no longer applied in state affairs.

What makes this particularly interesting and piquant in discussing Turing's biography, is that the interplay of rule-following Duty and creative Will was central to Turing's own theory of mind and machine. In his 1948 report on machine intelligence he described the problems to be addressed in terms of 'discipline' and 'initiative,' the latter being necessary to anything that could genuinely be called 'intelligence'. In his discussion of machine intelligence, a dialogue between mechanical and creative, obedience and surprise, runs throughout. It is a striking fact that he came down so strongly on the side of saying computers would be able to show intelligence, when he of all people knew the significance of creative originality. But this dialogue of compliance and rebellion – and resolving it by standing aside, withdrawing – ran through his life. In his 1950 paper, Turing made a joking reference to Casabianca, the boy on the burning deck who carries out his orders relentlessly as a computer, oblivious to common sense, and this image of military duty went back to his childhood learning of the poem at preparatory school, still within a First World War world. Even as a child he had known the limitations of rule-following, and had found the precepts of his class largely incomprehensible.

Yet Alan Turing had not been a rebel against his class or his school. In his upper-middle-class environment, dominated by the suffocating ideology of the Public School, he had neither complied nor rebelled; he had largely withdrawn into a scientific world of his own. There was, however, a moment in 1933 when the Anti-War student movement at Cambridge articulated the change that had taken place since the world of Duty plunged to disaster in 1914, and at twenty-one, Turing joined it and placed himself clearly on the modernist side. His contemporaries were insisting on deciding for themselves what to fight for, and the emergent left-liberal side decided not to fight for that old chant of Duty: 'King and Country.' 'Blatant militarist propaganda', Turing called a cinema recruiting film called 'Our Fighting Navy.' Ten years later, Alan Turing was to be seen by his WRNS assistants 'prancing' in his Bletchley Park hut at the news of the sinking of the Scharnhorst by that very Navy. Of course, the 1933 enlightenment had coincided with the one development that made war justifiable to the enlightened: the transformation of Europe's strongest industrial power into the aggressive engine of murderous fascism.

The young Alan Turing was well aware of the shock of Nazi Germany, and it was obvious where his sympathies lay. When visiting Germany in 1934, Turing's travelling companion was surprised to see how naturally a German socialist seemed to confide in him. Later, he was immediate in his response to the 1938 wave of persecution, and sponsored a young Jewish refugee. But in the intellectual context of the 1930s, dominated by the question of Communist party policy, Alan Turing's 1933 political engagement was short-lived; he was regarded as 'not a political person.' Alister Watson, who introduced him to Wittgenstein, was a Communist party member, as was also the young Robin Gandy who later became his student and closest friend. Others of Alan Turing's friends were fully engaged with political and economic issues. But for Alan Turing, it was the 'phoney' that attracted his scorn rather than political opposition; the compromises and alliances of political action were alien to him.

In temperament he was closer to his first lover, James Atkins, another mathematics student of his year, who was and remained a pacifist while becoming a professional musician. But Alan Turing found in the music of mathematics the means to undermine Nazi Germany. In 1936, just after he had completed his famous paper on computable numbers, and had arrived in Princeton, Alan Turing wrote to his mother, who was apt to ask him 'but what use is it':

> I have just discovered a possible application of the kind of thing I am working on at present. It answers the question 'What is the most general kind of code or cipher possible' and at the same time (rather naturally) enables me to construct a lot of particular and interesting codes. One of them is pretty well impossible to decode without the key, and very quick to encode. I expect I could sell them to H. M. Government for quite a substantial sum, but am rather doubtful about the morality of such things. What do you think? (Turing 1936)

What theory of a 'most general' cipher he had in mind, what was his theory of cipher security, and and what were the interesting codes, remain unknown. Nor, in Turing's reference to the question of 'morality of such things', is it clear whether he means the morality of selling his work, or the morality of doing military work. However, whatever moral consideration Turing found most perplexing, the fact is that at Princeton in 1937 he proceeded with a cryptographic idea embodied in an electromagnetic relay device, and saw it specifically as relevant to looming war with Germany. It is very possible that this device was indeed offered to the British government, and was the reason why he became the first scientific officer at the Government Code and Cypher School. He received a salary, and later some special payments, but what was more important, he paid a moral price: he sacrificed the freedom of his mind.

Freedom, to Alan Turing, was more important than questions of money. He waxed more eloquent over the Abdication crisis than over any other issue, expressing, perhaps naively, a support for the King's freedom to marry, and deploring the 'hypocrisy' of the Archbishop of Canterbury. But tellingly, he deprecated the indiscretion of the ex-King in allowing state papers to be seen by Mrs. Simpson. (Hodges 1983, p. 122) Soon afterwards, he had to promise to keep State secrets himself. But no doubt he found it exciting to be let into government's innermost secrets, and perhaps too the technical challenge of the Enigma problem became immediately and addictively fascinating.

The Abdication issue was an anticipation of his own quandary after his arrest as a homosexual in 1952. For in 1952 he showed himself the most devoted to personal freedom, but at the same time, as his friends did not know, he was the most tightly bound by secrecy at the highest level in the Anglo-American alliance. He was ahead of his time, as with all things, in an open insistence on his sexual identity, and his response to his trial and punishment (with hormone injections) in England was to travel over Europe. News of the Forbundet af 1948 organisation in Denmark and Norway, essentially the first open European gay movement, was a particular attraction.

It has yet to be disclosed what the security officials of Britain and the United States made of his priorities, but at a time when American commentators tended to place homosexuality on a level with communism as a danger to American interests it is not surprising that Turing was watched closely in 1953 when a young Norwegian tried to visit him. He was probably naive on such questions, for homosexuality had not been in itself a well-defined 'security' issue in 1939–1945. As Donald Michie has emphasised, there were gay men at Bletchley Park even more open than Alan Turing. The change towards explicit 'vetting' for homosexuality, and the explicit exclusion of homosexuals, came only after 1948, and possibly Turing was unaware of the position in which he had placed himself. So in 1952 he was shocked to be excluded from secret work, and in 1953 highly indignant to be the object of police surveillance. It was this 'security' development which for society generally, as individually for Turing, created a change of consciousness, politicising a hitherto 'personal' issue.

In assessing Alan Turing's place in mathematics and war, we cannot overlook the culture of mathematics itself, of which he was part. It was and is a reticent and quiet culture. The introduction to this article mentioned a double secrecy surrounding Alan Turing's life: the official secrecy surrounding cryptology, and the social taboo about his sexuality. But it is a general reality for mathematicians that the very nature of their subject attracts almost as great an enforced silence. Very recently, mathematical culture has dipped its toe in the business of fortune and celebrity, but in the long aftermath of the Second World War it had no such profile. The contrast with the prestige of physics after the open and visible atomic bomb is particularly notable. The Second World War lesson was well learnt, and the National Security Agency and its British counterpart GCHQ became major employers of mathematical graduates, so that the postwar flowering of mathematics has to some extent rested upon the demands of the most secret government work. But the near-invisibility of mathematics combines easily with the total secrecy to make this alliance one incapable of arousing public interest.

Gauss called mathematics the queen of the sciences, who sometimes condescends to serve. The priorities of self-effacing mathematical culture are well illustrated by Newman, who was faced with the very difficult task of writing a Biographical Memoir of Turing immediately after his dramatic suicide in June 1954. Turing had been elected a Fellow of the Royal Society in 1951 and as such called for such a detailed Memoir, which appeared as (Newman 1955).

Alan Turing had died at the height of his powers in 1954, wrote Newman, who interestingly did not say that his apex lay either before or in the Second World War. Because of its secrecy, Newman could say nothing of significance about the wartime work, but even allowing for this constraint, he severely understated Turing's contribution with bland expressions such as 'a mild routine' and 'congenial set of fellow-workers.' Newman also understated Turing's role in the emergence of the computer. He stated that Turing's theory of the universal machine was not known by the designers of post-war digital computers, and omitted Turing's own design (Turing 1946) from his list of Turing's works. Newman gave far more attention in his Memoir to Turing's very abstract and difficult work on ordinal logics (Turing 1939). Above all, he portrayed the Second World War not as allowing Turing to turn the logical theory into practical application, but as interrupting Turing in his work on the logic of the uncomputable and the Riemann Hypothesis, preventing him from settling into a 'serious' problem. Thus subtly, perhaps unconsciously, he put an Anti-War impetus into his assessment – discounting those subjects that the war had accelerated, emphasising those that it had interrupted. His priorities were essentially the reverse of those that modern computer science would expect.

Newman was perfectly aware of the importance of Turing's work to worldly affairs. Newman knew it at first hand, and in no way dissented from the common agreement that Turing was the leading figure in the wartime work. In 1946 Newman wanted to reject the formal decoration he was offered by the government because he considered the rank of OBE that had been awarded to Turing to be so absurd an undervaluation. But Newman had been repelled from the emergent world of the computer, and it seems that by 1955 he considered both war and computation, however important politically and economically, to be secondary, transitory, matters compared with the timeless mathematics that it was his task to assess. Perhaps he was right: the queen of the sciences sometimes sees far ahead of her subjects. It is a point of view that Turing would have respected: in his last period he probably drew as much strength as he could from the Platonic qualities of the mathematical sciences, as counter to the harsh ironies of the world.

In 1953/4 Turing wrote a last semi-popular article on undecidable problems (Turing 1954a) and studied the axioms of quantum mechanics: topics that owed nothing to his Second World War experience. It seems that he wanted to go back to what he had learnt in 1932 from von Neumann's axioms of quantum mechanics, and to think out for himself a new quantum mechanical theory, focussing on the problem of when and how wave-function 'reduction' is supposed to take place (Gandy 1954). Since the 1980s, Roger Penrose's views on logic, complex analysis and physics (Penrose 1989, 1994) have made this late development particularly intriguing. Penrose's views stand as the most radical contradiction of Turing's on the possibility of machine intelligence, but are based on the same materialist ground and focus on the very topics that perplexed Turing most: the interpretation of Gödel's theorem and the reality of a quantum-mechanical infrastructure to the brain. In particular (Penrose 1994) offers a detailed mathematical critique of Turing's 'mistakes' argument, as well as concentrating on the puzzle of the 'reduction' process. Had Turing followed his late interests further, and combined them with his knowledge of computability, he might have seen something much deeper connecting logic, space-time and quantum mechanics, than anything he did for computers. We cannot know.

The Second World War obviously stimulated progress in significant areas of mathematics and science. But maybe it stunted the growth of more subtle things that might have been. It is perhaps too soon to assess the balance. Many mathematicians might feel, though would perhaps not say openly, that the principle of the computer is a trivial matter compared with the serious problems of mathematics. Newman probably took this view in 1955; it is hard to imagine what Turing really thought, for he was no typical mathematician and had a vision that only partially overlapped with that of the classical discipline.

At the end, Alan Turing wrote in March 1954 a cryptic joke about quantum mechanics to his friend Robin Gandy, on a postcard (Turing 1954b) headed 'Messages from the Unseen World':

> The exclusion principle is laid down purely for benefit of the electrons themselves, who might be corrupted (and become dragons or demons) if allowed to associate too freely.

Free association was, of course, just what the State had forbidden to him. But this dialogue of acquiescence and dissent; of seriousness and humour; of self-assertion and withdrawal; abstraction and concreteness – these last words were typical of Alan Turing in mathematics and in war.

References

Bauer, F. L. (2000): *Decrypted Secrets* (new edition), New York: Springer 2000.

Davis, M. (2000): *The Universal Computer,* New York: Norton 2000.

Gandy, R. O. (1954): Letter to M. H. A. Newman, in King's College archive, published in volume 4 of The Collected Works of A. M. Turing, eds. R. O. Gandy and C. E. M. Yates, Amsterdam: North-Holland 2001.

Good, I. J., Michie, D., et al. (1945): General report on Tunny, p. 327. This document became available at the Public Record Office, London in 2000.

Good, I. J. (1979): Studies in the History of Probability and Statistics. XXXVII. A. M. Turing's statistical work in World War II, *Biometrika* 66 (1979), 2, pp. 393-396.

Good, I. J. (2000): Turing's anticipation of Empirical Bayes in connection with the cryptanalysis of the Naval Enigma, *J. Stats. Comp. & Simul.* 66, pp. 101-111, and in volume 4 of The Collected Works of A. M. Turing, eds. R. O. Gandy and C. E. M. Yates, Amsterdam: North-Holland 2001.

Hardy, G. H. (1940): *A Mathematician's Apology,* Cambridge: Cambridge University Press 1940.

Hodges, A. (1983): *Alan Turing: the Enigma,* London: Burnett, New York: Simon & Schuster; new editions London: Vintage 1992, New York: Walker 2000.

Hodges, A. (1997): *Turing, a natural philosopher,* London: Phoenix; also New York: Routledge 1999. Included in: *The Great Philosophers* (eds. R. Monk and F. Raphael), London: Weidenfeld and Nicolson 2000.

Hodges, A. (2002): Alan M. Turing, in: E. N. Zalta (ed.), *Stanford Encyclopedia of Philosophy,* http://plato.stanford.edu.

Mahon, A. P. (1945): *The History of Hut Eight,* Public Record Office, London 1945, HW 25/3.

Newman, M. H. A. (1955): *Alan M. Turing,* Biographical memoirs of the Royal Society 1955, p. 253.

Penrose, R. (1989): *The Emperor's New Mind,* Oxford, New York: Oxford University Press 1989.

Penrose, R. (1994): *Shadows of the Mind,* Oxford, New York: Oxford University Press 1994.

Turing, A. M. (1936): Letter to Sara Turing, in Kings College Archive; see also Hodges 1983, p. 120.

Turing, A. M. (1936-37): On computable numbers with an application to the Entscheidungsproblem, *Proc. London Maths. Soc.*, ser. 2, 42 (1936-37), pp. 230–265; also in volume 4 of The Collected Works of A. M. Turing, eds. R. O. Gandy and C. E. M. Yates, Amsterdam: North-Holland 2001.

Turing, A. M. (1939): Systems of Logic defined by Ordinals, *Proc. Lond. Math. Soc.*, ser. 2, 45 (1939), pp. 161–228; also in volume 4 of The Collected Works of A. M. Turing, eds. R. O. Gandy and C. E. M. Yates, Amsterdam: North-Holland 2001.

Turing, A. M. (1940): *Mathematical theory of ENIGMA machine*, Public Record Office, London 1940, HW25/3.

Turing, A. M. (1946): Proposed Electronic Calculator, report for National Physical Laboratory, Teddington; published in: *A. M. Turing's ACE Report of 1946 and Other Papers* (eds. B. E. Carpenter and R. W. Doran), Cambridge, Mass.: MIT Press 1986; and in Volume 3 of The Collected Works of A. M. Turing (ed. D. C. Ince), Amsterdam: North-Holland 1992.

Turing, A. M. (1947): Lecture to the London Mathematical Society, typescript in King's College, Cambridge, published in. *A. M. Turing's ACE Report of 1946 and Other Papers* (eds. B. E. Carpenter and R. W. Doran), Cambridge, Mass.: MIT Press 1986; and in Volume 3 of The Collected Works of A. M. Turing (ed. D. C. Ince), Amsterdam: North-Holland 1992.

Turing A. M. (1948): Intelligent machinery, unpublished report for National Physical Laboratory; published (ed. D. Michie) in *Machine Intelligence* 7 (1969), and in Volume 3 of The Collected Works of A. M. Turing (ed. D. C. Ince), Amsterdam: North-Holland 1992.

Turing A. M. (1950): Computing machinery and intelligence, *Mind* 49 (1950), pp. 433-460.

Turing A. M. (1954a): *Solvable and Unsolvable Problems*, Penguin Science News 31 (1954).

Turing, A. M. (1954b): Postcard, quoted in volume 4 of The Collected Works of A. M. Turing, eds. R. O. Gandy and C. E. M. Yates, Amsterdam: North-Holland 2001); reproduced in Hodges 1983, p. 513.

The Mathematician K. Ogura and the "Greater East Asia War"

TETU MAKINO*

Kinnosuke Ogura (1885–1962) was an excellent Japanese mathematician, struggling for the modernization and democratization of his home country. However, during the Sino-Japanese War, Ogura was trapped in the mobilization of the Tennoist aggression against China. This paper provides some evidence, seeks for an explanation, and draws some tentative conclusions.

1 Who Was K. Ogura?

Kinnosuke Ogura was born in 1885. He graduated from Tokyo College of Science (antecedent of Tokyo Science University) in 1905 where he lectured from 1910 to 1911. From 1911 to 1917, he was Assistant at the Department of Mathematics of the new Tohoku Imperial University. There in 1916 he received his Ph. D. with a thesis on *Trajectories in the conservative field of force*. From 1917 to 1937, he was employed as a Researcher at the Siomi Institute of Physical and Chemical Research (newly set up by Dr. Hantarō Nagaoka, antecedent of the Faculty of Science, Osaka Imperial University). During that period, he continued his research in France for two years, 1919–1922. In 1940, he was appointed Chief Director of Tokyo College of Science. In 1946, he was elected President of the Association of Democratic Scientists of Japan. He died in 1962. His Collected Works appeared in eight volumes in 1974 (see [Ogura 1–8, 1973–1975]).

2 Ogura's Struggle for the 'Scientific Spirit'

Ogura started his academic career studying applications of differential geometry to mathematical physics. He was the first mathematician to receive his Ph. D. degree as a graduate of a private college. In prewar Japan the gap between imperial universities and private colleges was so big that it caused journalistic gossip.

In Paris he studied European traditions of science and culture with some admiration. He attended the Congrès International des Mathématiciens 1920 at Stras-

* Faculty of Engineering, Yamaguchi University Ube 7558611, Japan.
Email: makino@yamaguchi-u.ac.jp

bourg, Alsace, read a paper, and talked with Hadamard, Fréchet, Eisenhart, D. E. Smith and others. At Collège de France he attended the seminar of Hadamard and the lectures by Langevin on the relativistic theory.

While in France he became convinced that Japan needed to popularize mathematics to working people and to raise the general public scientific level. Back in Japan, he gave lectures in many cities during 1922–1923. He declared in the Preface of his book on nomography [Ogura 1923, p. 9] that *Mathematics is a treasure not to be monopolized by mathematicians*. Also he published a book, *Fundamental Problems of Mathematics Education* [Ogura 1924], in which he emphasized the importance of spreading the scientific spirit. Ogura was the first Japanese to use the word "scientific spirit" as a translation of "ésprit scientifique" in an essay by Paul Appel.

His Japanese translation of *A History of Elementary Mathematics* by Cajori [Cajori 1917] triggered his own study of the history of mathematics. Under the influence of the book *Arts in Class Societies* by Russian marxist Plekhanov, Ogura wrote the articles *Arithmetic in class societies 1, 2* [Ogura 1929a], [Ogura 1929b] and *Mathematics in class societies* [Ogura 1930], which were published in the prestigious magazine, *Sisō*, in 1929, 1930.

These articles are pioneer works on the history of western mathematics from the view point of social, cultural, and economic history ("historical materialism"). They had a big influence on the Japanese world of thought.

Later he began studying Japanese and Chinese traditional mathematics. His article *Social nature of Chinese mathematics – Social situations in the Qin-Han era viewed from the 'Jiuzhang Suanshu'* [Ogura 1934], was highly appreciated by Chinese scholars: See the Chinese translation with comments by Yue Guang and the words of the editor Zhang Shenfu, Dagong-bao (Tianjin), July 12 and July 26, 1934. The editor Zhang Shenfu (Qinghua University) wrote:

> Ogura's work has opened a new phase of study of Chinese mathematics. Probably he is the first to observe social natures in Chinese mathematics. Moreover his work clarifies the closed relation between various sciences: mathematics and political economy. Studies in the same approach are expected to be done successively in China.

Ogura corresponded in a collaboration to exchange materials of old Japanese and Chinese mathematics with a Chinese scholar Li Yan (1892–1963), who is a pioneer of the modern study of the history of Chinese mathematics. (See the Preface of Li Yan, *History of Chinese Mathematics* (Zhong guo Suanxueshi), Shangwuyinshuguan, 1937, Shanghai.)

A recent review of the historical development of the history of mathematics in Japan (see [Sasaki 2002, p.292]) gives the following characterization:

> Ogura produced a number of influential books and papers, in particular on the history of mathematics and on mathematics education. Although not a strict Marxist historian, Ogura was certainly influenced by Marxist historiography in emphasizing the social aspects of the history of mathematics.

In prolongation of his criticism of the elitarian policy on mathematics education by the imperial Japanese government, he began to make efforts to guard against the upcoming racism and fascism in Europe and Japan. In order to criticize the Nazi philosophy of mathematics by Bieberbach [Bieberbach 1934], he wrote the article *Mathematics and racial characteristics* [Ogura 1935], published in *Tyūō Kōron*.

Ogura points out that Bieberbach's classification of mathematicians into Type S and Type J is very crude. Bieberbach assumes that the spirit is alienated from the real in mathematicians of Type S while the intuition and thinking are in harmonic unification for mathematicians of Type J. Bieberbach insists that French and Jewish are of Type S and German are of Type J. In such a way Bieberbach insists on the expulsion of French and Jewish mathematicians to purify German racial mathematics. K. Ogura acknowledged the existence of these two different types of mathematical mind, but criticized:

> Bieberbachs's racial characteristic concept is suprahistorical, rigidly psychologic. But the "racial" characteristics depend upon total history. The existence of mathematicians of Type S, so-called by Bieberbach, is not only French or Jewish. It is a common and universal fact in highly developed capitalist countries. It is a transparent deception that there do not exist representative mathematicians of German race of Type S. Now German capitalism is faced with a crisis and the despotism of the Nazis has appeared as a political form adopted by the bourgeoisie. On the other hand French and Jewish people are expelled, the working class is ordered to subordinate themselves, liberalism is suffocated. Bieberbach's troops stood up as ideologues of such a theory to protect the profits of the bourgeoisie. Especially their theory (?) serves not only for the expulsion of Jewish mathematicians but also for the rapid development and enlargement of industrial technology and military sciences, which is desired by the bourgeoisie. It is natural for them to better appreciate mathematics of Type J from the practical view points than mathematics of Type S, which is a formalism away from the real. Thus Naziism suffocates liberalism, and puts restrictions on the study of mathematics of Type S, which has been developed freely, by expelling French and Jewish mathematicians.

Thus Ogura warned readers about the possibility of appearance of "Bieber bachs of Japanese type".

This shows that Ogura opposed Nazi fascism and racism to protect the free development of mathematics at this time. In fact in the article *Task of scientists* [Ogura 1936], Ogura wrote:

> When the storm of fascism began to rage, European conscientious scientists gained selfawareness. Look! The science policy of Nazi Germany, which is the mother country of Jacobi who said "The purpose of science is the honour of the human mind", is now extolling the German spirit instead of the internationality of science, expelling many scientists, and strongly tinting science education with military colour.
>
> In Italy, natural science as a curriculum is illtreated in the name of respect for the classical spirit, with the curriculum of science being annihilated, and mathematics is taught in the spirit of classical education.

> This policy is to reduce people to stupidity and ignorance, isn't it? British and French conscientious scientists, who have a good long tradition of liberalism based on their deep scientific culture, are enemies of fascism instinctively. Now they have gained selfawareness. In France a group of scientists including colleagues of Poincaré, Hadamard, Langevin, Perrin and so on is fighting against this anticulturalism.
>
> The storm of fascism is not limited to foreign countries: in Japan anticulturalism is imminent, following a special Japanese path. In this situation scientists should devote themselves to the scientific spirit in their actions. The scientific spirit is the spirit which learns scientific heritages with due humility, examines them continually, discovers new and more detailed facts and creates more complete theories. It is quite the opposite of prejudice.
>
> Therefore the mind of a scientist should be kept in the condition of freedom. We scientists have freedom instinctively. We scientists have the courage to pursue the truth and to tell the truth. We scientists are essentially radicalists in this sense.

But Japan was entirely absorbed in the war against the Chinese people. In his *Reminiscences of a mathematician* [Ogura 1949], he wrote:

> Having arrived in Paris in 1920, I visited Professor Emile Borel, to whom I had been sending my papers and who cited my papers. Professor Borel told me that he stayed in Japan three days on the way to China. I asked his impression of Japan. His brief answer was "Japan is too militaristic". It was just after World War I and the Japanese people were enjoying " Taisyō Democracy". I had not imagined that a mathematician such as Professor Borel would point out the militarism of Japan in such a period of peace. I was so surprised at his keen insight.

3 What Was the "Greater East Asia War"?

"Greater East Asia War" is the name given by the Great Japan Empire Government in 1941 to the war which commenced in 1937 as the Sino-Japanese War (Marco Polo Bridge Incident). The war ended in 1945 just after atomic bombs were dropped on Hiroshima and Nagasaki. The prewar Japanese government called the war by this name in order to justify it as a war to ensure Japan's survival and selfdefence, to liberate Asia from Europe and the US, and to establish the "Greater East Asia Co-Prosperity Sphere".

But in fact this war cannot be justified. It was an imperialist war to invade Asian countries by mobilizing all Japanese people. As a result of the imperialist war and the brutality of Tennoist power, the people were forced to suffer great hardships with many people losing their lives and the country being devastated. The war of aggression caused the death of more than 20 million people in Asian countries. For example more than 200,000 Chinese civilians and war prisoners were killed by Japanese forces in Nanjing (Nanjing Massacre, December 13, 1937 – early in

January, 1938). Also in Singapore thousands of overseas Chinese were killed at the start of the Japanese occupation in 1942. Japan was the aggressor that used every means available to put Asian people under brutal oppression.

4 Ogura and the War

When the SinoJapanese War 1937–1945 began, 441 liberalists, leftists, anti-fascists in the academic and cultural area were arrested (Popular Front Incident, December 15, 1937 and February 1, 1938). The National Mobilization Law was set up in 1938 to mobilize all Japanese to the war. In 1940 Ogura engaged with the Imperial Rule Assistance Association ("Taisei Yokusankai"), which is an institutional embodiment of the new order intended to be a national organization for integrating and mobilizing people into a warlike state. This was a serious political mistake for him, but to us, it seems, his selfcriticism of this fact after the war (*Reminiscences* [Ogura 1949]) is insuffcient. Here is what really happened:

After some time, Ogura published an article *Rôle of scientists under the present emergent situation* [Ogura 1941]. He wrote:

> Now our Japan is faced with a very serious situation. The whole world is in a storm of historical changes. Here it is natural to advocate loudly the promotion of science through both the offcial and the private, in a time when we are pushing ahead with the realization of a high-degree state defence system.
>
> Stop discussions for the sake of discussion. The State needs the establishment of the following science policy immediately: Control strongly the study of science and technology for the purpose of the State! The special talents and knowledge of Japanese scientists should be sacrificed for the State, just like the soldiers sacrifice their blood for the State!
>
> Look at the emergent international situations. In the US, which is not yet at war, American mathematicians have organized a "War Preparedness Committee" in the last summer to start study of mathematics directly necessary for the war. Japan, however, which set a goal to establish a high-degree defence state and a self-suffcient economy of East Asia, has not yet established a war-time system of science. But five years have already passed since the beginning of the war. There are scientists who are not aware of this fact. It is so strange. In order to establish a high-degree defence state we must promote science and technology. For that there is no choice but to control science and technology.
>
> All wise scientists understand it. Under rational control by the State, scientists will do their best desperately, making up their mind to accomplish the highest task imposed by the State, with real delight. All scientists should pursue the highest task of the State and concentrate themselves to this.

The author cannot make the quotation without sorrow. An anti-fascist, dedicated democrat mathematician Ogura, even he could not resist and was trapped into the crazy spiritual mobilization of the Tennoist warlike power. How strong the mind control was!

After the war Ogura wrote in an essay *I am ashamed to be a scientist* [Ogura 1953]:

> I cannot bear talking about science and science education during the time when military fascism was raging. No. I should say briefly: Almost all of us surrendered to the pressure of the power; Of course, I was one of them.

5 Double-Edged Sword

In 1943 Ogura elaborated his bow for the Imperial War when he gave a lecture at a meeting of The National Science Association. He talked about École Polytechnique founded in the time of the French Revolution in Paris, which intended to cultivate good engineers for the increase of productive power and excellent offcers of artillery. He emphasized that the guiding principle of École Polytechnique is the synthesis of theory and application and the practice. His analysis is long but well done to explain the revolutionary character of the early École Polytechnique. This lecture was published as an article *School of science and technology in wartime – on early École Polytechnique* in the book *Mathematics in Wartime*, 1944, during the war, [Ogura 1944], see also Figure 1.

The beginning of the preface of this article is:

> Today our country needs a rapid advance of science and technology to win the Greater East Asia War. Today's war is a war of science and technology. Therefore the unification of science and technology under the political power and the innovation of science education, these important problems, are being discussed in every military, offcial and popular sector, and have been realized. But in the US, according to newspapers, a very clear and radical policy to make a thorough education of science and technology based on mathematics and physics is put in practice. Compared with the US our country is delayed much. I regret it.

This shows that Ogura's study on the history and philosophy of mathematics was devoted to the glory of the "Greater East Asia War". Ogura stands on the side of the militarist power.

It is interesting that Ogura published the same article after the war in 1948, by making some revisions, under the title *School of science and technology in the time of revolution – on early École Polytechnique* [Ogura 1948]. The above quoted phrases were replaced by

> Today our country needs a reconstruction of science to perform a Democratic Revolution and to construct a peaceful cultural nation. Therefore the innovation of the institutions of science and the innovation of science education are being discussed in all sectors.

The major portion of the article was not changed but he falsified the words "for the Greater East Asia War" to "for the Democratic Revolution". The same study of the history and philosophy of mathematics serves both war and peace! What does it mean?

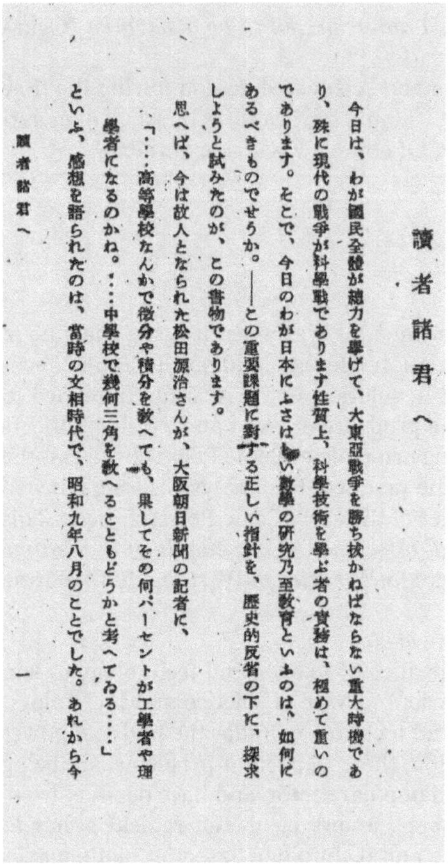

讀者諸君へ

今日は、わが國民全體が總力を擧げて、大東亞戰爭を勝ち拔かねばならない重大時機であり、殊に現代の戰爭が科學戰であります性質上、科學技術を學ぶ者の責務は、極めて重いのであります。そこで、今日のわが日本にふさはしい數學の研究乃至敎育といふのは、如何にあるべきものでせうか。――この重要課題に對する正しい指針を、歷史的反省の下に、探求しようと試みたのが、この書物であります。

思へば、今は故人となられた松田源治さんが、大阪朝日新聞の記者に、

「…高等學校なんかで微分や積分を敎へても、果してその何パーセントが工學者や理學者になるのかね。…中學校で幾何三角を敎へることもどうかと考へてゐる…」

といふ、感想を語られたのは、當時の文相時代で、昭和九年八月のことでした。あれから今

讀者諸君へ 一

Translation:
To the Readers
Today is a serious time to fight the Greater East Asia War with all the might of our nation. Especially contemporary war is a war of science, and the duty of those who study science and technology is very heavy. How should research and education in mathematics fit today's serious task for Japan? – I tried in this book to find a correct guiding principle for this serious task by reviewing history.
When Mr. G. Matuda said to a reporter of the OsakaAsahi (newspaper): "They teach calculus in high schools. But what percent of them will become scholars of technology or science? ... I doubt whether it is necessary to teach geometry and trigonometry in junior high schools"; he was the Minister of Education. It was in August 1933, at that time [next page] *less than ten years earlier. Today everyone recognizes that mathematics ia a very sharp-edged tool. If you lost the Japanese Spirit, then you could not construct Japanese mathematics. Therefore if we did not highly develop mathematics as a weapon, we would not know how to apologize to brave soldiers in the front line...*

Figure 1. *Left*: the first page of Ogura's 1944 book on *Mathematics in Wartime*; *right*: English translation. After 1945 lending of extant copies was banned and the book was excluded from Ogura's *Collected Works*.

Y. Kondō says in the commentary to [Ogura 1–8, 1973–1975, Vol. 1]:

> The fact that the lecture given during the war can be accepted after the war without major change shows the consistency of Ogura's thought. This is a rare case in Japan in that time when apostasies and disappointments were in vogue.

The author cannot agree with this affrmative estimation. Ogura's is not consistency but unprincipledness. T. Hirosige writes [Hirosige 1965, p. 123]:

> Here I do not intend to blame Ogura for unprincipledness. The problem I want to propose is that the contention described by Ogura with emphasis in that article can claim its raison d'être both as that which, during the war, gives a guiding policy for science and technology to win the war, and as that which, after the war, gives a guiding principle for science and technology to construct

> a peaceful nation and Democratic Revolution. This was done only by changing the adjective phrases.
>
> Therefore we must recognize that the general canons of the progress of science and the innovation of science education can be applied to any epoch and to any society if the epoch or the society needs science. In fact it is not the case that the development of science until now has been only combined with the ideals of peace and democracy.

T. Hirosige asserts that ignorance of the historical character of science leads to science-supremacism: if progress in science can be achieved, any thing, even military research, can be done.

The author thinks that this is a very important point in reconsidering the social responsibility of mathematicians. When a mathematician is doing his study motivated only by pure curiosity, can he claim that he is innocent of its social function? The author does not think so. Is this position too "ethical"?

6 "Consistency" of Ogura

Why did Ogura write such an article as *Rôle of Scientists under the present emergent situation*, [Ogura 1941], which claims the necessity of control of science and technology by the military state? That he charged himself with that responsibility as Chief Director of the Tokyo College of Science may be one of the reasons: *It is not the consciousness of men that determines their existence, but, on the contrary, their social existence determines their consciousness*, as K. Marx wrote 1845 in *Thesen über Feuerbach*. Perhaps there are other reasons.

Those who adore Ogura say that their teacher was consistent in his thought through his whole life: he was fighting for Japan's modernization in the field of mathematics and mathematics education. In fact in his article *Rôle of scientists...* [Ogura 1941], he wrote (to his fellow scientists):

> You are extremely afraid of the control of scientific study. But in fact there has been more terrible and worse control in your narrow field of science than control by bureaucrats. It is the "territories" in the scientific world, isn't it? Stupid guys are building their nests in universities, and remain safe until retirement age without any important work. They do so by dint of the "territories". You are afraid of the innovative control of scientific study which intends to get rid of this feudalistic guildship. Do you advance "the freedom of study" today? It might be a word for self-justification or defence of territories. Do you pass with territorialism under such an emergent situation of the State? "The freedom of study" or "the science for science", these sublime slogans, cannot justify your vulgar behaviour.

Here he opposes the feudal and closed character of the academic world in prewar Japan and insists on the necessity of modernization. One can say that in this sense he is consistent. But it should be pointed out that the innovation of science and technology should be done not by bureaucratic control of a warlike power but by democratic criticism by people in peace.

To this, T. Hirosige [Hirosige 1973, p.169] wrote:

> Unexpectedly K. Ogura wrote the article "Rôle of scientists...", in which he insists: "Control strongly the study of science and technology for the purpose of the State." But this does not mean his wholehearted conversion. Ogura says that scientists who are opposed to control of science are "gang leaders content with things as they are" and bigoted to the "feudalistic guildship", who want to protect their territories. Ogura censures them because they are afraid that the "innovative control of scientific study" will get rid of their feudalistic privileges. In other words Ogura is repeating here his longcherished advocacy of anti-feudalistic modernization.

Maybe this is true. But the author thinks that Ogura's strategy to achieve anti-feudalistic modernization of science and technology by dint of control by the brutal warlike Tennoist power was wrong. This shows the half-bakedness of Ogura's democratic thought.

Of course, it is easy to criticize Ogura's thought for us after the war, but his attitude was taken under serious situations. After the war Ogura wrote in *Reminiscences*, [Ogura 1949]:

> I was neither a fascist nor a governmentpatronized scholar, but I was neither a so-called defeatist. At that time, when the international situation was already progressing for, we should make preparations for war. What will happen if war commences without any preparation? I thought we should do the best rather than to keep silence. Thus I wrote that article, which should be totally criticized from today's view point.

To keep silence or to cooperate with the war of aggression? K. Ogura took the latter choice. But there were numbered scholars or literary persons who kept silence during the war. For example the famous writer Kahū Nagai (1879–1959), whose works Ogura loved in his youth, was one of them. Nagai was addicted to the nostalgia of good old Tokyo and writing his secret war-weary diary [Nagai 2001]. Conservative Nagai kept silence while progressive Ogura wrote a bellicose appeal. This is an irony of history. Ogura's choice of attitude was not an inevitable one. From the above passage of his *Reminiscences* [Ogura 1949] a tone of his self-justification can be read.

In fact it is very diffcult for pure mathematicians to choose a politically correct attitude when the situation is drastically changing. This is not irrelevant to contemporary mathematicians. How many mathematicians in the world are making efforts to stop the global, arrogant and "unilateral" military operations of US imperialism today? Ogura's case is a good lesson learned from Mathematics and Japanese history.

References

Bieberbach, L. (1934): Persönlichkeitsstruktur und mathematisches Schaffen, *Unterrichtsblätter Math. Naturw.* 40 (1934), pp. 236–243.

Cajori, F. (1917): *A History of Elementary Mathematics with hints on methods of teaching.* New York: Macmillan 1917. (Enlarged Japanese translation: K. Ogura and Y. Ide, *Syotō Sūgaku-si*. Tokyo: Sankaidō 1928).

Hirosige, T. (1965): *Science and History* (Kagaku to Rekisi). Tokyo: Misuzu Syobō 1965.

Hirosige, T. (1973): *Social History of Science - Institution of Science in Modern Japan* (Kagaku no Syakaisi - Kindai Nihon no Kagaku Taisei). Tokyo: Tyūō Kōron-sya 1973.

Nagai, K. (2001): *Diary at the Broken Heart Cottage* (Dantyō-tei Nitizyō), in 7 volumes. Tokyo: Iwanami Syoten 2001–2002. (Originally written from September 16, 1917 until his death).

Ogura, K. (1–8, 1973–1975): *Collected Works* (Ogura Kinnosuke Tyosakusyū), in 8 volumes. Tokyo: Keisō Syobō 1973–1975.

Ogura, K. (1923): *Graphical calculations and nomograms* (Zu-keisan oyobi Zuhyou). Tokyo: Sankaidō 1923.

Ogura, K. (1924): *Sūgaku Kyōiku no Konpon Mondai*, Idea Syoin 1924; [Ogura 1–8, 1973–1975, Vol. 3].

Ogura, K. (1929a): Kaikyū syakai no sanzyutu sono 1, *Sisō*, August 1929; [Ogura 1–8, 1973–1975, Vol. 1, pp. 58–76].

Ogura, K. (1929b): Kaikyū syakai no sanzyutu sono 2, *Sisō*, December 1929; [Ogura 1–8, 1973–1975, Vol. 1, pp. 77–111].

Ogura, K. (1930): Kaikyū syakai no sūgaku, *Sisō*, March, May, June 1930; [Ogura 1–8, 1973–1975, Vol. 1, pp. 112–182] .

Ogura, K. (1934): Sina sūgaku no syakaisei - Kyūsyō Sanzyutu o tūzite mitaru Sin-Kan zidai no syakai-zyōtai, *Kaizō*, January 1934; [Ogura 1–8, 1973–1975, Vol. 3, pp. 166–184].

Ogura, K. (1935): Sūgaku to minzokusei - Natisu sūgakuron no hihan, *Tyūō Kōron*, November 1935; [Ogura 1–8, 1973–1975, Vol. 1, pp. 224–244].

Ogura, K. (1936): Sizen-kagakusya no ninmu, *Tyūō Kōron*, December 1936; [Ogura 1–8, 1973–1975, Vol. 7, pp. 3–18].

Ogura, K. (1941): Genzikyokuka ni okeru kagakusya no sekimu, *Tyūō Kōron*, April 1941; [Ogura 1–8, 1973–1975, Vol. 7, pp. 32–47].

Ogura, K. (1944): Senzika ni okeru kagaku gizyutu gakkō - Syoki no Ekōru Poritekuniku, pp. 3–124 in: K. Ogura, *Senzika no Sūgaku*. Tokyo: Sōgen-sya 1944.

Ogura, K. (1948): Kakumei zidai ni okeru kagaku gizyutu gakkō - Syoki no Ekōru Poritekuniku ni tuite, in: K. Ogura, *Sūgaku-si Kenkyu*, Vol. 2, Iwanami Syoten, 1948, Tokyo; [Ogura 1–8, 1973–1975, Vol. 1, pp. 183–233].

Ogura, K. (1949): Iti sūgakusya no kaisō, *Kawade Hyōron*, October, November, December, 1949; [Ogura 1–8, 1973–1975, Vol. 7, pp. 151–344].

Ogura, K. (1953): Ware kagakusya taru o hazu, *Kaizō*, January 1953; [Ogura 1–8, 1973–1975, Vol. 7, pp. 124–144].

Sasaki, C. (2002): Japan, pp. 289–295, in: J. W. Dauben, Ch. J. Scriba (eds), Writing the History of Mathematics: Its Historical Development, Basel: Birkhäuser Verlag 2002.

Working within the System

WOLF GÖHRING*

This article is dedicated to my mother.

1 Introduction

"There is no homeostasis whatever. We are involved in the business cycles of boom and failure, in the successions of dictatorship and revolution, in the wars which everyone loses, which are so real a feature of modern times," writes Norbert Wiener in *Cybernetics* (The MIT press, Cambridge 1961, p. 159).

But war is not only "a feature of modern times", it also involves people who act and thus define "modern times". I will describe my involvement in warlike matters, as a child, as a student, as a scientist, as a trade unionist, as a member of a works council, and as a participant in an anti-war movement. Finally I ask: *Why* is war so real a feature of current times? And, could there be means for a modernizing of modern times such that war would be no longer so real a feature? I think that one has to study political economy and the reasons for economic non-cooperation in order to find an answer on how to defeat war.

2 Childhood and Youth

I remember a scene in summer 1942 when I was two and a half years old: I was sitting in the garden and eating rhubarb compote. Presumably this was before the first bombing night in these days which I also remember. From then on, the war was a mysterious fate that accompanied me in the following three years: The "others" fighting "us".

It took years to understand what really happened at that time. Last year, I took a look at my Latin schoolbook. The first phrase concerning Romans was: "*Romani bellum amabant.*" – The Romans loved war. All the following sentences in this short lesson concerned war (*Ludus Latinus, Lese- und Übungsbuch I.* 4th ed., Ernst Klett Verlag, Stuttgart 1950, p. 7). He who teached this stuff to boys of eleven, twelve years had been a officer in the nazi army. He understood nothing of what had happened. "*Vae victis*" he often commented on the situation six or seven years after this war.

* Fraunhofer Institute for Autonomous Intelligent Systems, D 53754 Sankt Augustin.
Email: wolf.goehring@bi.fhg.de

I would like to thank my colleagues Bernd Müller for translating the German quotations, and Stéphane Beauregard for totally reworking my awkward English. I would like to thank all my unnamed colleagues who also stood against military research.

Figure 1.
A feature of modern times. Why?

I lived in Saarland, which since 1948 had an autonomous government and an economic and customs union with France. There, people were very interested in the French colonial wars in Vietnam and in Algeria. A Frenchman who taught us a special French course became a soldier and was killed in Algeria. My father fell in the German attack on Moscow. He was a member of the "Propaganda Kompanie" and a photographer. A journal[1] published *His Last PK-Film* showing an exploding Soviet tank (see Figure 1) in the "midst of a battle". My stepfather was able to leave Stalingrad just before it was enclosed by the Soviet army. He often spoke of situations where he had luck, and he told of the German murderings he had seen in the Ukraine. Sometimes, on Sunday, I visited the cinema where the newsreels showed artillery duels from the Korean war. The war stories of the adults were mixed with radio news of the British-French bombing of Cairo when the Suez channel was nationalized. In those years, US Marines also invaded Lebanon. During the Hungarian revolt, I heard a man say triumphly: "Within one year I'll be back at Königsberg!" (Kaliningrad). A class excursion took us to the battlefield of Verdun in France where during World War I some hundred thousand people were killed. We had a look at the bones of ten thousands of soldiers assembled in the "bone house". Our French teacher accompanied us. When he was born his home town, Metz, was part of the German "Kaiserreich". It is now in France. I can't count the wars that assigned this region to France or else to Germany.

When writing essays in class, I preferred political themes. I wrote down without critique what I had read in the newspaper some days before. An example: Western Germany should have the Bomb because of its responsibility to defend itself and not to burden others with this task. During a sunny spring morning in 1957, our teacher of physics entered the classroom and, looking straight at us, he opened the

[1] I possess one page of this journal without title and date. Pictures are commented on in German and Spanish.

lesson: "Is it a big thing or not?" "It" was the "Göttinger Erklärung" (Goettingen Declaration) against the German Bomb which had been published a few days before. Our teacher knew some of those who had signed. They were his teachers at Göttingen University where he finished his studies in Physics and Mathematics a few years previously. Before he could start his studies after the war, he took part in submarine warfare including a long enclosure in his boat on the floor of the sea while being attacked by a submarine hunter. (This was dramatized in the film "Das Boot".)

In this lesson in April 1957 I learned that it was possible to stand against war, that scientists are responsible for what they are doing, that war should not be so real a feature of our times. This responsibility had as a result to stand *against* the Bomb. This contrasted to the responsibility reported by the newspaper which had as a result to stand *for* the Bomb.

3 Entering Computer Science

Many such episodes were mixed in my head when, shortly after my prediploma in Mathematics and in Physics, the University's Institute for Applied Mathematics offered me a job as a technical assistant. It was a job in computer science, in the field of artificial intelligence, pattern recognition, radar signal extraction (military and civilian), and cybernetics – all at the end of 1961. In this job, I wrote an excerpt on "prediction theory" (April 1962). I remarked that during the second world war Wiener developed a mathematical method yielding a high hit rate when firing on attacking planes. In the next lines, I mentioned weather forecast and the controlling of a moon rocket as similar tasks. I had no military ambitions or realistic military ideas in this context, and in the sequel I looked to understand the underlying mathematics.

The "logical" counterpart of Wiener's airplane hitting project were well developed airplanes based on mathematical computations. In *Cybernetics* (p. 5) Wiener writes:

> At the beginning of the war, the German prestige in aviation and the defending position of England turned the attention of many scientists to the improvement of anti-aircraft artillery. Even before the war, it had become clear that the speed of the airplane had rendered obsolete all classical methods of the direction of fire, and that it was necessary to build into the control apparatus all the computations necessary. (...) Accordingly, it is exceedingly important to shoot the missile, not at the target, but in such a way that missile and target may come together in space at some time in the future. We must find some method of predicting the future position of the plane.

Prediction theory and short tellings of the institute's director on computations of airplane wings showed to me in the early 1960s that mathematics could be used, like screws, for military as well as for civilian purposes. Only a few changes in methods and formulas would be needed. There were no frontiers.

4 My First Job

The assistant professor who supervised my diploma thesis (Vergleich von Lernprozessen. Universität des Saarlandes, January 1965) became director of an institute of computer science. He offered me a job in the field of learning and pattern recognition. When I arrived at my new job I joined a small team with two other colleagues. Within a year, we would get a PDP-7 with a display and a lightpen. We were to implement a time-sharing system and attach some 5-channel teleprinters disposed by the German telecom. In using the display and the lightpen we were to study man-machine dialog and pattern recognition – all in 36 kB of RAM. The project was financed with more than half a million Deutsch Marks from the German Federal Department of Defense.

By pattern recognition I understood the recognition of characters. The German DoD meant recognition of sonar echos: submarines, torpedos, even shoals of fish. As an alternative, they meant moving symbols around the screen and recognizing "good" or "bad" situations of tanks in a battlefield, "controlled" and "optimized" by an operator in front of a display in a command and control center. What does that say about my responsibility for what I was doing as a novice in science?

During the following months, we tried to figure out what to do with the PDP-7. At this time, the magazine *Der Spiegel* published an advertisement for ROLLS ROYCE showing tanks and another one showing BOING's "Chinook 2" over an extended jungle with a headline: "It passed its test in Vietnam!" The advertisement said: "This helicopter is suited for all your logistic tasks! Wherever you want to go with this helicopter it goes easily. You can transport two howitzers with their staffs." I wrote a reader's letter (May 5, 1966) to *Der Spiegel*. I asked what *Der Spiegel* wanted to suggest: Should I use the helicopter to solve logistic problems when trying to illegally transport people from the German Democratic Republic (GDR) to Western Germany or in trying to rob a bank? In a few lines, they answered:

> Quite surely, there are many reasons to refrain from accepting the ROLLS ROYCE or BOEING advertisements. But we think the right of freedom of speech which made us print the 'Call against the war in Vietnam' also includes the right of the companies ROLLS ROYCE and BOEING to offer their products using arguments which they hold to be adequate. We cannot see a violation of common law or common ethical norms when these companies refer, among other things, to the operation of their helicopters in Vietnam or in a case of conflict quite in the line of the political point-of-view of their governments.[2]
> [Letter from *Der Spiegel*, June 1, 1966, their sign: A/Wi/La. See figure 2].

[2] "Sicher: es gibt viel, was gegen die Aufnahme der ROLLS ROYCE- oder BOEING-Anzeigen sprechen könnte. Aber das Recht auf Meinungsfreiheit, das uns den 'Aufruf gegen den Krieg in Vietnam' drucken ließ, scheint uns auch das Recht der Firmen ROLLS ROYCE und BOEING einzuschließen, mit den Argumenten, die sie für gut halten, ihre Erzeugnisse anzubieten. Wenn sie dabei im Sinne der von ihren Regierungen verfolgten Politik unter anderem auf den Einsatz ihrer Hubschrauber in Vietnam oder im Konfliktfall hinweisen, erkennen wir darin keinen Verstoß gegen herrschendes Recht oder geltende Sittennormen." [Translations by Bernd Müller]

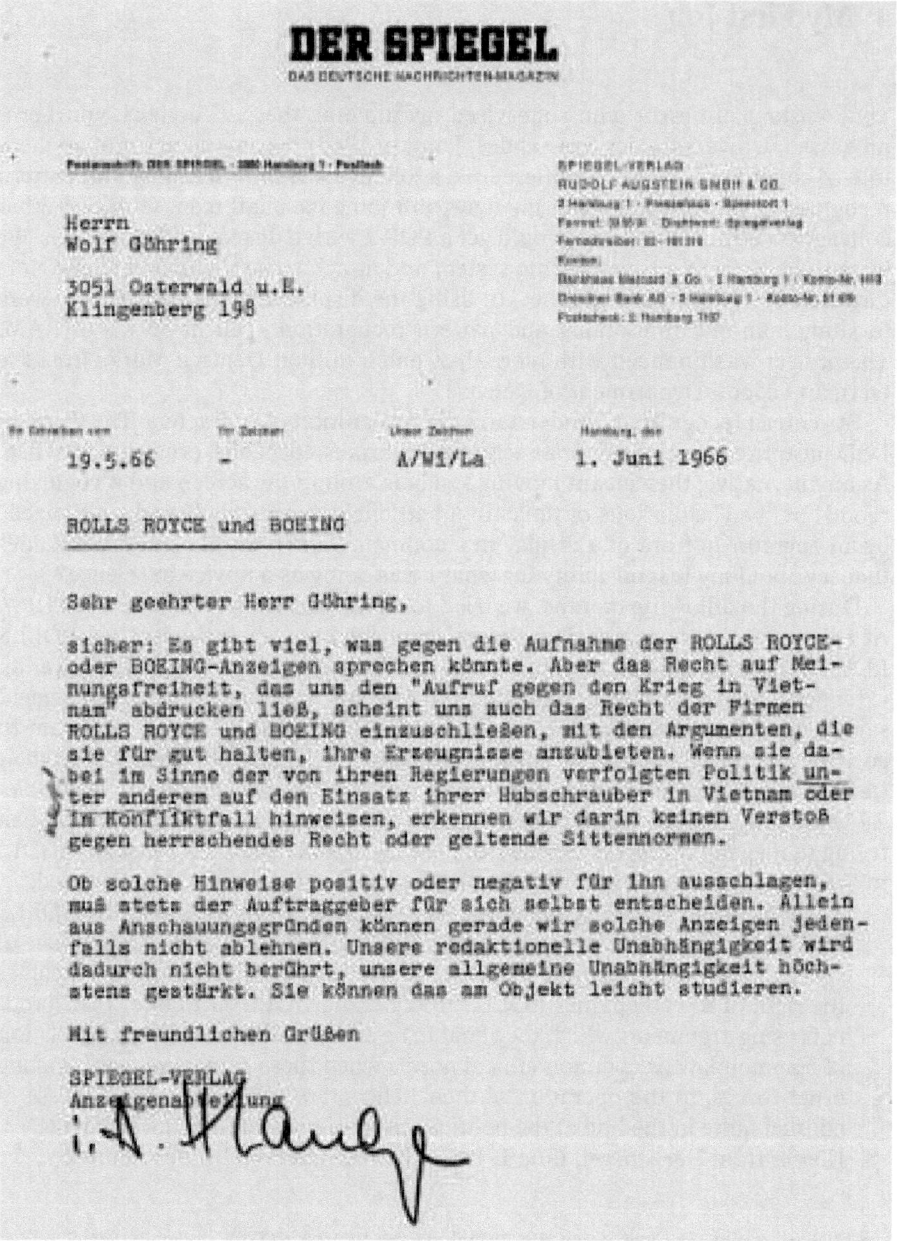

DER SPIEGEL

DAS DEUTSCHE NACHRICHTEN-MAGAZIN

SPIEGEL-VERLAG
RUDOLF AUGSTEIN GMBH & CO.

Herrn
Wolf Göhring

3051 Osterwald u.E.
Klingenberg 198

19.5.66 - A/W1/La 1. Juni 1966

ROLLS ROYCE und BOEING

Sehr geehrter Herr Göhring,

sicher: Es gibt viel, was gegen die Aufnahme der ROLLS ROYCE- oder BOEING-Anzeigen sprechen könnte. Aber das Recht auf Meinungsfreiheit, das uns den "Aufruf gegen den Krieg in Vietnam" abdrucken ließ, scheint uns auch das Recht der Firmen ROLLS ROYCE und BOEING einzuschließen, mit den Argumenten, die sie für gut halten, ihre Erzeugnisse anzubieten. Wenn sie dabei im Sinne der von ihren Regierungen verfolgten Politik unter anderem auf den Einsatz ihrer Hubschrauber in Vietnam oder im Konfliktfall hinweisen, erkennen wir darin keinen Verstoß gegen herrschendes Recht oder geltende Sittennormen.

Ob solche Hinweise positiv oder negativ für ihn ausschlagen, muß stets der Auftraggeber für sich selbst entscheiden. Allein aus Anschauungsgründen können gerade wir solche Anzeigen jedenfalls nicht ablehnen. Unsere redaktionelle Unabhängigkeit wird dadurch nicht berührt, unsere allgemeine Unabhängigkeit höchstens gestärkt. Sie können das am Objekt leicht studieren.

Mit freundlichen Grüßen

SPIEGEL-VERLAG
Anzeigenabteilung
i. A. Hange

Figure 2. No violation of ethic norms.

"Quite in the line of the political point-of-view of their governments" meant being in line with the faking of the Tonking incidence. I never more bought *Der Spiegel* though sometimes I read it.

My colleagues and I had no idea what to do with rectangles on the screen which should symbolize tanks. We had ideas of what to do if such symbols concerned flowchart elements or elements of electrical circuits. Thus I looked to implement some of *these* ideas (Input of geometric structures via CRT display and light-pen; experiences with program "FLUSS". In: W. Händler, J. Weizenbaum (eds.): *Display use for man-machine dialog.* Carl Hanser Verlag, München 1972, pp. 245–246). Submarine detection by sonar signals? This meant lengthy formal computations using the whole apparatus of Fourier transforms, correlations, and integration formulae. Some months after I started with my job and half a year before I wrote the letter to *Der Spiegel*, I had a meeting with my boss, a delegate from the Department of Defense and another from a firm that built electronic devices for the German Navy. I was the youngest in this meeting on technical and organisational aspects of a possible project on submarine detection by sonar signals. We talked in this meeting as if it was an everyday subject. Eventually I would have to work on this theme, in which case I would have to undergo a security check. I don't know if the check occurred. Later on, a colleague did this job instead of me.

5 In Touch with Command and Control Systems

In October 1966, I attended a NATO summer school on "Display Use for Man-Machine Interaction" in Paris. Participants came from universities, industry, and military organisations. Hardware, software and some applications were presented, such as touch screens and flight control. One talk concerned the ergonomics of the display workplace of a military ground operator. The system was connected to a ground network and flying airplanes. The better the ergonomics, the better the operator can control an air-attack. A few weeks later, I attended a 3-day NATO conference in Munich on "Systems for Command and Control". It concerned the presention of battlefield situations in command and control centres for groups of officers in order to make decisions for military operations in distant battlefields. The techniques presented were primitive – compared to today's video projection technology, but they were used in the war against Vietnam. In 1972, William J. Pomeroy described this in the *Daily World Weekend Magazine* (Dec. 16, 1972, pp. 6–7). I follow here the quotation in *Marxismus und Informatik* (by Jean Claude Quiniou. Pahl Rugenstein Verlag 1974, p. 149, orig. in French). Pomeroy tells us that already in 1968 there were leaks from Pentagon sources of electronic or automated battlefields. Great areas of Vietnam and Laos had been equipped with electronic detectors. This was the first stage of the electronic battlefield. The second one was a command and control centre equipped with computers that collected and processed information and sent commands. The third stage was a network of weapons. The command and control centre was connected with the Pentagon via satellite. An officer could be sitting in Washington, where he could react to electronic signals and operate weapon systems in almost any part of Indochina.

6 Responsibility for What I Do. But How Does this Work?

During the conference days at Paris, I met a Swiss girl and her German boyfriend, both selling paintings. We described what each of us was doing, and how we were living. I told of the conference and its military aspects. Then we had a long discussion through this night. Am I responsible for the application of the results of my research? "You are", he said. "Quit this job. I stay in Paris in order not to serve in the Western German army." My answer: "But I need a job, I can't quit it. And I try to find civilian applications. I can't stop others from using my results for military applications." "All you do in this field is for military applications", he replied. Research for the war which is "so real a feature of modern times" as an inevitable fate? At the institute, we also discussed these problems. One of my colleagues agreed to produce results for civilian instead of military use. It would be better to use the money of the Department of Defense in this way, otherwise it would be used to purchase military hardware. He supported me in developing a system for testing character recognition (Eingabe handschriftlicher Zeichen über einen Bildschirm und ihre Erkennung. In: *Elektronische Rechenanlagen* 12 (1970), pp. 188–193). Clearly, this very individualistic approach to avoid military research couldn't be generalized.

I often had my lunch at the cafeteria of a nearby university. Political groups distributed their flyers at the entrance to the cafeteria. In 1969, the basic group of the "Sozialistischer Deutscher Studentenbund" (SDS) at the mathematical-physical faculty offered a pamphlet entitled *Research and Destroy.* The authors critized the term "pure science" and showed how mathematics was involved in the real existing Vietnam war. They illustrated this point with copied extracts from diverse papers: Marvin B. Schaffer: Lancester Models of Guerilla Engagements, in: *Operations Research*, May–June 1968, p. 457f , Frank Mc Nolty: Kill Probability for Multiple Shots. An unnamed conference had a track entitled "Vietnam Applications" where Nigel Howard from the Management Science Center of the University of Pennsylvania presented "Applying Metagame Theory to the Vietnam Conflict", other speakers on "Optimizing Strike Force Planning", "Optimal Allocation of Air Strikes", and "Time Dependent Predictions of Guerilla Actions". In a job posting, the Stanford Research Institute (SRI) searched for "Weapon Systems Analysts and Operations Analysts [...] who desire to apply their experiences to systems problems". SRI offered them "a free investigative approach" and "competitive salaries and excellent fringe benefits."

7 Working in Industry

From late 1969 I worked for a German manufacturer of electronics, of both the civilian and military varieties. I joined a group which developed an operating system for commercial applications. In the hall where the computers were installed some were programmed by militaries: Civilian and military applications intermixed in the same hall. A colleague who helped me with a technical problem told me that he was developing a system for South Africa's Navy. It should serve to control for military reasons the sea traffic around the Cape. This cooperation with South Africa in 1970 took place while UN sanctions against this country were in effect.

I suspect that during the first year of my work at this firm, I was under surveillance since letters needed five days to arrive at my home. The military engagement of this firm may be illustrated by another affair. The secretary of our group married. Her husband studied computer science at a technical university. His studies were financed by this firm, and therefore he had to offer them his work services when he finished his studies. Shortly before his marriage he signed a contract. It seemed to be clear where he would begin to work after his honeymoon. But when he arrived on his first working day at 8 a.m. they refused, without any argument to allow him to work. What happened? Though the workplace was a civilian one, he failed the security check because his Austrian *father* had worked a dozen years ago as an engineer in East Germany.

8 Public and Collective Effort against Military Research

The period from 1965 to 1972 proved to me the difficulties in following one's responsibility when one is simultaneously subject to wage labor. By contract, I had sold my work services. The other side of the contract decided what work I had to do. Within the relationship of wage labor, I had no formal right to refuse military research. It was only by chance and in an individualistic manner that I could avoid an engagement in military research. But I had learned to engage myself in the following years *publicly and collectively* against military research.

In 1972, I entered the German national research institute for computer science (Gesellschaft für Mathematik und Datenverarbeitung GMD) which was owned by national and state Governments. At GMD, I joined the trade union. There, I found a forum of other people who discussed – among other things – the question of military research. In 1971, the Government offered a draft for new statutes of the institute, and in 1974 they sent a second draft. I looked for differences to the first one, and I found an omitted phrase, namely: The institute serves exclusively peaceful purposes.

What could this mean? *Publicly and collectively we stated that we are opposed to military research.* We initiated a broad movement among the 600 employees of the institute. We wrote a petition saying no to military research, signed by about 350 employees. Their reasons for signing the petition ranged from general ethical, political, economical or social considerations to personal experiences especially during World War II or to labor conditions.

There were a lot of public meetings in the institute, where directors also participated. Heavily discussed was whether "peaceful purposes" should be compatible with "military purposes". Some argued that military purposes could be peaceful, an argument similar to the old Roman one: *si vis pacem para bellum.* As common sense we found: Even if others outside the institute are willing to work in military projects, this is no reason that we should do it also. *But it would be impossible to us to control the usage of our results*. Mathematical formulae, methods, ideas, computer programs are products like butter or woollen socks: they can be used by civilians as well as by soldiers. When producing these things, it is essentially unknown who will buy and use them later on.

Our activities encouraged people in other national research labs to stand against military research; in return, they also encouraged us. Formally, our activities were unsuccessful, since the Government *refused* to include the peaceful purposes clause in the statutes of GMD. But practically we had become too suspicious for the military and we were no longer potential partners for them. We didn't reach a demilitarization of society but our activities had been an important step forward: In a collective effort, we made it publicly known that we didn't want to work on military projects. In the following years, we tried to consolidate our position. We informed the public on the problem of military research, not only at GMD but also via the trade union and via the union of works councils of public research institutes (Arbeitsgemeinschaft der Betriebsräte von Forschungseinrichtungen, AGBR). The Strategic Defense Initiative SDI, the German involvement in the second Gulf war, an intended cooperation of GMD with Bundeswehr by using a military multi-media lab were some occasions to inform the public on our position. Through these public and collective efforts, we could prevent the militarization of some part of public research. *But in general, we couldn't prevent military research in society.*

About eight years after the affair on "peaceful purposes" a governmental administrator - a physicist - proposed to the works council that GMD should develop intelligent mines. These mines would be placed along the border with East Germany. They would analyse the noise spectrum of an arriving tank and detect if it is a West-German or a Soviet one. In the latter case, the mine would explode. The administrator didn't mention that such mines would also be sold on the weapons' world market and used especially in civil war regions. Maybe he was thinking only to save some jobs in a munitions factory near GMD which produced mines. I participated in this meeting, and in the next works assembly at GMD I publicly stated that I was against this project in the field of pattern recognition and artificial intelligence. Later on, I found that producing ammunition had a bad tradition in this region. A few miles away about fifteen thousand workers produced most of the ammunitions for the German "West-Front" during the first World War (Oral information from town's archive in Siegburg).

9 Federal Government: Integration of Civilian and Military Research

It is worth giving some details of the dispute in 1974/75 with the Federal Government. Hans Matthöfer, Secretary of the Federal Department of Research and Technology (BMFT), answered the petition in accordance with the Federal Government's "Research Report IV" (Forschungsbericht IV) to the Bundestag:

> Political and factual reasons let the Federal Government emphasize that the defense research should not isolate itself. Furthermore, the Government holds the opinion that further integration of the Bundeswehr into society necessitates the transfer of defense relevant research work not only mostly to a defense internal organization [...] but also to existing scientific institutions.[...] Yet it is not intended to change the daily working circumstances of GMD in the future by offering GMD defense oriented tasks on a large scale.[3]
> [Matthöfer to GMD, November 6, 1974. All papers quoted in this chapter are to be found in the archive of the works council at Fraunhofer Gesellschaft Institutszentrum Birlinghoven, formerly GMD, Sankt Augustin. Folder §111 Gesellschaftsverträge]

In a meeting (October 30, 1975) with delegates of the petitioners, Matthöfer declared: Those who claim that GMD should serve only peaceful purposes are communists dependent on Moscow, or pacifists. For pacifists there would be no place at GMD. (Public report of GMD's works council. November 5, 1974). Within the next months some publications quoted this report which forced Matthöfer to deny his remark (Matthöfer to the works council, February 20, 1975). Based on some affidavits, the works council maintained its position (Works council to Matthöfer, March 11, 1975). Matthöfer answered that he had expressed his stand against pacifists, that anybody who didn't want to join a military research project would be disciplined. His "short remark on communistic efforts", he wrote however, "did not refer to those persons who have their very private conflict of conscience in mind, but to those who want to hinder the Government in following a majority decision and conviction in the Federal Republic and in looking for the proper defense abilities of this democracy. [...] To exclude GMD in principle from the legitimate task of defense research in a democratic state would contradict the basic decision for defense measures in the Federal Republic and therefore is not acceptable for me."[4,5] (Matthöfer to the works council, May 28, 1975) The newly elected works council (I now was member of it and its new president) published in the institute a short comment which also was communicated to Matthöfer. It was

[3] "Die Bundesregierung legt aus politischen und sachlichen Gründen Wert darauf, daß sich die Verteidigungsforschung nicht isoliert. Sie ist vielmehr der Ansicht, daß eine weitere Integration der Bundeswehr in die Gesellschaft es auch erforderlich macht, die für die Verteidigung notwendigen Forschungsarbeiten überwiegend nicht einer verteidigungsinternen Organisation, [...] sondern den vorhandenen wissenschaftlichen Einrichtungen zu übertragen. [...] Es ist jedoch nicht beabsichtigt, die Betriebswirklichkeit in der GMD dadurch zu verändern, daß in Zukunft in großem Umfang verteidigunsbezogene Aufgaben an die GMD herangetragen werden sollen."

[4] "(Meine) kurze Bemerkung über kommunistische Bestrebungen (stand in einem ganz anderen Zusammenhang. Sie) bezog sich auf diejenigen, denen es nicht um ihre ganz persönliche Gewissensentscheidung geht, sondern die die Bundesregierung daran hindern wollen, der Mehrheitsentscheidung und -überzeugung in der Bundesrepublik zu folgen und für die Verteidigungsbereitschaft dieser Demokratie zu sorgen. [...] Die GMD prinzipiell von der Verteidigungsforschung als legitimer Aufgabe des demokratischen Staats auszuschließen würde im Widerspruch zu(r) Grundentscheidung für die Verteidigungsbereitschaft der Bundesrepublik stehen und kann daher von mir nicht akzeptiert werden."

[5] This was stated during the time of the "Berufsverbote".

stated as a joint view "that quite ostensibly research for the defense sector, which means military research, cannot be reconciled with peaceful purposes."[6] (Works council, June 30, 1975. See figure 3) Matthöfer reacted with a new letter: "I cannot understand your position which discriminates in principle and impedes all support of the manifold efforts to keep up the outer security of the Federal Republic by involving GMD in respective projects and tries to make it impossible. (...) Also, I cannot share your position that defense oriented research would contradict tasks which serve peaceful purposes."[7] (Matthöfer to the works council, August 13, 1975. See figure 4).

Eleven years later, this debate had an echo. On August 27, 1986, Josef Rembser from BMFT talked on "Actual questions on international politics in science and technology"[8] before an assembly at DESY (Deutsches Elektronensynchroton Hamburg). DESY's works council had invoked this assembly of the employees, and it communicated Rembser's paper to AGBR. Concerning military research outside the nuclear sector "there exists an interesting exchange of letters from the year 1974 between the Gesellschaft für Mathematik und Datenverarbeitung (GMD) in Birlinghoven/Bonn and the Federal Secretary of Research at that time, Hans Matthöfer"[9], Rembser said. He quoted in full length Matthöfer's letters of November 6, 1974 (pp. 11–12), and of August 13, 1975 (p. 13). He commented: "The statements of the Federal Secretary of Research more than 10 years ago, a social-democratic Secretary of Research by the way, point in the right direction concerning the question if military research is allowed in the non-nuclear fields of work of our Institutions of Major Research"[10] (p. 13). Rembser asked how far statutes constituting *peaceful purposes* would allow *research for defense*. "This question is to be seen [...] in the context of the foundational tasks of most of the Institutions of Major Research in the nuclear research and nuclear technology sectors. Here, in May 1955 the Federal Republic of Germany declared in the context of the Declaration of Sovereignty of the Three Occupation Forces bound by international law that research and use of nuclear energy is conducted only for peaceful purposes and that the development, production and possession of nuclear weapons are renounced on a voluntary basis."[11] (pp. 10–11)

[6] "[...] daß sich offensichtlich Forschung für den Verteidigungsbereich, militärische Forschung also, mit friedlichen Zwecken nicht vereinbaren läßt."

[7] "Ich kann Ihre Haltung, jegliche Unterstützung der vielfältigen Bemühungen um die Aufrechterhaltung der äußeren Sicherheit der Bundesrepublik Deutschland durch Mitarbeit der GMD an entsprechenden Projekten grundsätzlich zu diskriminieren und zu verhindern, nicht verstehen. [...] Ich kann auch Ihrer Auffassung nicht folgen, verteidigungsbezogene Forschung stehe in einem Widerspruch zu friedlichen Zwecken dienenden Aufgaben."

[8] "Aktuelle Fragen internationaler Wissenschafts- und Technologiepolitik"

[9] "... gibt es einen interessanten Schriftwechsel aus dem Jahre 1974 zwischen der Gesellschaft für Mathematik und Datenverarbeitung (GMD) in Birlinghoven/Bonn und dem damaligen Bundesforschungsminister Hans Matthöfer."

[10] "Die Aussagen des Bundesforschungsministers vor mehr als 10 Jahren, eines sozialdemokratischen Forschungsministers im übrigen, sind auch heute richtungsweisend für die Frage der Zulässigkeit militärischer Forschung in den nicht-nuklearen Arbeitsgebieten unserer Großforschungseinrichtungen."

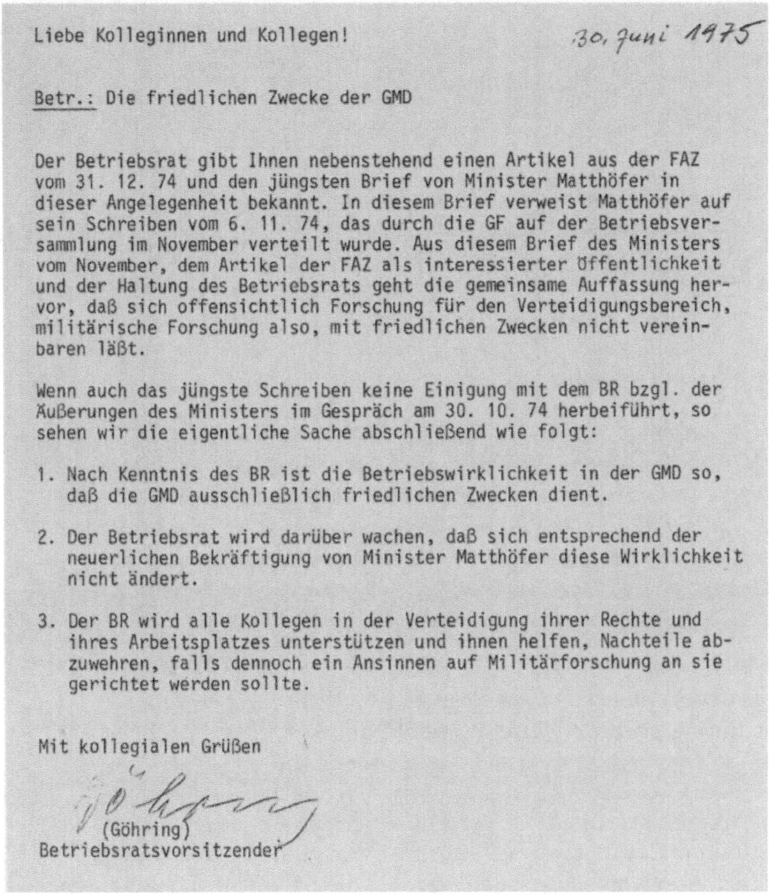

Liebe Kolleginnen und Kollegen! 30. Juni 1975

Betr.: Die friedlichen Zwecke der GMD

Der Betriebsrat gibt Ihnen nebenstehend einen Artikel aus der FAZ vom 31. 12. 74 und den jüngsten Brief von Minister Matthöfer in dieser Angelegenheit bekannt. In diesem Brief verweist Matthöfer auf sein Schreiben vom 6. 11. 74, das durch die GF auf der Betriebsversammlung im November verteilt wurde. Aus diesem Brief des Ministers vom November, dem Artikel der FAZ als interessierter Öffentlichkeit und der Haltung des Betriebsrats geht die gemeinsame Auffassung hervor, daß sich offensichtlich Forschung für den Verteidigungsbereich, militärische Forschung also, mit friedlichen Zwecken nicht vereinbaren läßt.

Wenn auch das jüngste Schreiben keine Einigung mit dem BR bzgl. der Äußerungen des Ministers im Gespräch am 30. 10. 74 herbeiführt, so sehen wir die eigentliche Sache abschließend wie folgt:

1. Nach Kenntnis des BR ist die Betriebswirklichkeit in der GMD so, daß die GMD ausschließlich friedlichen Zwecken dient.

2. Der Betriebsrat wird darüber wachen, daß sich entsprechend der neuerlichen Bekräftigung von Minister Matthöfer diese Wirklichkeit nicht ändert.

3. Der BR wird alle Kollegen in der Verteidigung ihrer Rechte und ihres Arbeitsplatzes unterstützen und ihnen helfen, Nachteile abzuwehren, falls dennoch ein Ansinnen auf Militärforschung an sie gerichtet werden sollte.

Mit kollegialen Grüßen

(Göhring)
Betriebsratsvorsitzender

Figure 3. Peaceful purposes *irreconcilable* with military research.

Matthöfer *refused* to fix "peaceful purposes" since he didn't want to exclude the possibility of military research at GMD. Fixing peaceful purposes in the statutes would exclude this possibility as in the political declaration of 1955. When the works council confronted Matthöfer with the logical consequence of his position, he did not agree "that defense oriented research would contradict tasks which serve peaceful purposes." He interpreted refusing military research as an impeding of "the manifold efforts to keep up the outer security of the Federal Republic."

[11] "Die Frage hängt [...] mit den Gründungsaufgaben der meisten Großforschungseinrichtungen auf dem Sektor Kernforschung und Kerntechnik (zusammen). Hier hat die Bundesrepublik Deutschland im Mai 1955 im Zusammenhang mit der Souveränitätserklärung der 3 westlichen Besatzungsmächte völkerrechtlich verbindlich erklärt, sie betreibe die Erforschung und Nutzung von Kernenergie ausschließlich zu friedlichen Zwecken und verzichte freiwillig auf die Entwicklung, die Herstellung und den Besitz von Kernwaffen."

HANS MATTHÖFER
BUNDESMINISTER
FÜR FORSCHUNG UND TECHNOLOGIE

53 BONN-BAD GODESBERG, den 13.8.1975
POSTFACH 120370
STRESEMANNSTRASSE 2
RUF (02221) 591

An den

Betriebsrat der
Gesellschaft für Mathematik
und Datenverarbeitung mbH

- z.Hd. Herrn Göhring -

5205 St. Augustin 1
Postfach 1240

Eingegangen
DA 25.8.75
20. AUG. 1975
C 15.10.75
BR-Nr. 1464

Sehr geehrter Herr Göhring!

Ich habe Ihren Brief vom 17. Juli 1975 mit Erstaunen gelesen. In meinem Schreiben vom 6. November 1974 hatte ich meine Auffassung zum Ausdruck gebracht, daß verteidigungsbezogene Forschung nicht in größerem Umfange an die GMD herangetragen werden sollte und daß in der GMD nur solche Projekte aus den einzelnen Bundesressorts bearbeitet werden sollen, die mit der Zielsetzung und Struktur des Forschungs- und Entwicklungsprogramms der GMD übereinstimmen.

Damit habe ich eine Mitwirkung der GMD an verteidigungsbezogenen Aufgaben in einem sinnvollen und zweckmäßigen Rahmen nicht ausgeschlossen. Ich kann Ihre Haltung, jegliche Unterstützung der vielfältigen Bemühungen um die Aufrechterhaltung der äußeren Sicherheit der Bundesrepublik Deutschland durch Mitarbeit der GMD an entsprechenden Projekten grundsätzlich zu diskriminieren und zu verhindern, nicht verstehen. Sie überschreitet nach meiner Meinung auch Ihren Auftrag, nach dem BetrVG die Interessen der Arbeitnehmer der GMD zu vertreten.

Ich kann auch Ihrer Auffassung nicht folgen, verteidigungsbezogene Forschung stehe in einem Widerspruch zu friedlichen Zwecken dienenden Aufgaben. Mir ist daher Ihre Interpretation meines Schreibens vom 6. November 1974 unverständlich. Eine "gemeinsame Auffassung" in diesem Punkte, die Sie gegenüber den Mitarbeitern der GMD in Ihrem Rundschreiben vom 30. Juni feststellen, kann ich aus Ihrer Haltung zu dieser Frage nicht herauslesen.

Mit freundlichen Grüßen

Ihr Hans Matthöfer

(M a t t h ö f e r)

Figure 4. Peaceful purposes *reconcilable* with military research.

10 Military Research, Warfare, and War as Part of Political Economy

In his talk at DESY Rembser generalized the intertwining of military research and security politics. He underlined that "science and technology determine international politics not only since our days" and he named the nuclear bomb, the radar and the Haber-Bosch synthesis. Science and technology would also be used as instruments of foreign politics he explained. "This is increasingly true in our days where we can observe an exacerbating race for technological leadership, for the early utilization of new technologies in civil markets"[12] he stated (p. 2) The amount of military R&D in the Federal Republic, he mentioned, was still low as compared to other countries, even if it was then increasing. (pp. 5–6). He regretted that "it is not always easy for our colleagues in the military research and technology sector to declare publicly that they work for military customers. Also in our country, things are approaching the situation in other Western countries because of an increase of objectivity in the attitudes of the public and the scientific community. Nevertheless, R&D for defense purposes still have a shadowy existence within scientific life."[13] (p. 9) As an example for "ideologies", "incrustations", "one-sided political valuations"[14] he pointed to an advertisement in *Die Zeit* (May 23, 1986) where scientists declared that they would engage themselves in impeding SDI-research in German national research labs.

With respect to SDI, Rembser stated: "The SDI discussion, as it was and still is conducted in a small part of the German scientific community, gives me cause for worry because of its narrow-minded fundamentalism and because of the one-dimensional earnestness of the discussants. Must we as Germans again and again be in such a way thoroughly and without compromise, even intolerant when it comes to political and social positions? Why don't we have an eye on our European neighbors, Scandinavia or Great Britain? We did not operate always with luck in our recent history when we wanted to be the schoolmasters or forerunners of Europe, or even of the whole world. Do we have to back out, to refuse technologies as the first, acting alone and without consultation or synchronization with our Western and Eastern neighbour countries while around us and in the farther

[12] "Dies gilt in wachsendem Maße in unseren Tagen, in denen wir einen sich verschärfenden Wettlauf um technologische Führung, um die Erst-Nutzung neuer Technologien für zivile Märkte beobachten [...]"

[13] "Es ist für Kolleginnen und Kollegen aus der militärischen Forschung und Technik nicht immer einfach, dazu zu stehen, daß sie für militärische Auftraggeber arbeiten. Zwar nähern sich auch bei uns die Verhältnisse dank objektiver werdender Haltung von Gesellschaft und wissenschaftlicher Community der Situation in andern westlichen Staaten. Dennoch führen FuE für Verteidigungszwecke im wissenschaftlichen Leben noch ein gewisses Schattendasein."

[14] "Ideologien", "Verkrustungen", "einseitige politische Bewertungen"

world a thoroughly reflected 'getting in on' these technologies takes place?"[15] (p. 18) In this philippic on the opponents to the *military* SDI, Rembser didn't name the technological research areas of SDI. Two pages later he lists these areas, yet he doesn't give any remark on responsibility if one would do research on "kill assessment", "directed- or kinetic-energy weapons", on "lethality" or on "battle management" (p. 20). He called such research "a thoroughly reflected 'getting in on' these technologies" that would happen abroad. Then he discussed the US policy on research and development: "The strengthening of their own technological capacity by enforcing defense research in important sectors of basic research and leading-edge technologies is part of the openly-stated U. S. politics. These political goals are primarily coined according to geostrategic goals, they are not motivated scientifically or technologically. Technology certainly is always a specific factor.

Many U.S. companies finance [...] a large part of their R&D costs via military orders. [...] It goes without saying [...] that the transfer of technologies from the military into the civilian sector - and clearly vice versa - will be most smooth when taking place in the management and in the organisation of one and the same group of companies. In this context, American companies have great advantages over their European and Japanese competitors."[16] (p. 22)

Let's summarize: Rembser called the position of those who are opposed to military research to be of "narrow-minded fundamentalism", of "one-dimensional earnestness", and "intolerant". He claimed to orient "our engagement" on the positions of our European neighbours. He called it an "objective attitude" if one accepts military research as it is done abroad. He followed Matthöfer in the "further integration of the Bundeswehr into society". With "exacerbating race for technological leadership, for the early utilization of new technologies in civil markets", Rembser stated the indissoluble intertwining of warfare, economy and competition. In sum, he showed in his talk before this assembly of wage laborers

[15] "Die SDI-Diskussion, wie sie in einem kleinen Teil der deutschen wissenschaftlichen Gemeinschaft geführt wurde und wird, macht mich besorgt durch ihre enge Grundsätzlichkeit und den eindimensionalen Ernst der Diskutanten. Müssen gerade wir Deutschen immer wieder besonders gründlich und kompromißlos, ja intolerant sein, wenn es um politische und gesellschaftliche Standpunkte geht? Warum orientieren wir uns nicht an europäischen Nachbarn, etwa an Skandinavien oder an Großbritannien? Wir haben in unserer jüngeren Geschichte selten glücklich operiert, wenn wir Schulmeister oder Vorreiter in Europa, ja sogar in der Welt sein wollten. Müssen wir als erste, allein und ohne Abstimmung bzw. ohne Gleichtakt mit unseren westlichen und östlichen Nachbarländern Ausstiege aus bestimmten Technologien vollziehen, bestimmte Technologien verweigern, während um uns herum und in der entfernteren Welt ein wohlüberlegter Einstieg in diese Technologien erfolgt?"

[16] "Die Stärkung des eigenen Technologiepotentials durch forcierte Verteidigungsforschung in wichtigen Sektoren der Grundlagenforschung und der Hochtechnologien ist erklärte US-Politik. Es ist eine Politik, die primär von geostrategischen Zielsetzungen geprägt wird, nicht aus wissenschaftlich-technischen Motiven. Technologie ist dabei immer ein besonderer Faktor.
Zahlreiche US-Unternehmen finanzieren [...] einen großen Teil ihrer FuE-Aufwendungen aus militärischen Aufträgen. [...] Es ist unstreitig, [...], daß der Transfer von Technologien aus dem militärischen in den zivilen Sektor - und natürlich auch umgekehrt - dort am besten funktioniert, wo er sich im Management und in der Organisation ein und derselben Unternehmensgruppe abspielt. Hier gibt es große Vorteile amerikanischer Unternehmen gegenüber der ausländischen, europäischen und japanischen Konkurrenz."

- scientific, technical and others, who depend on public funds and on funds from third parties - that military research, warfare, and war are an integral part of political economy which, indeed, has been a fact for a long time and which has not changed during the few years since Rembser's talk.

11 Conclusion

A moral, ethical or pacifistic view on military matters and a consideration of personal responsibility are helpful and worthy. Empirically, one can see that war has never been overcome by pure moral or ethical arguments. This approach has failed for thousands of years. Responsibility is claimed from those opposing military research as well as from those supporting it. Also, war was never overcome by an argument like *Si vis pacem para bellum*. This argument also has failed for at least two thousand years. Also modernized versions as we find them in our newspapers will fail.[17] *Parare bellum* - preparing war, or, in Matthöfer's words, "the manifold efforts to keep up the outer security" means to produce weapons. This production is an integral part of the social system of producing things.

In answering why war is so real a feature of our days one has to analyze the *economic system*, or, in other words, the *political economy*, and *how* war is embedded in it. Or, if one denies the need for this analysis, then the answer has to be found in separating this question from political economy (like separating variables in solving some differential equation). The claimed analysis would include how war could be overcome.

The solution lies - as I see it - in a superseding ("aufheben" in the dialectical sense) of the existing system of exchange. In some way, things are produced *independently* from each other. This independence leads to that these products become merchandise which are to be sold and bought on a market where competing sellers and competing buyers meet each other. *As an absolute contrary to war and competition, mankind needs cooperation and an instrument that facilitates it.* The instrument may be found in information and communication technology. Its real purpose lies in connection and cooperation instead of competition and waging war *despite the fact* that it is used today to "optimize" one's position in competition and hence in war. This contradiction indicates the driving dialectic of the future development[18] which, step by step, will enhance cooperation. Through an extremly enhanced cooperation mankind could supersede the system of merchandise production.

[17] A typical example is the political comment by Martin Winter in *Frankfurter Rundschau* (August 1, 2002, p. 3): "Only a Europe which is able to defend itself and which is willing and able to apply military force outside of its borders will be powerful enough to be able to prevent wars." ("Nur ein Europa, das sich selbst schützen kann und bereit und fähig ist, militärische Macht auch jenseits seiner Grenzen einzusetzen, wird genug Gewicht auf die Waage bringen, um Kriege verhindern zu können.")

[18] Concerning civilian economy this dialectic is explained to some extent in: Wolf Göhring: Was kommt nach E-Commerce? - Eine Perspektive für die Informationsgesellschaft. In: *UTOPIEkreativ* 137 (March 2002), pp. 233–243.

Ethics and Military Research

On the Moral Responsibility of Scientists

Jesper Ryberg*

This paper discusses the personal moral responsibility of scientists contributing to military research. A number of arguments defending the view that scientists do not carry any responsibility (or only a marginal responsibility) for the way in which their work is used are evaluated. It is argued that none of the arguments are convincing. Furthermore, some of the difficulties related to ascribing moral responsibility to scientists are considered.

In the time shortly after the end of the Second World War an interesting correspondence took place between Albert Einstein and the American Quaker A. J. Muste. The background for this correspondence was a public appeal in which Einstein's *Emergency Committee of Atomic Scientists* urged the public for donations to support scientists in their attempts to develop controls limiting the development and use of nuclear weapons. The point made by Muste in his letters was not that the content of the appeal was wrong, rather it was that if the appeal were to be taken seriously, then what Einstein and other scientists should do would be to renounce any involvement in constructing such weapons in the first place. He concluded by asking: "As for the masses, how can they be expected to believe that atomic weapons are as worthless and horrible as the scientists say they are, when the scientists continue to make the things [...]" and, furthermore, he declared that this "cannot make sense to ordinary human beings".[1]

The history of science reveals several examples of scientists who have followed Muste's prescriptions by renouncing involvement in military research. To take two examples within mathematics from very different epochs: Around 1600 the British mathematician John Napier – the founder of the theory of logarithms – experimented with a new form of artillery design. According to reports of witnesses, his invention cleared a very large area, destroying "all living creatures exceeding a foot in height". Napier was forced to take great pains to conceal the work-

* Department of Philosophy and Science Studies, Roskilde University, Denmark.
Email: ryberg@ruc.dk

[1] Muste, A. J., *Letter to A. Einstein* dated September 15, 1947. Located at: Records of the fellowship of reconciliation, Swarthmore College Peace Collection, Swarthmore, PA. My attention was directed to this correspondence by Chalk, R., Drawing the line: an examination of conscientious objection in science, in: Mircham, C. and Siekevitz, P. (eds.): *Ethical Issues Associated with Scientific and Technological Research for the Military, Annals of the New York Academy of Science*, vol. 577, New York 1989. I here quote from and draw on Chalk, p. 61.

ings of his invention and, in the latter part of his life, spoke specifically against the construction of new weapons.[2] Another more recent example is that of the mathematician Norbert Wiener – the founder of the field of cybernetics research – who had been working on guided missiles during the Second World War but, in the aftermath of the Hiroshima bombing publicly announced his decision not to "publish any future work [...] which may do damage in the hands of irresponsible militarists".[3] Though there are thus notable examples of scientists who have acted and made decisions which accord well with Muste's recommendations, it seems to be a fact that most scientists who are offered an opportunity to engage in – or who have in their careers carried out – military research do not take such radical stands. Many scientists are probably less concerned with moral implications than with the scientific attractions of their work or, as J. Robert Oppenheimer once put it, the scientific "sweetness" of the problems.

What I shall be concerned with here, however, is not the question of what motivates scientists in their work nor the question of how scientists throughout history have perceived their role in relation to military inventions and development but, rather, the ethical question which constituted the core of the Einstein-Muste correspondence – namely, whether scientists can be held to be acting rightly or wrongly in carrying out military research or, as it is often put, whether scientists in this respect have a moral responsibility. Should we, for instance, regard it as ethically illegitimate when scientists contribute to the construction of weapon systems or to the development of other sorts of military technology? Due to the comprehensive attention which over the last decades has been directed to this sort of question, and owing to the limited length of this article, several aspects of the question will have to be left untouched. What I shall do is to consider, in the following five sections, a number of arguments which have been presented in the debate; all arguments are to the effect that scientists either do not have a moral responsibility or that if they do bear such a responsibility this is only to a very marginal extent. I shall suggest that neither of these arguments are convincing. In the final section, I shall briefly argue that – in so far as scientists do have a moral responsibility – the relation between such a responsibility and the justification of warfare is pretty complex, which means that it is a very difficult task to make evaluations of responsibility for the individual scientist.

1 The Nature-of-scientific-inquiry Argument

The first argument I shall consider basically holds that the idea of the moral responsibility of scientists rests on misrepresentation of the nature of scientific work. The argument has been presented most forcefully by Robert Hoffman who

[2] See Chalk, R., Drawing the line: an examination of consientious objection in science, op. cit. note 1, p. 65.

[3] Quoted from Chalk, R., Drawing the line: an examination of conscientious objection, op. cit. note 1, p. 69. Original in Wiener, N., A scientist rebels, *Atlantic Monthly* **179** (1946), p. 46; *Bull. Atomic Scientists* **3** (1947), p. 31.

contends that the claim that the freedom of scientists to pursue the truth should be withheld only if the scientist rightly recognizes his moral responsibility, and the claim that the scientist may sometimes have a positive duty to limit his research, are both mistaken. These views are simply based upon a "defective theory" of scientific inquiry. According to Hoffman, inquiry, rightly conceived, is an "activity that seeks to eliminate an undesirable characteristic of a situation", namely, "its doubtfulness or indeterminateness". Thus, the starting point of inquiry is the "recognition of the indeterminateness of a certain situation with respect to a specific issue". Now, on the ground of this somewhat meagre characteristic of scientific inquiry, he goes on to claim that we can "see how odd it is to hold a scientist responsible for the specific truth he finds". Hoffman's point is that to hold a scientist responsible for what he discovers is "tantamount to demanding that he rightly foresee the outcome of his inquiry before initiating it". But if that were possible, that is, if he could foresee the specific truth it would yield, "there would be no need for inquiry".[4] In short, since the essential nature of inquiry is that one cannot foresee what one will discover, the scientist cannot properly be held responsible for the results that follow from his research.

Hoffman's argument can be perceived as related to the traditional "moral luck argument" which has received wide attention in the modern discussion of ethical theory. According to this argument, it is absurd to hold someone morally responsible for consequences which follow from acts as a result of mere good or bad luck. However, in my view it is hard to find any interpretation of Hoffman's argument which does not leave it highly implausible. The first thing worth noticing is that, even if the sketched description of what characterizes scientific inquiry is correct, it does not follow that the scientist is exempted from responsibility in all parts of his work. Carrying out an inquiry is only one part of the scientific work; another is the communication of the results of an inquiry. Thus, even if it is correct, an inquiry is a process in which the scientist discovers something which at the beginning was unforeseeable, then there is no reason why he should not be held responsible for communicating the results when doing so can be expected to have terrible consequences. As mentioned, abstaining from publishing certain results was exactly the device adopted by Wiener. To this Hoffman might, of course, respond that all he is concerned with is the discussion of responsibility in relation to scientific inquiry. However, even if this is so, there is a second point which makes it reasonable to regard the argument as defective, namely, that it rests on a misrepresentation of the nature of scientific research. Admittedly, there exists scientific research which is *pure* or *fundamental* in the sense that it, from the outset, is not clear what the precise purpose of the inquiry consists in. Moreover, it is obviously also correct that scientific inquiries now and then produce results which were totally unexpected, that is, which almost pop up as luck would have it. However, most scientific research does not fit this description. On the contrary, typical research is better described as "mission-oriented", in the sense that the

[4] All quotations are from Hoffman, R., Scientific research and moral rectitude, *Philosophy* **50** (1975), pp. 475–477.

scientist seeks to solve pretty well-defined problems.[5] When Harvard Professor Louis Fieser during the Second World War was presented with the task of developing a means for burning out Japanese from bunkers, it did not make much sense to hold that he, at the beginning of his research, had no idea what he was looking for. Even though it is trivially true that Fieser did not from the start know the precise outcome that would follow from his inquiry - which in this case turned out to be napalm - he certainly had an idea of what would constitute a satisfactory solution to the problem. The same is the case, to a larger or minor extent, with much other research. That is not to say that it may be clear how the results of the research will in the end be used by those who make that kind of decision - but I shall return to this question below. What is important is that scientists may have qualified expectations with regard to what may follow from their work if it turns out successfully. But, in that case, the nature of scientific inquiry does not itself rule out the idea of ascribing a moral responsibility to the scientist. If such a responsibility does not exist it must be due to some other reason.

2 The Morality-is-handled-by-the-state Argument

Another argument which has now and then been presented in the discussion of the moral responsibility of scientists seeks to reject such a responsibility on the ground of an appeal to the contention that questions of right and wrong do not lie in the hands of individual scientists. For instance, some of the atomic scientists who responded to Muste's appeal not to take part in the construction of weapons did so by objecting that it is simply not the responsibility of scientists to judge whether research projects should be initiated or continued. Hans Bethe held that a "strike" by the scientists "would only antagonize the public of the United States who would rightly accuse us of trying to dictate the policies of the country".[6] And Willie Higinbotham even more emphatically proclaimed that "We believe in government by the people [...] if scientists were to walk out on all military projects they would be taking the law into their own hands just as surely as the Ku Klux Klan".[7]

Put in a more general form, the argument which can be reconstructed on the ground of these quotations may go the following way: As long as a scientific research project does not in any way violate the law and is perhaps in one way or the other endorsed by the government of the society, the individual scientist should not concern himself with the ethical aspects of the research and cannot either be held responsible in the sense that he, for instance, may be regarded as having acted

[5] See also the criticism of Hoffman's view in Belsey, A., The moral responsibility of the scientist, *Philosophy* **53** (1978).

[6] Bethe, H., *Letter to A. J. Muste* dated December 16, 1946, location see op.cit, note 1. I here quote from Clark, R., Drawing the line: an examination of conscientious objection in science, op.cit. note 1, p. 62.

[7] Higinbotham, W. A., *Letter to A. J. Muste* dated August 5, 1946, Located ibid., quoted from ibid. p. 62.

wrongly by carrying out his work. Morality, so to speak, is handled by the state and by the state only. Hence, as long as these conditions are satisfied it cannot be wrong for a scientist to take part in military research. In fact, it might perhaps even be held to be wrong not to take part in such projects. Now, is this argument convincing?

In order to answer this question, clarification is required. It is simply not obvious what the precise content of the argument amounts to. At least two overall interpretations can be distinguished, each of which seems to call for further elaboration in order to be conclusive. The first interpretation would be to hold that the claim that "the state handles morality" means that whatever the government judges right or wrong simply *is* right or wrong. What makes a certain policy right is that it is sanctioned by the state. The second interpretation might be that the claim means that governmental decisions and regulations function as some sort of filtering system in the sense that if some kind of project is wrong, and thus should not be undertaken, it would be filtered out by the government in the first place. Thus, on this view it is not the governmental decisions themselves which make a certain policy or project right or wrong – they are, so to speak, not right- and wrong-producing – but the government sees to it that only what is morally right makes it through the filter.

Now, it should be noticed that if the intention of the morality-is-handled-by-the-state argument is to show that the scientists do not carry any moral responsibility, that is, that they should be regarded as *amoral* in their professional roles, then the argument is flawed. On neither interpretation does it follow that scientists do not have a moral responsibility. No matter whether governmental decisions produce right and wrong decisions or merely filter out the right from the wrong, there would still seem to remain a moral responsibility for the scientist, namely, to act in accordance with what is right and wrong, that is, to follow the policies which are endorsed by the government. In that sense, the scientist is not left genuinely amoral in his work. However, it might still be held that the sort of cases which have typically led to discussions of the responsibility of scientists such as, for instance, scientists' contributions to state-mandated development of mass-destruction weapon systems, would be ruled out by the argument. There would be no room for conflict between what is morally right and what is endorsed by the state. Be that as it may, however, there are still reasons to doubt the argument in either interpretation.

The question which the first interpretation naturally gives rise to is: why should we accept the position in the first place? That is, why should we accept that some policy is morally right just because it is endorsed by the state? Probably the most appealing answer would go along the lines of Higinbotham's quoted subscription to "government by the people"; that is, to hold that if a policy is endorsed by society through a representative government, then it is right precisely because it reflects the voice of the people. However, even though we in a democratic society would apparently in many situations regard such a view as reasonable, it is nevertheless a position which is morally highly controversial and which would thus need further support to be convincing. Firstly, it is not hard to construct (or find) examples of decisions which are made by representative governments but which might still strike us as morally dubious. Secondly, if we accept that what is right and wrong is in fact determined relative to what a certain decision-maker decides, why should the decision-maker then be a representative government and not another particular group of people or perhaps certain individuals? This is a question that needs to be answered in order to bring support to the view. And notice that the view cannot be supported by the sort of moral arguments by which the legitimacy of democratic decisions are often sustained. If something is right if and only if it is endorsed by a representative government, then the claim that this position is morally right becomes circular.[8] Finally, the claim that what is right is what is endorsed by a representative government seems dubious if the notion of endorsement is not qualified. The obvious question is whether governmentally endorsed projects are right even if they are endorsed on the ground of factually mistaken assumptions. Suppose that a government endorses a certain project on the ground of the assumption that the project will not have any environmentally damaging consequences, but that a scientist knows that this assumption is not correct. In that case, it is certainly a dubious position to hold that what the government decided is right. However, suppose that the conception of endorsement is qualified by holding that what is right is what is endorsed by a government if it is not endorsed on the ground of false assumptions. In that case there might be cases where certain projects are endorsed by the government but where a scientist would be right in not acting in accordance with the endorsement. And there might be cases in which a scientist could rightly be morally accused for having participated in a military research project, even when the project was actually endorsed by the government. Thus, there would not be much left of the initial idea that morality is only a governmental affair.

Let us then turn to the second interpretation of the argument. Is it more reasonable to hold that a representative government sees to it that only what is right passes through the filter of decisions? As in the case of the first interpretation, neither does this version of the argument justify the conclusion that the scientist is amoral in his professional role. It seems that there would still be something right which the scientists ought to act in accordance with, namely, the decisions made by the government (and *vice versa* with regard to policies which are governmentally rejected). Be that as it may, there is an obvious problem that confronts the position, namely, that it rests on an assumption which is highly dubious. In order to reach the conclusion that there cannot be situations in which a scientist acts wrongly by following governmental decisions or rightly by acting against such decisions, it will have to be presupposed that the government always correctly manages to distinguish right from wrong. However, it is certainly hard to see how such an assumption of infallibility can be supported. On the contrary, it is reasonable to assume that even the most well-functioning governments may sometimes fail in their decisions with regard to what should and what should not be done. Which means that there may still be something to be said by others, such as scientists engaged in military research.

Thus in sum, the idea that "the state handles morality", understood in the sense that once a certain scientific project is endorsed by the government there would

[8] The position that "something is right because it is decided by the government" is right because it is decided by the government.

be no moral questions left on the shoulders of scientists, is not plausible. Neither the interpretation that governmental decisions create right and wrong, nor the – in one way – more modest interpretation that the government filters out right from wrong, succeeded in bringing support to the picture of a scientist who is basically amoral in his professional life. And both interpretations on closer scrutiny turn out to be highly dubious. However, perhaps the important thing is not whether military research is governmentally endorsed in the first place but rather how a government or whoever is politically responsible decides to apply the outcome of the scientific work. This is the point of the argument to which we shall now turn.

3 The Scientists-do-not-decide-on-application Argument

The argument which, I believe, most naturally comes to mind in the discussion of the moral responsibility of scientists who perform military research is that the ascription of responsibility to scientists is misplaced because it rests on a conflation of research and application. The argument has been presented forcefully by Hoffman who, as an example, considers a virus inquiry which leads to the discovery of chemical compounds that are toxic to humans. Now, as Hoffman puts it, one of the "logical implications of the statement that a certain chemical compound is toxic to human beings is the statement that *if* the compound were to be used effectively in warfare, it would harm people". However, as he emphasizes, this statement is hypothetical, that is, it does not enjoin anyone to use the chemical compound in waging warfare. The decision to wage chemical warfare is simply not an implication of the scientific inquiry nor the truths it discovers. It is, Hoffman contends, therefore "senseless to ascribe responsibility to him [the scientist] for the use of his discovery. Responsibility for that use is rightly ascribed to whoever formulates the policy and whoever makes the decisions".[9] If the result of military research is wrongly applied it is the government or another possible decision-maker who should be held responsible. One should not shoot at the scientists.

Put in a more general form, the argument is based on the premise that the people who make fatal decisions or acts are those who should be held responsible: they are the ones who may be deciding or acting rightly or wrongly. Responsibility should be ascribed to the latest mind which enters a stream of events. The problem with this view simply is that, even though there are situations in which it seems reasonable, there are certainly also cases in which it seems highly problematic. To illustrate this we can use an example made up by Immanuel Kant in a short famous essay titled "On the Supposed Right to Lie from Altruistic Motives".[10]

[9] Hoffman, R., Scientific research and moral rectitude, op.cit. note 4, p. 476.

[10] Kant's "On the supposed right to lie from altruistic motives" can be found in *Critique of Practical Reason and other Writings in Moral Philosophy*, University of Chicago Press, Chicago 1949.

Kant imagines a person who is fleeing from a murderer and who tells you where he is going to hide. When the murderer comes along and asks where the fleeing person went you have pretty good reasons to believe that, if you tell the truth, that person will be found and killed by the murderer. What should you do – tell the truth or lie? The reason that Kant's essay is famous is not that it constitutes the place where he provides the most elaborate insights of his moral philosophy but rather that it clearly illustrates what most people regard as an implausible rigorism of his position: according to Kant it would be wrong to lie even if this were the only way to save a person's life. However, if one agrees that Kant's answer is morally unacceptable, that is, that you may act wrongly by telling the truth, then one cannot at the same time maintain the view that it is only the last person in a serious of events to whom responsibility may be ascribed. Put in another way, Hoffman is wrong when he holds that one should only be held responsible for what follows logically from descriptions of one's acts. If one has good reasons to believe that something fatal will follow if one acts in a certain way and that this could be prevented if one does not do so, then this provides a sufficient base for the ascription of responsibility to the moral agent. If this view is applied to the discussion of military research it implies that, if a scientist has reasonable grounds for regarding it as likely that the results of his research will be used, by the government (or whoever makes the relevant decisions), in a way that will adversely affect mankind, then he may be acting wrongly in carrying out the research or by communicating the results. Thus, the scientists-do-not-decide-on-application argument cannot, I believe, stand closer scrutiny. If the argument at first sight seems convincing, this may be because one imagines situations in which one has no idea how the outcomes of one's research will be used by others. In such a case, it may strike us as much more reasonable to exempt the researcher from any moral responsibility. However, just as this lack of insight into future consequences did not exist in the murderer example, neither is there general reason to believe that it correctly represents what is at stake in military research. That more may be said on the possibility of making such estimations on the application of the results of scientific research is an issue I shall return to below. However, enough has been said to reject the argument that has been considered in this section.

4 The Replaceability Argument

An answer which is frequently met in contexts in which someone is blamed for having acted in a way which had undesirable consequences, is that the act did not really make any difference because the consequences would have occurred anyway, namely, as a result of someone else's act. More plainly put, the argument says that "If I did not do it, someone else would" therefore "I did not really do anything wrong". It might be suggested that this well-known argument can be transferred directly to our discussion of the moral responsibility of scientists. If a particular scientist had not taken part in a military research project someone else would simply have replaced him. Therefore, there is not much sense in holding the scientist responsible for what he did. After all, it did not make any difference.

The argument clearly has some appeal. We are often inclined to believe that the rightness or wrongness of an act should be evaluated on the ground of the consequences of the act. For instance, in many cases we find it reasonable to hold that an act was wrong because it led to a result which was worse than what would have been the case had the act not been performed. Therefore, if one can show that a scientist's work did not really make any difference because his work would have been carried out no matter whether or not he did it, there would no longer be a ground for regarding carrying out his work as wrong. Nevertheless, there are good reasons to be sceptical with regard to this argument.

It is obvious that in order to be sound it must be presupposed that the empirical premise of the argument is correct. That is, that the scientist actually would have been fully replaceable had he not undertaken his research. The premise would be wrong if other scientists, for personal or moral reasons, would not have taken on the work, or if other scientists did not possess the sufficient skills to do the work. However, even if we assume – which is probably often the case – that the replaceability premise is correct, there is another more basic reason to reject the argument. It is simply not plausible to hold that the possibility of replacement exempts one from moral responsibility. As Douglas P. Lackey has pointed out there are probably only a few people who would be willing to accept this significance of replaceability if the focus were to be on right or praiseworthy actions. As he puts it "I would be surprised to find a scientist working on a project who would accept the argument that he does not deserve to be paid for his work because, if he had not done it, someone else would have stepped in and done the same work".[11] Be that as it may, I believe that the implausibility of the alleged moral significance of replaceability can be demonstrated straightforwardly by a simple example which is far from the work of scientists.

Suppose that Smith and Jones both intend to kill an innocent person before them, and that Smith knows that if he does not kill the person Jones certainly will. Now, will it be wrong if Smith kills the innocent person? The answer must certainly be in the affirmative. However, if Smith does not kill the person it will not make any difference: the person will be killed by Jones. Thus, it seems that there might be cases in which an individual act may be judged wrong even though it did not make any difference in the final result.[12] To object that one can never be absolutely certain that a particular result would have been produced anyway – that Smith could never be certain that Jones would have done the killing – and that this makes an important moral difference, is not a way out of the problem if one wishes to maintain that scientists do not carry a moral responsibility: neither in their case would there be certainty with regard to replaceability. Thus, I believe that the replaceability argument should not be accepted.

[11] Lackey, D. P., Military funds, moral demands: personal responsibility of the individual scientist, in Mitcham, C. and Siekevitz, P. (eds.), op.cit. note 1, p. 128.

[12] See also, Parfit, D., *Reasons and Persons*, Clarendon Press, Oxford 1984, pp. 70–73.

5 The Share-of-the-total Argument

The earlier discussion of Kant's murder example showed that it is not plausible to maintain that one is exempted from responsibility merely because another person's decisions are also involved in the production of a bad result. However, it might be held that the example may bring support to another argument which does not seek to establish that scientists do not have any moral responsibility but rather that even if they do bear a responsibility it is only a very marginal one: a responsibility that may almost be ignored.

The argument is pretty simple. On the one hand, it does not seem reasonable to hold that one does not bear any moral responsibility if one told the murderer where the fleeing person was hiding and this, as expected, led to the death of the person. On the other hand, it seems counter-intuitive to hold that one bears the total responsibility for the death of this person: after all, what one did was only to tell the truth to someone who then carried out the murder. The fatal result would not have occurred had the murderer not acted as he did. Thus, the truth-telling person bears some but not the full responsibility for what happened. Put in more general terms, the view is that in situations in which many people contribute to a certain outcome, each contributor is only responsible for a small share of this outcome. Now, if this apparently appealing view is correct it has obvious implications for the discussion of scientific responsibility. Suppose that the work which a scientist has done is in the end applied in a weapon system and that this system is used in a way that has fatal consequences. Obviously the fatal result is then not only a consequence of the scientist's decision to carry out his research. Several other people's decisions and acts have been involved in the production of the fatal outcome, for instance, other scientists, engineers, politicians, military personnel, and many others. There is usually a very long distance from the scientist's writing desk to the final consequences of the application of his work. Therefore, if what might be called the "share-of-the-total view" is correct, that is, if each contributor to a project does not bear responsibility except for a small share of the final outcome, it would seem that the individual scientist only makes a morally almost negligible contribution to the bad result.

Sometimes it might, of course, be the case that the distance between the result of the scientific research and the consequences of the application is not so long. Moreover, there might also be cases in which the fatal consequences are of such a high magnitude that even a small share of the total responsibility would amount to something which was far from negligible. However, even if we leave these possibilities aside there is a more general reason to reject the argument, namely, that the share-of-the-total view on closer inspection turns out to not be morally acceptable. In cases where the persons who are involved in a certain project each makes a contribution which is necessary for the production of the final outcome, it simply does not seem morally plausible to maintain that each person bears responsibility only for a share of this outcome. This can be illustrated relatively easily by adopting an example presented by the Oxford philosopher Derek Parfit.[13] Suppose that, due to an accident, the lives of a hundred people are in danger. These people can be saved if I and three other people join in a rescue mission. If any of us fails to join, all of the 100 people will die. However, if I fail to join, I could save, single-handedly, 50 people. What should I do?

If one adopts the share-of-the-total view then, by joining the rescue mission with the three others, I would be responsible for saving 25 lives. Compared with the 50 lives I would save by acting single-handedly, this view therefore implies that I ought to operate on my own. I should not join the rescue mission. But this clearly is an unacceptable conclusion. It is indisputably better if 100 lives are saved than if 50 lives are.[14] However, in order to reach this conclusion, that is, in order to make my participating in the rescue mission the rational answer, we will have to give up the share-of-the-total view in favour of a view which ascribes full responsibility – that is, responsibility for the whole outcome – to each person whose work is a necessary condition for the success of a project.[15] Admittedly, the claim that we should be held responsible for having benefited (or harmed) someone in the sense this principle suggests, does not correspond fully with our ordinary idea of when we benefit (or harm) someone. We are often inclined to believe that we have benefited (or harmed) someone only when the act we perform is the chief or immediate cause of the benefit (or harm). However, as the example indicated, we may benefit (or harm) someone even when an act is a more remote part of the cause that led to the benefit (or harm). The share-of-the-total view simply constitutes what Parfit rightly refers to as a "mistake in moral mathematics".

This leaves us with a pretty simple conclusion. If the work of a scientist who contributes to military research is a necessary condition for the final application of, say, a certain weapon system or some other piece of military technology, then the fact that the scientist is far distant from the application of the weapon system, in the sense that many others' work and decisions are also necessary for the application, does not reduce the responsibility of the scientist. The scientist may – along with many others, such as politicians and military decision-makers – be held totally responsible for a bad result.

6 Concluding Remarks

In the previous sections, I have evaluated a number of arguments to the effect that scientists cannot rightly be said to bear a moral responsibility in their pro-

[13] Ibid. p. 67-70. The argument is also discussed by Lackey, D. C., Military funds, moral demands: personal responsibility of the individual scientist, op. cit. note 10, p. 129.

[14] It is, in the example, assumed that the other three persons would not be able to save any lives if they were acting single-handedly.

[15] Strictly speaking, the view that full responsibility should be ascribed to all who contribute with something that is necessary for the production of a certain outcome, does not follow from a mere rejection of the share-of-the-total view. But the former principle seems to provide the most plausible route out of the problems that face the latter view.

fessional life as military researchers. According to the first argument, the idea of ascribing moral responsibility to scientists was based on a lack of understanding of the nature of scientific research. The open-endedness of the scientific inquiry, it was held, makes it absurd to ascribe moral responsibility for the results. As was argued, however, this argument basically ignored the mission-orientedness of most military research. The second argument sought to establish the lack of responsibility on the ground of the view that morality is an issue for the state, not for the individual scientist. As was made clear, the argument did not succeed in showing that scientists are exempted from any responsibility and neither of the considered interpretations of the view on closer examination turned out as plausible. According to the third argument, the amoral nature of the scientist's work was a simple implication of the fact that it is not scientists who decide whether or how the results of their work should be applied. Responsibility should be ascribed to politicians and other decision-makers, not to scientists. However, as the example of the murderer and the fleeing person demonstrated, it is on closer scrutiny not reasonable to hold that only the final agent or decision-makers in a series of events leading to a bad result should be held responsible. The fourth argument suggested that due to the fact that a scientist may have been replaced by another if he had not carried out his research, the ascription of responsibility was misplaced. As a response, I argued that replaceability, even if it is correct, does not exempt one from responsibility. Finally, it was argued that the argument, that a scientist is only one part in a series of agents whose work and decisions may lead to a bad or good outcome, does not imply that responsibility is correspondingly reduced. Anyone who makes necessary contributions to a project should be ascribed responsibility for the total result of the project. In sum, neither of the arguments I have examined has succeeding in showing that scientists are exempted from responsibility in their work as military researchers.

To me it seems much more plausible to maintain the opposite view, namely, that military researchers may be held to bear a responsibility. Though I shall not here engage in a closer discussion of the precise foundation of such a responsibility, I wish to make one further comment on the matter. Acting in accordance with a moral responsibility may be a very complicated matter. To take a few examples. The term military research – which is often used somewhat loosely – covers an enormous range of activities. For some of these activities it is a fact that both military and civilian applications are apparent. It is a well-known fact that not only nuclear power, but also drugs, pesticides, aircraft, radar, processed food, satellites, computers, transistors, lasers and many other technologies have been developed for military purposes but have also had obvious civilian applications. There is no reason why a possible civilian spin-off should not be counted in when the scientist is considering the pros and cons for engaging in a research project. But this clearly complicates the ethical evaluation.

Furthermore, when considering whether it would be right or wrong to take part in a certain military research project it is often assumed that the answer to this question depends on whether a war or some other kind of military operation, in which the result of the research is applied, would be morally justified. Roughly put, it is assumed that if the military operation is justified it would not be wrong to carry out the military research, or perhaps it would even be right to do so and, oppositely, if the military operation is morally unjustified it would be wrong to undertake the military research. For instance, pacifists – like Muste – who believe that all sorts of warfare is unjustified usually hold that military research is wrong. And advocates of the just-war tradition also often seem to defend this sort of relation between whether wars are just or unjust and whether military research is right or wrong. However, this relation is clearly much too simplistic. Even if a military operation is just, a certain military technology may lead to an increase in the number of victims and, oppositely, if a military operation is not justified the application of a certain technology may nevertheless decrease the number of victims. In fact, if there is something to the old saying *si vis pacem para bellum,*[16] then even a pacifist may regard military research as right. Thus, there is no simple relation between the legitimacy of war and the rightness or wrongness of military research. This clearly underlines the complexity of an ethical assessment of this sort of research.

The reason for mentioning these examples is, on the one hand, to underline that it may in practice be very difficult for the individual scientist to assess whether it would be right or wrong to engage in miliary research. But also, on the other, to emphasize that this complexity of bearing and carrying out responsibility should not be conflated with the view that scientists do not have any responsibility. The former view makes it vital to engage in and seek to sort out the practical complications – the latter view makes this work irrelevant.

[16] "If you want peace prepare for war".

IV
Enlightenment Perspectives

Mathematical Thinking and International Law

A Questionable Relationship

IB MARTIN JARVAD*

We examine the hypothesis that natural law philosophy in general and modern international law, in particular from the 17th century on, was based on the successful application of mathematical methods to modern physics and to modern war. However, it was not a case of direct application, apart from a few scattered references, rather modern international law was part of the same broad intellectual movement and change of all the sciences. It is shown that the natural law foundation of international law (by Hugo Grotius) was characterized by a strict method, where deductions from minimalist axioms produced natural laws of universal validity as the necessary relations among normative phenomena, including the important part of the law of war and peace. The power of contracting, axiomatically included among the natural powers of man and human societies, justified the historically contingent positive laws while the natural law made and still is the basis for the validity and obligation of the natural law as well as the positive treaty law.

1 Introductory Remarks

The vast changes, which we now refer to as modernity, gained momentum during the later half of the 16th century. The breakdown of the Respublica Christiana resulted in chaos out of which the European Nation State System slowly emerged (roughly speaking from the Peace Treaty of Westphalia 1648) eventually putting an end to the armed conflict between Catholics and Protestants.

During the very same period of time, the foundations were laid for the revolution of, or rather the establishment of the natural sciences, as it gradually liberated itself from the tutelage of theology. Simultaneously, the quest for a secular theoretical basis for law and justice began.

This was a period of time in which interdisciplinary interests were pronounced. The application of mathematics was particularly important in the process of changing from the medieval Aristotelian observation-based conception of the physical

* Department of Philosophy and Science Studies, Roskilde University, Denmark.
Email: jarvad@ruc.dk

world, to that of a theory-based conception. Galileo's physics has been interpreted as a vindication of Platonism and pure theory against empiricist Aristotelianism[1].

2 A Theoretical Point of Departure in Neo-Platonism

Historians of science disagree considerably as to the relative importance of the different factors contributing to this development. However, there is general agreement that it was within the field of mathematized physics that the most important advances were made. For purposes of the present discussion[2] I will depart from Koyre's theory which I take to hold for Galileo, Descartes and in part Newton, though not with respect to Bacon. Koyré's claim was that a theory-based Platonism was vindicated against an empiricist Aristotelianism. Plato claimed the objective cognition of both mathematics and justice to be possible. This led to his well-known idealist philosophy that the forms contemplated by the mind in mathematics and in justice are real, and what we experience as discoveries of new forms, are only reminiscences - a regained memory. Being one with God before birth and after death was to see and grasp the forms in their entirety.

However, the new Platonism where mathematics came to be seen as the 'language of nature' was different. It did not necessarily object to observation and experimentation. Rather theoretically determined experiments become meaningful when we assume the mathematization of nature and discard miracles. In Koyré's point of view, it is the theoretically designed experiments, which make the decisive difference between the skill (the techne) of the craftsman, and the ingenuity of the designer of scientific instruments and machine technology. The vindication of Platonic idealism in physics is therefore only partial.

This line of thought, however, included the idea that what had been established by logical necessity also had to hold for God. Mathematics was particularly important for separating the understanding of nature from the domination of religion. It took more than three centuries to gradually diminish the role of God as the creator, the regulator, the upholder, and the governor of nature before, finally, Laplace could boldly answer Napoleon that he had no need for this hypothesis in his model. The medieval notion of Thomas of Aquino, which had assimilated Aristotle into Christianity, stated that contemplating the laws of nature (classical Greek natural philosophy) was one way of worship. Integrating these notions allowed natural philosophy to have a niche in an otherwise religious universe. However, this way of thinking was eventually replaced by the separation of the physical world that

[1] Koyré, A. *Etudes d'histoire de la pensée scientifique*, ed. Galimard 1973, also in English: Galileo and the Scientific Revolution of the XVIIth Century, *Philosophical Review* 1943, pp. 333–348.
Federico Cesi (1585–1630) founder of l'Accademia dei Lincei wrote in a letter to Galileo Nov.1612 that those to be admitted to the academy should 'neither be slaves of Aristotle nor any other philosopher but noble and free intellects in physical matters' and the common programme should be to 'eradicate the principal dogmas of the presently ruling doctrine' [my transl.], see Montalenti, G. et al, *Federico Cesi e l'Accademia dei Lincei,* Napoli 1988, p. 68.

followed the laws of logical necessity or the mathematical method. Mathematicians, natural philosophers, and engineers could be, and most often were, pious believers, however, religious beliefs had ceased to play an important role in their professional lives.

So it is an attractive hypothesis that mathematics, in this wide sense, in a similar way may have contributed to the establishment of a secular natural and rational law, including the law of the new- nation state system. Below we will produce some textual support for this hypothesis.

The significance of mathematics for the establishment of international law could be based on statements by Grotius himself. For instance in the *Prolegomena*[3] which is a brief exposé of the theory on which the treaty on the law of war and peace is based, Grotius explicitly refers to the method of mathematicians (Prol.58); '*Vere enim profiteor, sicut mathematici figuras a corporibus semotas considerant, ita me in iure tractando ab omni singulari facto abduxisse animum*' which in Kelsey's translation runs as follows: 'just as mathematicians treat their figures as abstracted from bodies, so in treating law I have withdrawn my mind from every particular fact'.

Van Eikema Hommes[4] maintains that the entire treatise is based on the application of the mathematical method, and states as a fact that Grotius knew Descartes. Tuck[5] holds that Grotius cited mathematics as the methodological model for the human sciences and bases his theory on similar programmatic statements from an earlier unpublished work by Grotius, *De iure Praedae*, see below.

Programmatic statements are not in themselves convincing, rather we shall look for ourselves. Furthermore we do not find any important traces of directly applied mathematics in the manner of Hobbes or Spinoza. Grotius probably understood mathematics too well to do that.

It is also a controversial hypothesis. Ethical philosophy of the last century has been almost unanimously negative towards the platonic notion that any objec-

[2] I may stress that this choice is for this particular purpose and that one reason for choosing it is that it leads to falsifiable hypotheses. Other contributions are fruitfull in other respects also for the revolutionary development of political and legal philosophy of the epoch for instance Needham, J.: *Science and Civilization in China*, Cambr. 1954, and following esp. Vol 2, Sect. 18 Human law and the laws of nature in China and the West, pp. 518–583; Kelsen, H.: Causality and Retribution, *Philosophy of Science* 8/4 (Oct. 1941), pp. 533–556; Zilsel, E.: Physics and the problem of historico-sociological laws, same issue, pp. 564–579, and Zilsel, E.: The genesis of the concept of physical law, *The Philosophical Review* 51/3 (May 1942), pp. 245–279.

[3] The *Prolegomena* is quoted from Hugo Grotius: *Prolegomena to the Law of War and Peace*, tr. Francis W. Kelsey, introd. by Edward Dumbauld, The library of Liberal Arts, N. Y. 1957. Quotations are cited by paragraph in arabic.

[4] van Eikema Hommes, J. H.: *Major Trends in the History of Legal Philosophy*, North Holland, Amsterdam, New York, Oxf. 1979, pp. 83–87.

[5] Tuck, R.: Grotius and Selden, ch. 17, in part IV, The end of Aristotelianism, in: Burns, J. H. with the assistance of M. Goldie (eds.): *The Cambridge History of Political Thought, 1450–1700*, Cambridge Univ. Press 1991, pp. 499–523; see esp. p. 505. See also Tuck, R.: *Natural Rights Theories, Their Origin and Development*, Cambridge University Press 1979.

tive basis for justice exists, towards the idea that justice is somehow a reality and towards the idea that right and wrong is within the grasp of human cognition. Non-cognitivism[6] in different versions has ruled supremely.

Natural law theory has been the target of often abusive polemics. Instead of making 'speculative' assumptions about what are natural points of departure for deductive reasoning about law, moral philosophers and legal theorists[7] have pointed to emotions, to the authority of absolutist or democratic lawmakers, to popular traditions or case law as the only reality.

According to this contemporary trend in moral and legal philosophy there is nothing beyond the whims and fancies or the emotions of the citizen or the prince or the democratic lawgiver or the courts and their traditions, customs and habits.

Although only few have admitted it, not even the standard and seemingly uncontroversial demands for universality, consistency and coherence can be upheld in non-cognitivist philosophy. Thus we are deprived of a platform on which to take a critical stance, and simply have to accept the given as it is. Legal non-cognitivism slides easily into the decisionism of the leading legal theorist of the Third Reich, Carl Schmitt[8].

Being discontented with this state of affairs, I think that it is necessary to investigate again the natural law that Hugo Grotius claimed was a possible object for human intellectual cognition. And I expect that the subsequent presentation will suffice to convince that the natural law conceived by Grotius was not the free-wheeling speculative invention and disguise of the prejudices of an age and a culture. Rather, it was a sober and disciplined exposition of necessary relations among legal phenomena characterised by adequate definitions, *les relations necessaire de les choses* as Montesqueu called them. Further it will become clear, I hope, that Grotius' theory of law like his theory of religion has a markedly sociological twist and that it is far from narrow-minded prejudice.

Investigating the significance of mathematics in the formation of international law of war and peace is markedly different to investigating the role of mathematics in warfare, where mathematical science based technologies were successfully employed.

The secular natural law enterprise seems only to bear some resemblance to mathematics in so far as both seek to establish something that has no physical tangible existence whatsoever by means of logical necessity.

Recently, some attention has been paid to those phenomena in human society that have no physical factual existence. Modern language philosophers like Austin[9], and successors like Searle and Habermas have called attention to the cre-

[6] Couture, J. and Nielsen, K.: Introduction and afterword to (and eds. of) The Ages of Metaethics, *Canadian Journal of Philosophy Supplementa*, Vol. 21, 1995.

[7] See for instance Ross, A.: *On Law and Justice*, London 1958.

[8] Schmitt, C.: *Die Diktatur*, Berlin 1927, *Politische Theologie, Vier Kapitel zur Lehre von der Souveränität*, München und Leipzig 1934, and *Der Nomos der Erde im Völkerrecht des Jus Publicum*, Köln 1950.

[9] Austin said in his Harvard lectures from 1965 that his first inspiration came from law, see Austin, J.: *How to Do Things with Words*, Oxford 1962, 1975, 1980, footnotes on p. 2 and 7.

ation of institutional 'facts' by verbal acts. Concepts like 'person', 'state', 'republic' etc. are fictitious and not tangible, yet real. Legal fictions are real was the claim of the medieval Roman lawyer Bartolo of Saxoferrato. The point boastfully underlined by Galileo is that, even if the law of free fall only holds in vacuo and never in real life conditions, it holds as a logical necessity, and that is the centre of Koyré's theory.

Likewise the theory of natural law seeks to both establish its own objects as the normative objects derived by pure reason and to investigate and analyse critically the norms established in tradition, by volitions and accepted by authority.

The timelessness of the platonic universe is replaced by an ambiguity with respect to the difference between inventions and discoveries; do we invent or do we discover in mathematics and in law?

Since I am not a mathematician, I am reluctant to propose definitive answers to the question of the possible relationship between mathematics and international law. Rather I shall attempt to present the theory developed by Grotius and his methodological principles for the reader's consideration.

3 From Medieval to Early Modern Warfare and the Application of Arithmetic and Calculus

The applications of mathematics in warfare during the transition from medieval to early modern times were important. However, it did not involve the application of advanced mathematics, only simple methods of calculation and bookkeeping. The need to develop new techniques in warfare, fortification, artillery and navigation may have served as an incentive to pursue mathematics and natural philosophy in its own right. The really decisive changes were in social organisation, especially the operational analysis facilitating the introduction of firearms.

In the early renaissance, Machiavelli revived interest in Roman republicanism and military organisation, mainly drawing upon Titus Livius and through him on Polybius. His Art of War[10] together with works of later humanists such as Lipsius[11] served to stress warfare as an instrument for political ends rather as an end in itself. The Catholic Church had for long advocated the same[12] against the norms of chivalry that tended to see fighting in duels, tournaments, and war as the truest test of male virtue, bravery and valour. The order and discipline of the army or the man of war served to enhance the fighting power of a limited number of men and was ensured by calculated order while marching, in encampment and in battle for-

[10] *L'Arte della Guerra* 1521 , *The Art of War*, tr. Ellis Farneworth, revised ed. and introduction by Neal Wood, Da Capo Press 1965, p. 90.

[11] Lipsius, J.: *Sixe Bookes of Politickes or Civil Doctrine*, tr. William Jones, 1594, facsimile reprint Da Capo Press, 1970.

[12] See Thomas of Aquino: *Summa Theologica,* Quaestio 40.

mation. All three aspects involved the application of arithmetic and calculus, but at a very basic level, namely that of the drill sergeant or the quartermaster.

In turns, the order and discipline of the army and the navy served to facilitate the adoption of technical innovations such as firearms. In both defence and attack on land, the rational architecture that was employed was based largely upon the Roman model described and praised by Polybius. In naval warfare, the mode of fighting changed from close encounters and man-to-man fighting to artillery exchanges between movable naval forts.

These changes in warfare vastly increased the costs of waging war, and military laws were drawn up, not only to ensure discipline, but also to prevent destructive pillaging and marauding. In fact it became part of military discipline to abstain from molesting and plundering civilians.

This was not a particular European discovery. Also the Mongol conquerors of China under Genghis Khan's successor Kublai Khan realised that more revenue was gained from a conquest by letting the civilian production continue and the civilian administration and tax collection proceed.

But purely instrumental restraint tends to break down into anarchy, chaos, and general destruction as in the Thirty Years War, because utilitarian calculus depends on what counts as utilities. If the utter destruction of an enemy and the devastation of his lands and means of subsistence is the object of warfare, restraint is not rational. Polybius records that the final destruction of Carthage included the killing of all, men, women, and children, as definitive destruction of the only competing power was the object of Rome's war. Therefore a general norm of restraint must have another basis.

4 Modern International Law of War and Peace: Hugo Grotius

The establishment of modern law of nations was the achievement of one man Hugo Grotius (1583–1645). There were some before him and many after, but his contribution was unique in that he laid the theoretical normative foundation for the new nation state system firstly in Europe, later universally. Grotius made great efforts to credit his predecessors even if he tended to overdo it with copious quotations. But many of those who came after him lifted long passages out his works as if it were the common heritage of mankind.

His life and political career was intimately connected to his theoretical achievements.

He was born into a family that belonged to the commercial-administrative nobility of the Netherlands. His father Jan was the mayor of Delft and curator of the leading protestant university in Leyden. His uncle, Jan's older brother, had renounced the position as head of the family to become a law professor there. Hugo was a child prodigy, who, from the age of 11, studied at Leyden with Joseph Scaliger, a leading humanist and philologist at the time. As a child and a youth, he wrote poetry and religious drama in Latin and translated some of it into Dutch. He was renowned as a Latinist and poet, and corresponded with a wide circle of fellow humanists.

Following the family tradition he studied law after his humanistic studies and took up legal practice following the ideals of ciceronian rhetoric[13]. He had immediate success and in 1604 he was engaged by the Dutch East Asia Company, one of the biggest commercial enterprises in Europe at the time, to defend the seizure by admiral Hemskeerk, a relative of the de Groot family, of a richly loaded Portuguese carrack in the Straight of Malacca. The legality of the seizure was duly tried before the Dutch Court of Admiralty in order to determine whether it was a case of piracy, or a lawful prize and booty.

His brief was a lengthy treatise on the law of war and peace, and the freedom of the seas to peaceful passage and commerce. The title is *De Iure Praedae.* Except for the chapter on the freedom of the sea it remained unpublished until it was found in 1868.

The issue of the freedom of the sea was of the first importance in those days. The Portuguese monarchy had extended their slave raids on Western Africa around the South of Africa and with help from Arabic navigators da Gama could open the sea route to the Far East as the first European. The Spanish monarchy sponsored the attempt to open a Westward route to the Far East and discovered the New World. In 1493 pope Alexander Borgia arbitrated between Portugal and Spain and a line was drawn dividing the monopoly on the sea routes and the new foreign lands to be colonized and Christianised. This papal bull was fiercely contested by all others than the two beneficiaries, and also criticized in the Catholic Church. The Dominican Victoria and his school of students in Salamanca openly disputed the competence of the pope to make this decision.

The skirmish between a flotilla of three small armed merchant vessels of the Netherlands East Asia Company and the huge Portuguese armed merchant vessel, a galleon or carrack named Catharina[14] was instigated by a local Malay prince who wanted to break the monopoly of the Portuguese in the spice trade. The Portuguese tried to forcibly block the passage of the Dutch traders, but failed, and had to surrender. The vessel and its rich cargo were brought to the Netherlands where action was initiated at the Court of Admiralty to have it condemned and sold as lawful prize and booty. The case was very tricky on the Dutch side because a number of shareholders in the company were Mennonites and determined pacifists on religious grounds. They threatened to pull out their capital, and set up a competing ship owners and trading company in France to avoid being involved in what seemed to be piracy.

[13] Ciceronian rhetorics was markedly different from the rhetorics of the sophists that were the opponents of the Greek philosophers, depicted in caricature in many Platonic dialogues. The ciceronian ideal was a fusion of philosophy, history, and the art of forensic oratory. Cicero himself was both noted as a jurist and pupil of Scaevola, a famous and courageous orator, and an ardent lover of Greek philosophy. Philosophers nowadays, however, tend to regard him as unoriginal.

[14] According to Grotius' own account in his Annales the prisoners taken on the ship numbered more than 700.

First of all, the defence of the seizure involved the point that the Dutch had a right to sail in the East Indies, in casu the Straight of Malacca. Secondly, if the Portuguese would prohibit them from doing so, they had the right to proceed right away without awaiting a legal ruling. Thirdly, the Dutch had the right to resort to war in order to enforce their right of passage and commerce. Fourthly, also a private person or company had a right to wage a lawful war and finally, that any booty and prize taken in the course of a lawful war would be a lawful booty and prize.

Grotius brief probably settled the matter, and the 76 page long chapter on the *Freedom of the Sea*[15] published anonymously and without his knowledge in 1608, provoked Selden, a leading English lawyer to write a two volume rebuttal *De Mare Clausum* in 1635 to defend the English position of claiming access to sea routes, lands, and waters monopolised by others, while still retaining English monopoly on English routes and waters.

Grotius rose to the highest civil and political positions in the Netherlands, second only to Oldenbarneveldt, the elder statesman, who had by and large shaped the federal constitution of the United Provinces of the Netherlands, the Union of Utrecht. Oldenbarneveldt had taken Hugo under his wing when Hugo was still a boy and used to refer to him as 'my Grotius'.

Grotius then served as ambassador for the Netherlands to England and approached King James on a double mission. One was to resolve the disputes between the Dutch and the English over the herring fisheries. Herring was a key commodity in the Baltic trade. Furthermore, he was to respond to King James' intervention to the estates general of the United Provinces against toleration. King James sided with the orthodox party in the reformed church and warned against more lenient, latitudinarian views in the conflict between so-called remonstrants (latitudinarians, tolerant) and counter remonstrants (orthodox, intolerant).

The educated bourgeois-aristocratic environment of Grotius was not favourable to the primitive strict orthodoxy of the ultra-Calvinists. Arminius, who was the leading theologian of the tolerant party, was a friend of the family. Grotius had already been asked to defend Vorstius when he applied to succeed Arminius in his chair of theology at Leyden University, and was attacked by the orthodox professor Gomarus. Vorstius was successful in obtaining the chair.

The core issue of the struggle was predestination versus free will. The orthodox party held that the perfection of God implied not only God's complete foreknowledge of who were to be saved and who were to be condemned, but also His omnipotence and that consequently whatever happened, was His will. Furthermore, the perfection of God implied that it was perfectly just that some were predestined to be the vessels of evil while others were destined to salvation. Man's free will did not count in any way[16]. The Arminists favoured another interpretation

[15] Grotius: *The Freedom of the Seas or The Right which Belongs to the Dutch to Take Part in the East Indian Trade*, tr. Magoffin, R. New York 1916.

[16] It is understandable how a vulgar dogmatic protestant orthodoxy may be favorable to the callous bigotry and to the propensity to saving and industry of puritan capitalists as Max Weber presented it, but it is beyond my comprehension how Weber could imagine that a religion of

of St.Augustine, namely that man had a free will, for instance to embrace God and thus be saved, and further that predestination did not suspend the duty to do good, since man had the free will to do right or wrong. The orthodox accused the Arminists of the pelagian heresy[17] while the Arminists preached toleration.

Luther, Melanchton and Calvin had preached against toleration fearing sedition and disunity, and King James followed suit as a pious orthodox protestant. Grotius did not succeed in either mission with King James.

Back in the Netherlands the religious conflict intensified. Orthodox preachers incited the lower classes of the townsmen to march against tolerant preachers and their congregations[18]. The Arminists were the leading faction in the province of Holland but not in the other provinces. The Stadtholder, prince Mauritz of Oranje Nassau[19] saw the spreading riots as a chance of establishing himself as absolute sovereign by siding with the orthodox. By a coup d'etat he had Oldenbarneveldt, Grotius and Hoogerbeets arrested and charged before a hastily convened ad hoc synod. The charges were unclear and finally in the final sentence formulated as high treason. The process was irregular and the competence of a church synod dubious, to say it mildly. Oldenbarneveldt was sentenced to death and executed, Grotius to imprisonment for life in the Loevenstein Castle, and Hoogerbeets to imprisonment in his own home for life.

selfrighteousness could be advantageous to the advancement of science, see Weber, M.: *Die protestantische Ethik und der Geist des Kapitalismus*, Tübingen 1920. Weber states in note 48 that the pronounced preference of protestant asceticism for mathematized, rationalised empiricism is well known and need not be discussed further and refers to Windelband, W.: *Lehrbuch der Geschichte der Philosophie*, 1892, pp. 305–307. I have only had access to the 2. ed. of this work and have not found the evidence. It seems that Windelband has omitted this "well-known" fact.

[17] Pelagius, a theologian of British origin, was attacked by St. Augustine for the doctrine that salvation was a just retribution for doing good. Knowing pelagianism only through St. Augustine is to know only the one side of a debate. Revival of the teachings of St.Augustine, especially the attack on pelagianism, was a major concern of Luther, Calvin and the Protestant break with the Catholic Church. Antipelagianism remained a central concern of protestant theology and was the reason behind the break of Søren Kierkegaard with the official Danish church.

[18] Tex, J .den: *Oldenbarnevelt*, 2 Vols., Cambr. 1973, describes the start of controversies between liberal and strict calvinists already in 1608 in Alkmaar (Vol 2, p. 511). He mentions that Oldenbarnevelt attempted to even out matters by letting preachers of the conflicting factions debate in public, presumably hoping that the uneducated audience would tire of the often unintelligible dogmatic details. He was not succesful, and it seems that the lower classes of townspeople positively craved for orthodoxy, dogmatism and strict discipline.

[19] Prince Mauritz was not only an outstandingly successful general but also the author of the rationally designed new model army later developed by Gustav Adolf and the English Puritans. It was based on the Roman conscript army as described by Livius, Tacitus and Polybius and given a renaissance by Machiavelli and Lipsius. Mauritz had studied with Lipsius in Leyden. Mauritz' preoccupation with morale, uniformity of belief , unity of command and suppression of dissent is typical of the military way of thinking advocated by Lipsius. It is worth noting that Gustav Adolf in the last part of his career was a much less dogmatic lutheran protestant that in his youth and consequently much more disposed for reconciliation of the warring churches, see Ahnlund, Nils: *Gustav Adolf Den Store*, Stockholm 3. opl. 1932.

A year later, Grotius was helped to escape by his courageous and determined wife. He fled to Paris, where his wife and children joined him. There he wrote in one year the Law of War and Peace, which immediately became universally famous. Still, they suffered economically difficult circumstances until he was appointed ambassador for Sweden at the French Court. France was to make a treaty with protestant Sweden to oppose Austria in the Thirty Years War.

Once the treaty was concluded, Grotius became more attracted to scholarly pursuits, especially religious conciliation, which many thought of as an obsession. At the centre of his teaching – for we may call them so – he placed the Christian commandment of brotherly love as a continuation of the stand, which he had taken when supporting the Arminists. Gradually it became a theology of his own, where he sought reconciliation between not only the various Protestant Churches but also the Catholic Church, which led to him being eyed with suspicion by all camps.

Through the late 1630s and early 1640s he published a series of irenical works in which he saw the church as the visible social institution, as a body, not in any mystic sense, but as a social reality. The *Votum pro pace ecclesiastica*, the vote for peace among the churches (1643) should be concord between Christian states and concord as the soul of the church that Christ wanted to be one and united. Most explicitly he exhorted all who sought to establish peace among Christians to demolish all dogmas that destroy public peace, *demolire dogmata, qua pacem civilem perturbant. Prius est, bonum civem esse, quam bonum Christianum*[20] (Being a good citizen is prior to being a good Christian). But this statement lead an otherwise sympathetic theologian as Posthumus Meyes to conclude that Grotius was no theologian, for in his view no theologian could say so, only a statesman and jurist, "For the church will never be confined to tangible or ethical visibility, and that it is not concordia either which represent its profoundest being, but the Holy Spirit"[21].

He renounced his post as ambassador for Sweden and went to the court of Queen Christina, heir and successor of Gustavus Adolphus to collect his rear pay, and on the way back to the continent in 1645 he suffered shipwreck, fell ill, and died on the coast near Rostock.

5 The New Law of War and Peace

Its significance was first and foremost its generality and the *formal equality of all* before the law. This was highlighted recently by a group of Japanese scholars in a joint volume on the Grotian normative theory of war[22]. They added a precise critique against the (otherwise progressive) advances made by the Spanish school

[20] Quoted from Meyes, G. H. Posthumus: Hugo Grotius as an Irenicist, in: Feenstra, R. (ed.) *The World of Hugo Grotius*, Amsterdam & Maarsen 1984, pp. 43–63. Meyes made a minor error in referring to *Via ad Pacem Ecclesiasticam*. The quote is taken from *Votum pro Pace Ecclesiastica* from 1643.

[21] Ibid., p.63.

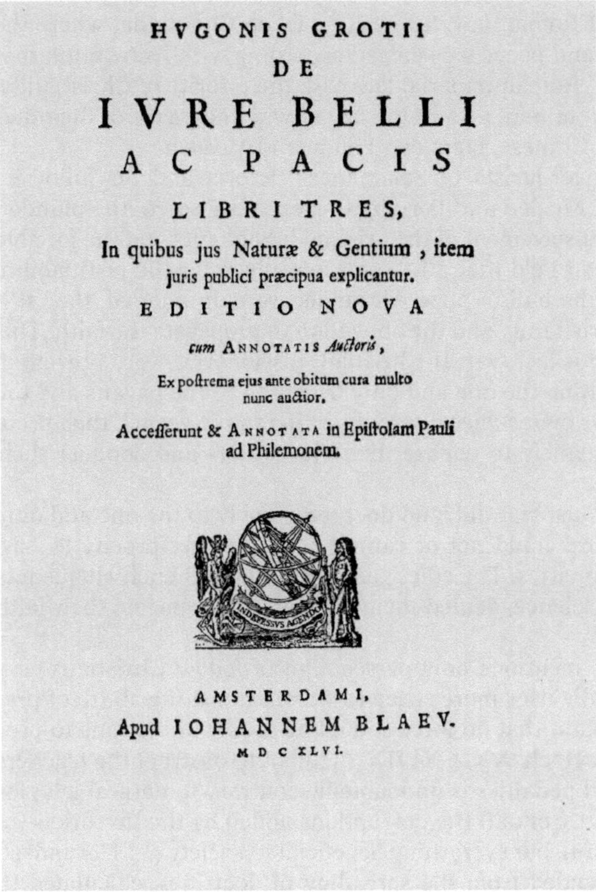
HVGONIS GROTII
DE
IVRE BELLI
AC PACIS
LIBRI TRES,
In quibus jus Naturæ & Gentium, item
juris publici præcipua explicantur.
EDITIO NOVA
cum ANNOTATIS Auctoris,
Ex poſtrema ejus ante obitum cura multo
nunc auctior.
Acceſſerunt & ANNOTATA in Epiſtolam Pauli
ad Philemonem.

AMSTERDAMI,
Apud IOHANNEM BLAEV.
M D C XLVI.

Figure 1.
Title-page of the second edition of Grotius' *De iure belli ac pacis*. As indicated by the armillary sphere, the publisher was also engaged in mathematical publishing. Historians of international law credit Hugo Grotius with the creation of modern international law, as in particular established in the Peace Treaty of Westphalia of 1648 and the Charter of the United Nations of 1945, and trace the origins of it back to patterns of mathematical thinking of striking public appeal in Grotius' time. Military analysts of our time blame the striking public appeal of mathematics supported modern warfare for undermining international law.

of Salamanca prior to Grotius for still being christocentric and thus subjecting formal equality of peoples, creeds, and states to the priority of Christianity.

When Grotius started out on his major opus, the treatise *On The Law of War and Peace*, the full title of which runs *De iure belli ac pacis libri tres, in quibus ius naturæ & gentium, item iuris publici præcipua explicantur*[23], he looked back upon the hostile relations of city states of antiquity, where each judged according to its own particular code of justice, to the imperial Pax Romana which judged

[22] Yasuaki, Onuma (ed.): *A Normative Approach to War, Peace, War and Justice in Hugo Grotius*, Oxford Clarendon Press 1993; see also the review Gordon, E., *American Journal of International Law* 1995, pp. 461–463.

[23] Amsterdam 1625 and several later editions which until the 1646 edition were revised by the author. Here is used the English translation by A. C. Campbell, London & New York 1901, in the Hyperion reprint ed. 1979. Quotations are cited by book, chapter and paragraph in roman numerals and with arabic numerals for sections within paragraphs.

according to the priority of Roman law, to the Respublica Christiana, where the normative ordering of war and peace was judged according with its own internal fusion of Greek philosophy, Roman imperial law with the priority of Christianity. This ethnocentric bias is prominent also in the law of war and peace of the other great cultures, the Hebrew, Chinese, Japanese, Hindu and Moslem.

Even if the great theologer-jurists of Salamanca Victoria and his followers denounced the conquest of Mexico and Peru by Cortez and Pisarro, the plunder, indiscriminate killing, the enslavement of the original inhabitants and the forcible baptism of them, and instead held that all the reasons given for the justifiability of the conquest, including the bull of pope Alexander were unfounded, they still maintained the truth of Christianity and the obligation to propagate the faith. This clause provided the cause for just war. If Christian missionaries were prevented by pagan rulers from preaching the one and only true faith to the pagans and the pagan rulers prevented their own subjects from listening to the gospel, then there was a just cause for war, namely to depose the unjust rulers and conquer their lands.

This particular cause for just war did, and does, only apply to the one and only true faith of Christianity and could not or cannot admit any reciprocity to, say, Moslem, or Buddhist missionaries. The edict guaranteeing the French Huguenots their right to liberty of conscience, denied them the right to propagate their faith in catholic France.

Grotius, very cautiously, mentions only persecution of fellow Christians as a cause of just war, but actually uses more space to warn against the abuse of pretexts. He states that "it is plain that no force should be used with nations to promote its acceptance" [Book II, ch. XX, § XLIIX] "[...] but to obstruct the teachers of Christianity by pains and penalties is undoubtedly contrary to natural law; for the doctrine of Christ, apart from all the corruptions added by the inventions of men, contains nothing hurtful, but everything beneficial to society [...] Nor indeed can any danger be apprehended from the spreading of doctrines, calculated to inspire greater sanctity of manners and the purest principles of obedience to lawful sovereigns" (Book II, ch. XX, § XLIX).

Grotius did not exclude the commands of God from the law of war and peace, but he included them in a secular system of law. All others Christian jurists preceding him included secular norms into an overarching religious system. So his way of circumventing the preoccupation with the dogmatically correct Christianity is to include it into a system of laws of nature, where God is the creator of nature, but explicitly also bound by the laws of nature.

Thus the solution offered by Grotius belongs on the level of the system of law rather than on the level of particular laws. This may be why Grotius proudly and correctly states "That body of law, however, which is concerned with the mutual relations among states...few have touched upon. Up to the present time no one has treated it in a comprehensive and systematic manner; yet the welfare of mankind demands that this task be accomplished" (Prol.l) .

According to the Japanese scholars formal equality in the eyes of law was the unique achievement of the modern European international law as shaped and developed by Grotius.

The method of arguments *a priori* and *a posteriori*

The arguments a priori are pure reason, while the arguments a posteriori are factual evidence. Natural right – which seems to be used as synonymous with natural law (Book I, ch. I, § IX) – "Natural right is the dictate of right reason, shewing the moral turpitude, or moral necessity of any act from its agreement or disagreement with rational nature, and consequently that such an act is either forbidden or commanded by God, the author of nature." (Book I, ch. I, § X).

"We are said to reason a priori, when we show the agreement or disagreement of any thing with a reasonable and social nature; but a posteriori, when without absolute proof, but only on probability, anything is inferred to accord with the law of nature, because it is received as such among all, or at least the more civilised nations." (Book I, ch. I, § XII).

The assumption about human nature is the stoic term *sociableness* "Among the traits characteristic of man is an impelling desire for society, that is, for the social life – not of all and every sort, but peaceful, and organised according to the measure of his intelligence, with those who are of his own kind" (Prol. 6).

Concerning the inclination to be sociable Grotius quotes Seneca, *On Benefits*, Book IV, ch. XVIII on ingratitude, where Seneca argues that men are weak compared to animals, but superior because of reason and society – "he who in isolation could not be the equal of any creature, is to become master of the world [...] It was society which gave man dominion over all other living creatures [...] it can be invoked against Fortune" [chance, IMJ].

To the appetitus socialis is added the "power of discrimination" between what is "agreeable or harmful" and thus "to follow the direction of a well tempered judgment [...]. Whatever is clearly at variance with such judgment is understood to be contrary to the law of nature, that is, to the nature of man" (Prol. 9).

The reason underlying this judgment is not to be understood as the dictates of appetite and instinct[25]. Anticipating the analysis of the concepts of law and freedom in Rousseau and Kant[26], Grotius declared appetite and instinctive drives to be subordinated; "agreement with reason, which is the basis of propriety, should have more weight that the impulse of appetite; because the principles of nature recommend right reason as a rule that ought to be of higher value than bare instinct" (Book I, ch. II, § l, 2).

"To this exercise of judgment belongs moreover the *rational allotment* to each man, or to each social group of those things which are properly theirs [...] leaving to another that which belongs to him, or in fulfilling our obligations to him"

[24] Tuck in his study of rights theories naturally stress the natural rights theory of Grotius. Sometimes Grotius distinguishes between rights and laws. Property is a right and so is the right to wage just war to recover property or even to inflict punishment. But the duty not to steal is a law.

[25] Thomas Hobbes conceived of freedom as the unimpeeded movement towards satisfaction of appetites and other needs.

[26] The concept that freedom consists in subjecting one's appetites and instinctive drives to the norm one has set for oneself.

thereby drawing a sharp demarcation between law and other principles of distribution (Prol. 10).

To the human power of discrimination is added the general human *power of contracting* "[...] since it is a rule of the law of nature to abide by pacts (for it was necessary that among men there be some method of obliging themselves one to another, and no other natural method can be imagined) out of this source the bodies of municipal law have arisen. For those who have associated themselves [...] or subjected themselves [...] had either expressly promised [...] or impliedly to have promised [...]" (Prol. 15).

I will stress not the social-contract theory of the origin of society which had its roots in the conciliar movement in the Catholic Church in the Middle Ages and was later lifted out of Grotius' writings by Locke, Rousseau and others, but rather call attention to the legal power of all men to create obligations with everyone else.

In the course of the advent of bourgeois society, certain nuclei of the preceding society survived, in particular the obligation of children to parents, and wives and servants to the master of the household. Otherwise all obligations, including the entire system of municipal laws, were to be based upon or construed as being based upon the legally accepted exercise of human will as the one and only legal power. All other social obligations were discarded. All considerations of what was due to holders of status and rank, or due to the deserving poor, or to the church and the holy Christian societies, were subjected to the human will. The magnitude of this conceptual change may be the reason for the awkward yet unmistakeable way of expressing it.

Thus, the elements of the natural law, as expressed by Grotius, are only the assumptions about sociableness, the duty to honour pacts, and the powers or reason, of discrimination between right and wrong and of contracting.

The position of the Christian God in the system of natural law

God is, as has been stated in the quotes above, the author of nature, thus we obey God by following natural law; but God is unable to change it. "Now the *Law of Nature is so unalterable*, that it cannot be changed even by God himself. For although the powers of God are infinite, yet there are some things, to which it does not extend. Because the things so expressed would have no true meaning, but imply a contradiction. Thus two and two make four, nor is it possible to be otherwise; nor again, can what is really evil not be evil." (Book I, ch. I, § X, 3; author's italics). Furthermore "God himself suffers his actions to be judged by this rule" (ibid. with Biblical references).

And finally the famous *etiamsi daremus* "What we have been saying would have a degree of validity, even if we would concede that which cannot be conceded without the utmost wickedness, that there is no God, or that the affairs of men are of no concern to him." (Prol. 11).

An outline of the classification of types of law

The statements above lead to define Gods volitional law as a special class of law. They are directives binding for those to whom they are addressed. The platonic Greek Zeus, and the stoic Roman Jupiter may direct them to all mankind. Gods directives may be applicable to some, for instance to Jews and to Christians together. Certain prescriptions of the Old Testament are binding on Jews only. Other prescriptions of the New Testament are obliging for Christians only (Book I, ch. I, §§ XV–XVII incl.).

Another class of law is that which is "allowed by the law of nature, not absolutely, but according to a certain state of affairs. Thus, by the law of nature, before property was introduced, everyone had a right to the use of whatever he found unoccupied; and before laws were enacted, to avenge his personal injuries by force" (Book I, ch. I, § X in fine). Here we find the notion of original communism and the notion of original appropriation of the fruits of nature, and the notion of the natural right to punish others from primitive revenge.

Human volitional right is either a civil right or a right that is more or less extensive. "The civil right is that which is derived from the civil power. The civil power is the sovereign body of free men, united together in order to enjoy common rights and advantages. The less extensive right, and not derived from the civil power itself, although subject to it, is various, comprehending the authority of parent over children, masters over servants, and the like. But the law of nations is a more extensive right, deriving its authority from the consent of all, or at least of many nations." (Book I, ch. I, § XIV) and "by mutual consent it has become possible that certain laws should originate as between all states or a great many states; and it is apparent that the laws thus originating had in view the advantage, not of particular states, but of the great society of states. And this is what is called the law of nations, whenever we distinguish that term from the law of nature." (Prol. 17).

The concept of the sovereign state is defined thus "That power is called sovereign, whose actions are not subject to the control of any other power, so as to be annulled at the pleasure of any other human will [...] the common subject of sovereign power is the state" (Book I., ch. III, § VII).

The law of nature relates not only to what exists independently of human will, but also to what follows from the exercise of human will. "Thus property, as now in use, was at first a creature of the human will. But after it was established, one man was prohibited by the law of nature from the seizing the property of another against his will" (Book I, ch. 1, § X, 2).

Thus international law, sovereign states, municipal law, and law of property are defined as created by man, as acts of will. This act of creation is made possible and meaningful by the law of nature and the law of nature consists of the relations of the concepts so created.

Grotius assumes, without stating it, a parallelism between law among men and law among nations. But the latter is incomplete in the sense that the enforcement of justice is by means of self-help. He therefore proceeds to investigate the justice of war, the ultimate means of self-help.

Investigations a posteriori

The detailed and meticulous record of empirical evidence for the doctrines and practices of international law is by far the bulk of the text. It displays vast erudition, it is not easy to read, and at times it seems self-indulgent. But though he frequently uses the term authorities his method is critical in two ways.

Although the hurried writings, which took him only one year, made him make many errors, his early humanistic training in critical examination and edition of classical texts served him well. In the many editions in his own lifetime, he made thousands of corrections.

The abstract natural-law theory served as a method of criticism of the logical consistency of the received wisdom, where even his favourites, such as Cicero, do not escape criticism (Book I, ch. III, § III in fine).

A modern reader, however, will find the lack of attention to the historical socio-economic and cultural context a grave shortcoming. Like other humanists he quoted presocratic Greek philosophers alongside post-socratic, with Biblical texts and Christian philosophers and historical experiences from any age in one pell mell.

His answer to the first question 'is war at all justifiable', is affirmative (Book 1, ch. II)

Hence, a just war by natural law as well as tradition is war waged for the following reasons: defence, recovery of property or debts or punishment of offences committed. This is a decisive break with the age of feudalism where the lay or monastic orders of the knights existed for the purpose of warfare. Not only does Grotius repeat Lipsius' saying that war must be waged in order to restore peace (Book I, ch. I), he also circumscribes the causes that can justify it.

With respect to defence the concept of pre-emptive strike is expressly rejected, it may be expedient but is not just.

With respect to recovery of property he dedicated nine chapters to a detailed analysis of property law.

With respect to obligations he dedicated the next six chapters to promises, contracts, and treaties and one chapter to damages out of contract.

The most difficult subject by far is punishment as a just cause for war

At first it seems that he bases it on historical evidence (Book II, ch. XX, § XXXVIII), but then he also claims that it is a right resulting entirely from the law of nature (§ XL in fine) He expressly rejects Victoria and the Salamanca school who "suppose punishment to be an effect purely arising from the authority of civil law" (ibid). The natural law basis is "the liberty of inflicting punishment for the peace and welfare of society, which belonged to individuals in the early ages of the world" (ibid).

There seem to be some conceptual and theoretical confusion. When Grotius defines war as a test of force or as a duel (Book I, ch. I), the parties are obliged to abide by the law of war, as the law of duel obliges duellists. They are to treat each other with respect, and certain deceptions are dishonest and unlawful, while others are part of the game. If these rules are followed, there is no kind of reproach involved in the struggle[27]. Punishment is different, involving a relation of superiority and inferiority and involves a notion of the end of "punishment, which some philosophers have called correction, some chastisement, and others admonition" (Book II, ch. XX, § VI). "The power of inflicting the punishment, subservient to this end, is allowed by the law of nature to any one of competent judgment, and not implicated in similar or equal offences." (Book II, ch. XX, § VII). After a long discussion "it may be inferred how unsafe it is for a private Christian, whether from motives of personal interest, or from those of the public good, to take upon himself the punishment of an offender, and particularly to inflict death." (Book II, ch. XX, § XIV). With these provisos, few, if any, would ever be justified in punishing others, and what comes out of the discussion is the cautious "pronouncing all wars to be just, that are made upon pirates, general robbers, and enemies of the human race [...] those who have renounced the ties and law of nature" (Book II, ch. XX, § XL).

Offences against God are considered as primarily a relation between the soul of one man and his maker (Book II, ch. XX, § XLIV) and it is held to be "plain that no force should be used with nations to promote its acceptance" (Book II, ch. XX, § XLVIII). "But to obstruct the teachers of Christianity by pains and penalties is undoubtedly contrary to natural law and reason", the reason why being that "Nor indeed can any danger be apprehended from the spreading of doctrines, calculated to inspire greater sanctity of manners, and the purest principles of obedience to lawful sovereigns" (Book II, ch. XX, § XLIX).

It seems that Grotius' position is very close to that of the theologians and jurists of Salamanca, but still it is definitely different. It was no longer the obstruction of missionaries per se that was contrary to law of nature, rather it was the obstruction by pains and penalties, and further this obstruction was not declared a just cause of war, it was only contrary to natural law. And the reason for declaring it contrary to natural law was not the unique truth of Christianity, but the inoffensiveness of its teachings.

Neither popular sovereignty nor absolutism from natural law

Although Grotius held that civil law, also called municipal law within states is based upon the social contract, he vehemently rejects the theory that "sovereign power is vested in the people, so that they have a right to restrain and punish kings for an abuse of their power" (Book I, ch. III, § VIII).

Rather he claims that this is solely a matter of the positive arrangement of a constitution. Some people subject themselves unconditionally and "entirely relinquish their rights" (ibid) while in the Athenian Republic "The city is not governed by one man, but in popular form" (ibid). The estates general may "In some places [...] serve as a greater council to the King [...] leaving him [...] full liberty to exercise his own discretion [...]. But in other places they form a body with power to inquire

[27] Hampshire, S.: *Justice is Conflict*, Princeton University Press, Princeton New Jersey, USA 2000.

into the prince's measure and to make laws" (Book I, ch. III, § X). His arguments in this respect are purely positivist. Direct or indirect democracy, republicanism, unlimited monarchy, limited monarchy all depend upon the actual and particular constitutional arrangements, and not on natural law of either the inalienable sovereignty of the people as maintained by the Spanish theologians and jurists and later Rousseau and Kant or the absolute royal sovereignty as maintained by Bodin, Hobbes, and Filmer.

Newer doctrines of natural law hold democracy to be a natural right and to follow from reason.

Human rights

To a modern reader interested in human rights Grotius seems passé. In his analysis men can sell themselves to slavery as well as to servitude.

He wrote before the civil rights of the Magna Carta were declared to be universal rights of man[28]. In his analysis they were positive volitional law and his position was close to that of the common lawyers. They were fiercely contested in England in the first half of the 17th century. The civil rights and freedoms were based on the pact between King John and the magnates and popular representatives of the Realm in the two houses of Parliament. Interpretation thereof was to be agreed by both parties to the pact. Only after the Glorious Revolution the new King William of Oranje Nassau and his Queen Mary had to accept the sovereign power of Parliament to pass the Bill of Rights as legislation, as statute law. In the American Colonies the settlers also wanted a constitutional Bill of Right, even if they were setting up a Republican Constitution with a popularly elected president and a representative legislator. In response to the popular wish for something even stronger and more binding upon future rulers, the civil rights and liberties were formulated as inviolable and inalienable universal rights in line with the new philosophies of John Locke. The US declaration was exported back to Europe and further elaborated in the Declaration de les Droits des Hommes and further in the 2nd half of the last century in the UN Declaration of 1948 and the European treaty on Human Rights of which the latter definitely is positive treaty law.

The ambiguity between discovery and invention persists. Human rights are both established by human will of what shall be and claim to be discoveries of what is.

6 In Conclusion

The critique of natural law as freewheeling dogmatic speculation is clearly unreasonable in this instance. But another more refined standard critique by modern positivist legal theorists is that natural law is simply tautological. All deductions must be contained in the assumptions. But this is not a serious objection, however.

Tautology is circular, but there is nothing surprising or objectionable in circularity when you analyse complex concepts by means of their relations to other concepts. Not infrequently you may go in a circle and arrive at one of the concepts that you have touched upon before. Circularity is empty and objectionable only if you move in small circles[29].

The method of progressive complexity in abstract moral and legal reasoning was characterised by Grotius in the above mentioned discussion of war as punishment where he repeats and sums up his warnings against being carried away by the opinion that custom is a part of natural law. The third warning is "to make an accurate distinction between general principles, such as the duty of living according to the dictates of reason, and those of a more particular though not less obvious meaning; as the duty of forbearing to take what belongs to others. To which many truths may be added though not quite so easy of apprehension: among which may be named the cruelty of that kind of punishment, which consists in revenge, delighting in the pain of another. This a method of proof similar to that which occurs in mathematics, the process of which rises from self-evident truth to demonstrations, the latter of which, though not intelligible to all alike, upon due examination obtain assent" (Book II, ch. XX, § XLI–XLIII).

The question raised above concerning the character of the natural law developed by Grotius as the system of the necessary relations between moral legal phenomena has been answered, I hope.

The distinction drawn by Grotius between natural unchangeable law and volitional and diverse law is tenable, I think, and resolves the problem of the blurred border between discovery and invention.

But new developments create new ambiguities between discovery of what is and invention of something new. Among these I have only touched upon the issue of human rights. Another is so-called humanitarian intervention[30]. Both are exceedingly complex, yet most often reduced to simplistic slogans.

Grotius' reasoning about international law may in some aspects be similar to the reasoning in mathematics. But the lean style of writing is not that of Grotius. He turns instead every stone on his path and wanders often far off the track and demonstrates his superior classical humanistic learning.

In his writings on international law Grotius was first and foremost a man of practical affairs, a statesman and a jurist, who presented his case as convincingly

[28] I have presented a more detailed examination of the struggles from the petition of rights to the bill of rights and the transformation of the rights from particular rights to universal in Jarvad, I. M.: *Fra sædvaneret til konstruktivisme – Om menneskerettighedernes retsfilosofiske grundlag, Filosofi og videnskabsteori*, RUC, 3.række Preprints and reprints No. 3, 1999. For the present purpose I will make the presentation very brief.

[29] Strawson, P. F.: *Analysis and Metaphysics: An Introduction to Philosophy*, Oxford University Press 1992. I have attempted to apply Strawson's analytic philosophy to legal theory in Jarvad, I. M.: *Refleksivitet og selvreferens i retten – eller magt*, Blume, Ketscher, Rønsholdt (eds.) *Liv Arbejde og Forvaltning, Festskrift til Ole Krarup*, København 1995, pp. 55–68.

[30] I have tried to discuss it in another paper: Enforcing human rights and democracy and European consciousness, in: Mongardini, C. (ed.): *La Nascita di una conscienza europea*, Roma 2001, pp. 81–89.

and persuasively as possible. When not engaged in such pursuits he was attracted to literature and to theology.

Above I have tried to lay bare the foundations of his natural law and shown in some respects how it was applied as a critical standard towards what was received from state practice, history and authorities. I do think these foundations of his natural law theory deserve attention for they became the point of departure for lawyers and philosophers for centuries to come.

The first point to stress concerning Grotius' *basic concepts of natural law* is the fact that they are narrow. In comparison to the rich Aristotelian, Thomist and neo-Thomist notions of the good life[31], they are lean. What constitutes the good life is for everyman to decide.

The second is characterizing *man as a being with a social appetite.* It is not clarified directly in the text what it really means to repeat this stoic commonplace. However, I suggest that we must infer that is not an appetite or a drive as in Hobbes' mechanical concept of man, Nor is it an idea in man's mind. It must be concluded from the presented arguments, that it is innate, but unlike the instinctive sociablenes of – some – animals, it seems to be to be an orientation of the rational mind, a *directedness of reason.*

The powers of man are twofold

The *power of discrimination* is presented as the power to discriminate between beneficial and harmful. This seems conceptually unclear. Grazing animals know instinctively to avoid certain poisonous vegetables, and may learn by experience to avoid others. Carnivorous animals learn that toads taste bad. This is not the sense of good and bad, which Grotius had in mind. The issue for the power of discrimination is not the utility or the satisfaction of appetites or instinctive drives as in Hobbes and those moral philosophers and economists who propagated the self-love theory. It might be an ability to discriminate between what is good and what is bad for society. But even if we assume a functional usefulness for society as the (phenomenological) orientation of reason, the concept of justice still escapes us. For it may be functional, yet unjust, for society to punish someone innocent as a scapegoat.

From the text even in the brief excerpts here presented it is clear that Grotius is not concerned with survival as the quote from Seneca could suggest, nor with the agreeable, or the good life, *eudaimon*, but with justice tempered with mercy and compassion. This capacity must therefore be the *human intuition to grasp what is right and wrong and therefore just and unjust.*

The second power of man is the *power of obliging oneself* by explicit declaration of one's will or by tacit consent. All that is not natural law or God's volitional law is human law based on contract and therefore obligations derived from explicit or

31 See for instance the neo-thomists Maritain, J.: *Natural Law,* South Bend, Indiana 2001, originally articles from 1943, 1951, 1952 and Finnis, J.: *Natural Law and Natural Rights,* Oxford 1980, For a non catholic neo-aristotelianism see Nussbaum, M.: *The Theraphy of Desire: Theory and Practice in Hellenistic Ethics,* Princeton Univ.Press 1994, and *Loves' Knowledge: Essays on Philosophy and Literature,* New York Oxford Univ. Press 1990.

tacit *human will to be obliged*. This is not a commonplace tautology but a total repudiation of the medieval notion of justice as based on consuetudines and/or authority. Consuetudines in themselves constituted law and justice, just as authority per se implied the justice of lawmaking, *rex ius in pectore habet*; the king has justice in his breast. Instead Grotius based all human law on consent as auto-obligation. In order to do so he stated the power of *auto-obligation* as a natural capacity of man only in the restricted sense of a self evident truth or tautology, namely that otherwise the term of promise and contract would be without meaning, and that the role of authority is to reinforce what is obligatory. The contract theory is not the path for pre-social man to enter society individually or to constitute it collectively; it *is the meaning of the society of the legally competent citizens.* Alongside this revolution in the general theory of law, Grotius also had the daring to create a normative theory of international relations as a proto-society of the legally competent states.

His method was not *induction* or the search for general principles in a mass of empirical data, such as the work of Gratian from around 1140 on the collected mass of the decrees of popes and ecclesiastical authorities to establish some order and system in the canon law called by himself *Concordia disconcordantium canonum* (bringing consistency among the inconsistent canons) to name an illustrious example.

Nor is his method *axiomatic deductive* as the systems of law of Hobbes or of ethics by Spinoza.

But his method is the *confrontation between deductively produced theorems with empirical observations stressing repeatedly that natural law is the law that make volitional or customary positive law valid and obliging.*

7 Perspectives: Is the Grotian Approach Still Relevant?

The application of mathematics to physics and civil and military engineering in the 16th and 17th centuries revolutionised warfare. In war between equals it was imperative to mobilise all available resources, scientists included. It is not unreasonable to point out that the application of mathematics to population statistics, health, education, productive capabilities and resources, and means of taxation was necessary to meet the rising costs of war. Alongside new technologies of administration appeared a growing awareness of nationalism and religion as means to obtain ideological support for bellicose activities.

In this paper I have examined the hypothesis about the relation between mathematics and international law of war and peace, leaving Grotius' theology and its significance for the law of war and peace for another occasion. The international law of war and peace presented an alternative to the mobilisation of societies to defence. Grotius performed this theoretical breakthrough amidst the 30 Years War between Protestants and Catholics.

Grotius broke with the medieval fusion of Aristotelianism and Christianity and like his contemporaries in mathematics, he grounded his theory on minimalist axioms. He did not reject the modern way of thinking; rather he embraced it and put it to constructive use as foundation for his theory about the normative relations between states.

His theory was and is controversial. The opposed theory of international relations is the so-called 'realist' theory of international relations. Realists see his theory as naïve and idealistic. At the end of the day only crude physical power counts.

Hobbes formulated the realist theory a few years after Grotius. Hobbes held that the state of nature is a-moral and a-legal and only self-interest reigns. Everyone is free to grab everything and only fear of retaliation by others will lead the prudent to restraint. There is really no difference between means of defence and of attack in so far as both may be seen as threatening; countermeasures are imperative; armaments must spiral and armament costs will frustrate the satisfaction of appetites. Hobbes held that this logic must lead rational agents to create by contract a sovereign monopoly on physical force. Hobbes held further and in opposition to Grotius that international relations must remain an a-moral state of nature. It is not very clear how or why Hobbes arrived at this position.

Most often the relevance or irrelevance of the Grotian approach is argued in terms of whether international relations are best described as a lawless Hobbesian world or as a moral-legal Grotian world. This debate, however, misses the point of the axiomatic-deductive method of international law.

The debate over whether the world of international relations is best described as a Hobbesian lawless anarchy or as a Grotian moral legal society is in itself an application of the Grotian approach. The empirical observations of shortcomings do not disprove the Grotian approach. Instead they demonstrate the various moral and legal shortcomings of the actors in international relations.

If we consider European international relations, including North America, the Grotian approach has by and large been a success. This underscores the soundness of his assumption of the natural moral-legal capacity of man and human societies. Norms of war have been established between belligerents and their uniformed armed forces and towards third parties. Since the WW II the two main pacts held each other in check by mutual fear (MAD) and even if the Western alliance threatened with pre-emptive strike on enemy territory, which is outlawed as aggression under the UN-Pact this has been on a verbal level only.

If we consider the global international relations, however, asymmetrical warfare has been going on all the time since the great discoveries in the 15th century. Even if the extension of nationhood to old and new states in the formal organisations of the League of Nations and the UN was a relative success for the building of a Grotian *world community*, the national sovereignty of states on the periphery of the global economy remains precarious.

Intervention and staging of coups d'etat in states in the 'spheres of influence' of great powers has been and still is the disorder of the day.

Technologies of aerial bombardments, rockets and steering systems nourish fascinating and dangerous illusions about the possibility of conducting war to punish wrongdoers with minimal risks to the executioner. It seems reasonable that neither airplanes and their crews nor the home bases can be hit back immediately. The targets of modern asymmetrical war are like sitting ducks.

Such warfare may be conducted for humanitarian reasons as former colonial wars were fought to spread Christianity and civilisation, but motives were and are mixed. Conquest and free access to the valuable resources of other nations were and are the tempting spoils of war.

The weakness of the Grotian approach in this respect is the notion of punitive war to which he expressed numerous reservations, but which he was not ready to dismiss from his system of law. Punitive war may easily be a pretext and it unavoidably hits the innocent, and is therefore unjust. War as proper self-help and duel is not.

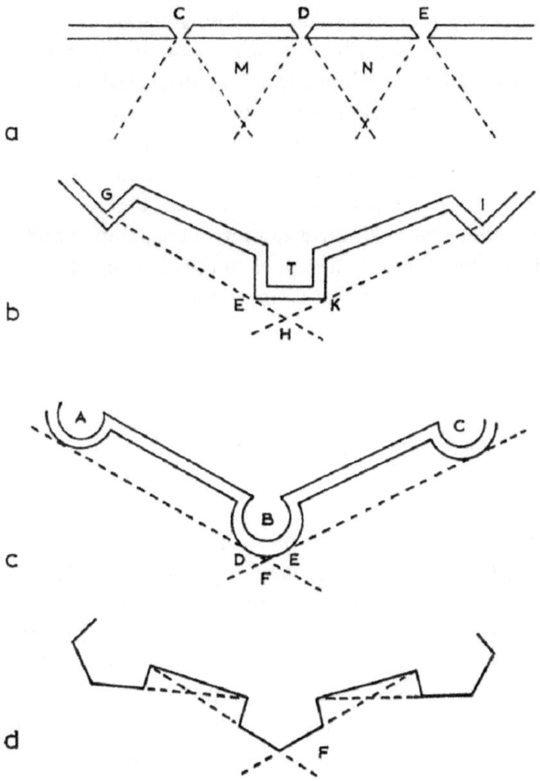

Figure 2. Genuine mathematical rationality was important in 16th century's changes of military practices. Around 1500 a new efficiency artillery called for re-design of fortifications. Simon Stevin's *Art of Fortification* from 1594 shows the evolution of the mathematical principles inherent in fortification, beginning with the high-medieval pre-fire-arms military castle (a) through the introduction of square protruding towers reducing dead angles (b), how this was improved by making the towers circular (c), and finally the system which lasted from Stevin's to Napoleon's times (d). Such innovations, together with the bookkeeping of Dutch commerce provides the background for Grotius' appeal to almost-mathematical reason. It might be interesting that the Latin translation of Stevin's *De havenvinding* of 1599 was by the young Grotius. [Acknowledgement: Simon Stevin's *De Stercktenbouwingh* (The Art of Fortification), 1594. Reproduced from Dijksterhuis, Simon Stevin - Science in the Netherlands around 1600, Martinus Nijhoff, The Hague 1970, p. 107]

Calculated Security? Mathematical Modelling of Conflict and Cooperation

JÜRGEN SCHEFFRAN*

The pioneering work of Lewis Fry Richardson on modelling the arms race raised expectations that mathematics can contribute to peace and conflict resolution. Based upon Richardson's model, various extensions are discussed, with a focus on time-discrete nonlinear models showing chaotic behavior. As a general framework for the modelling of conflict and cooperation in international security a multi-actor dynamic game is introduced, with mathematical conditions for unstable interaction, potentially leading to violent conflict, arms race and war. On the other hand, the approach provides a basis for the evolution of cooperation and coalition formation. In more detail, the case of instabilities in the offense-defense competition is discussed and the role of uncertainties in complex international relations.

1 Conflict Modelling in a Complex World

While the simple two-player arms race became the paradigm of conflict studies during the Cold War [Rapoport 1960], many of the world's current conflicts are more adequately described by complex multi-actor dynamic games. Decisions and interactions are shaped by a variety of actors and factors, which could provoke instability, rapidly changing security conditions and the outbreak of violence, making conflict-resolution more difficult. Not only the arsenals of armament are relevant for security, but also economic and technological interconnections as well as social and ecological factors, on global and regional levels. Security and sustainability are increasingly linked.

Understanding the emergence of collective behavior and the evolution of cooperation is a dynamic field of current interdisciplinary research, and knowledge transfer between the natural and social sciences can be highly fruitful. Mathematical modelling and its computer-based implementation can contribute to a deeper understanding and the development of new instruments for conflict-resolution and cooperation, disarmament, security and peace-building. Different mathemati-

* Potsdam Institute for Climate Impact Research, Germany.
Email: scheffran@pik-potsdam.de

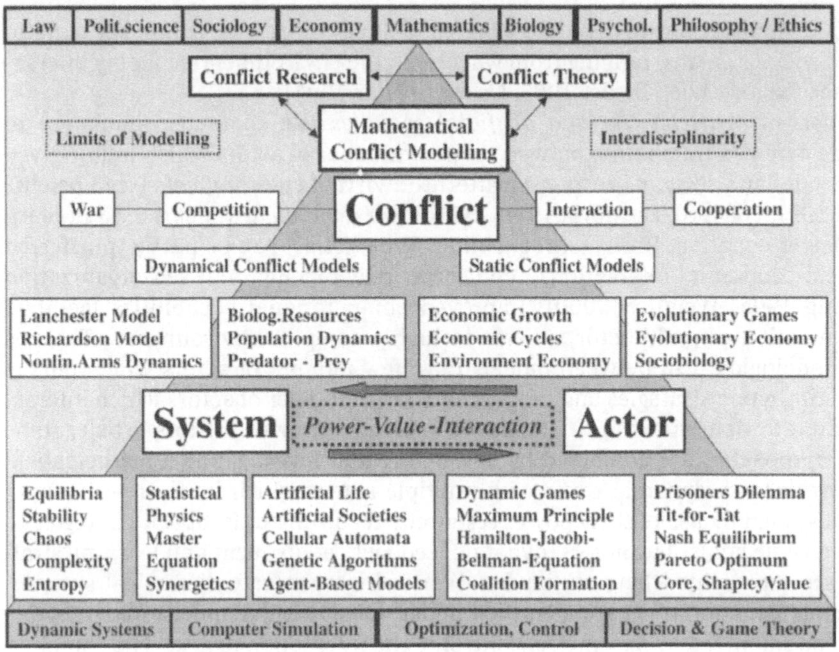

Figure 1. Landscape of mathematical conflict models.

cal methods have been applied to this field (on some see [Neuneck 1995]; a survey is given in Figure 1). The theory of *dynamical systems* determines equilibria and stability of differential and difference equations, as well as chaos, self-organization and phase transitions [Krabs 1998; for applications see Braun et al. 1983]. *Control theory* helps to determine optimal paths in dynamical systems. *Game theory*, based on the epochal work of von Neumann and Morgenstern 1944, analyzes rational choice among different options depending on decisions by other players.

Dynamic games deal with repeated game situations in which players interactively adapt their behavior to their environment, according to their own incentives, preferences and expected outcomes as well as the information sets and decisions taken by other players.

Differential games extend Pontryagin's maximum principle to the optimal control of dynamical systems by a few number of players, seeking to optimize their individual payoff functions for a given time-period (see for instance Dockner et al. 2000). The *repeated prisoners' dilemma* has been used to analyze the evolution of cooperation in experimental games [Axelrod 1984, 1997]. Increasing attention is focused on the link between *cooperative dynamic games* and *coalition formation*. *Evolutionary games* analyze the competition among populations of game-strategies depending on their fitness in the replica equation (for a survey see [Hofbauer/Sigmund 1998]). In economic *oligopoly theory* the competition of firms can be represented by the adaptation towards reaction functions according to Nash-Cournot strategies, the so-called "tatonnement process" [Simaan/Cruz 1976, Szidarovsky/

Li 2000]. Applications of dynamic-game models comprise a large number of natural and social systems, ranging from warfare to the environment-economy interaction (see for instance [Olsder 1995, Carraro/Filar 1995]).

Models of *artificial life and artificial societies* use computer simulation to analyze complex interaction between a great number of actors which follow given action rules and stimulus-response patterns in virtual environments [see Epstein/Axtell 1997, Gaylord/D'Andria 1998]. *Nonlinear dynamical systems* used in physics (Master equation, Boltzmann equation, Synergetics) can be partly transferred to socio-economic interactions, collective phenomena and self-organization [Helbing 1995, Weidlich 2000]. Improved computational capabilities facilitate the simulation of multi-actor systems in highly complex environments. There is a methodological gap between models for a few number of actors which pursue optimizing game strategies and models for a large number of actors which interact according to deterministic or stochastic rules of behavior. In this article, some of the approaches are combined by developing and investigating a mathematical framework for a dynamic game with multiple actors which in discrete time act upon a system by use of their power resources according to feedback strategies to achieve value goals. The actors' dynamic feedback, adjustment and learning strategies are represented by reaction functions which generate dynamical systems. Mathematical conditions are identified under which the dynamic system is stable or can be stabilized by parametric control of the interaction matrix. The combination of cooperative and dynamic game theory with computer-based multi-actor modelling provides a basis to further develop instruments of decision-support that can be used in international relations and economy as well as in other fields.

2 Weather, Arms and Mathematics – Richardson's World

Lewis Fry Richardson (1881–1953), a british physicist, psychologist and pacifist, applied mathematics to better understand arms race and war. He also made important contributions to weather forecasting, but the main drawback of his mathematical technique was the time necessary to produce a weather forecast. It usually took three months to predict the weather of the following day, but with the electronic computers after World War II, Richardson's method of weather prediction became more practical.[1]

Less known during his lifetime remained Richardson's work on mathematical conflict modelling [Hess 1995]. He tried to predict and prevent war by finding general laws, common to all nations. Being concerned that the arms race between the major powers in Europe during the 1930s could lead to another major war, he derived a set of differential equations that would describe the arms buildup, using empirical data from the First World War to fit the curves [Richardson 1960ab]. He was aware of the strengths as well as the weaknesses of applying mathematics to social phenomena:

[1] On Richardson's life, work and publications see Ashford 1985, 1993.

Figure 2. Lewis Fry Richardson (1881–1953).

> To have to translate one's own verbal statements into mathematical formulae compels one carefully to scrutinize the ideas therein expressed. Next the possession of formulae makes it much easier to deduce the consequences. In this way absurd implications, which might have passed unnoticed in a verbal statement, are brought clearly into view and stimulate one to amend the formula. An additional advantage of a mathematical model is its brevity, which greatly diminishes the labor of memorizing the idea expressed. If the statements of an individual become the subject of a controversy, this definiteness and brevity lead to a speeding up of discussions over disputable points, so that obscurities can be cleared away, errors refuted, and truth found and expressed more quickly. [...] Mathematical expressions have, however, their special tendencies to pervert thought: the definiteness may be spurious, existing in the equations but not in the phenomena to be described; and the brevity may be due to the omission of the more important things, simply because they cannot be mathematized. [Richardson 1960a, p. xvii]

The basic philosophy in Richardson's model is the assumption that for two arbitrary countries each country increases its own armament level x_i proportional to the armament of an opponent x_j and reduces it proportional to its own armament $(i, j = 1, 2)$. These three terms drive the increase or decrease of armament in a linear and additive way:

$$\dot{x}_1 = \frac{dx_1}{dt} = k_1 x_2 - a_1 x_1 + g_1,$$

$$\dot{x}_2 = \frac{dx_2}{dt} = k_2 x_1 - a_2 x_2 + g_2 .$$

The *defense coefficients* k_i measure the degree to which the ith country reacts to the opponent's armament (military threat) x_j, *the fatigue coefficients* $a_i > 0$ take into account the effect of economic constraints, reducing own armament proportional to x_i . The *grievance term* g_i measures political and strategic objectives and perceptions, the "*outward attitude of threatening or cooperation*" [Richardson 1960a, p. 13]. $g_i > 0$ represents aggressive intentions, promoting an arms buildup, while $g_i < 0$ corresponds to good-will and the tendency to cooperate for an arms builddown.

The equilibrium conditions $\dot{x}_1 = \dot{x}_2 = 0$, where the armament levels of both sides do not change, leads to straight lines in the (x_1, x_2)-space of armament levels, whose intersection point (x_1^*, x_2^*) is the so-called "*balance of power*"

$$x_i^* = \frac{k_i g_j + a_j g_i}{a_1 a_2 - k_1 k_2}, \quad (i, j = 1,2),$$

provided that $a_1 a_2 - k_1 k_2 \neq 0$. For $g_1 = g_2 = 0$ the only equilibrium for all coefficients is the point (0, 0). Stability of the equilibria is determined by the eigenvalues λ of the matrix A of coefficients a_i and k_i which are the solutions of the characteristic equation $|A - \lambda I| = 0$. Asymptotic stability is guaranteed for all eigenvalues having negative real parts.[2] Stability can be characterized by the stability index

$$\sigma \equiv \frac{a_1 a_2}{k_1 k_2} > 1$$

for positive a_i, k_i. If the product of the defense coefficients k_i exceeds the product of the fatigue coefficients a_i ($\sigma < 1$) the arms race escalates (unstable arms race), while for $\sigma > 1$ the armament level approaches the equilibrium asymptotically.

Richardson extended the equations to several nations for the arms races 1909–1914 and 1933–1939, using military expenditures as armament variables x_i. These calculations supported his view that "*foreign policy had then a rather machine-like quality*" [Richardson 1960a, p. 33] and led him to conclude that increasing armaments could lead to war breakout, while a constant level of armament corresponds to a steady state without war.

3 Critique and Extensions of the Richardson Model

Richardson's model initiated a flood of publications on the armament dynamics and a debate about its applicability to real-world phenomena which raised several critical issues. Describing countries as structureless entities by one single "catch all variable" x_i was seen as questionable. Expenditures are a weak indicator of military threat and its security impact which depends on the systems procured.

[2] The eigenvalues λ describe how the trajectories approach to or escape from the equilibrium once they are slightly displaced. The equilibrium is stable if nearby solutions stay nearby for all future times. If all eigenvalues have negative real part (Re $\lambda < 0$), all nearby solutions will be attracted by the equilibrium (asymptotic stability), requiring trace$(A) < 0$ and det$(A) > 0$. For Re $\lambda > 0$ the equilibrium solution is unstable and the trajectory goes off.

The variables and coefficients of the Richardson model are difficult to measure. Expenditures are not easily available, verifiable and reliable for all countries, and difficult to compare for different countries.

An arms race does not only have quantitative aspects, but also qualitative aspects, which are important for new strategies and doctrines. Not only objective and measurable quantities are relevant, but also subjective and irrational factors, like expectations and anticipations, instincts and traditions of political leaders. The Richardson model describes "politics without personalities", where state authorities are black boxes and decisions are hidden in the budget. The Richardson coefficients are fixed parameters, decoupled from strategy and security interests. The armament dynamics is reduced to a mechanistic and deterministic interaction, where by the choice of the initial conditions $x_i(0)$ and the coefficients a_i, k_i, g_i the future is determined, leaving no room for political decisions or control. In reality, the coefficients are time dependent and influenced by strategic objectives.

Both sides are assumed to have complete knowledge of the armament levels and react instantaneously. In reality, each side has limited information about the other side's strength, and worst-case assumptions provoke higher reactions towards arms buildup. Additionally, time lags exist for information gathering, decision making and weapons acquisition. Another critical time constant is the long lead time between technological innovations and deployed weapons systems.

The linearity and simplicity of the Richardson equations leads to a few types of system behavior (oscillations, asymptotic decay, exponential increase). Reactions of real systems may be disproportionate and highly nonlinear, showing a variety of qualitatively different modes of behavior. To describe the decisionmaking on the armament dynamics, time-discrete difference equations may be more adequate than continuous differential equations. The arms buildup is not only an action-reaction process, driven by the mutual stimulation of the opponents' armament (positive k_i), but is also stimulated by a bureaucratic and budgetary eigendynamics (negative a_i), guided by a competition of domestic influences and interests. Mutual stimulation and self-stimulation are limiting cases, which are coupled and can enforce each other.

Although an arms race may provoke crisis unstable situations, it does not necessarily lead to war, either because both sides want to avoid warlike situations by building more or less weapons or because one side reaches economic, technological or political upper limits of armament. In the latter case, which excludes a permanent and unlimited arms race, the economic damage to a society by additional armament exceeds its security gain.

Such critical remarks do not only apply to the Richardson model, but to any arms race model. Several extensions and adaptations have been proposed to improve the deficiencies with regard to the following points: using the number of weapons or their lethality as a measure of the armament level; derivation of the Richardson coefficients from strategic considerations; nonlinear (economic) constraints; the application of optimal control theory and differential games; time-discretization and difference equations; the influence of the bureaucratic process; stochastic Markov processes; decision rules.

Michael D. Intriligator developed a framework for the strategic armament dynamics, discussing "*decision rules with regard to weapons procurement and their implications for stability and war initiation in the context of a dynamic model of an arms race*" [see Intriligator 1975]. Critizing the simple action-reaction scheme, Intriligator included strategic considerations, institutional aspects and decision rules [Intriligator/Brito 1985, pp. 133–134]:

> Modelling the arms race as either a mechanistic or an optimizing process, however, fails to account for the institutions of defense decision making. To the extent that these institutions are large, complex, bureaucratic organizations, as indeed they are, they tend to rely on neither passive mechanical responses nor on explicit optimization rules but rather on rules of thumb or heuristic decision rules with regard to weapons procurement.

Both sides build or destroy weapons (missiles) x_i based on a model of a strategic missile duel according to decision rules $\dot{x}_i = F_i(x_i, x_j)$ for countries $i, j = 1, 2$, matching the strategic objectives. If the acquisition of missiles is proportional to the gap between desired levels x_i^* and actual levels x_i of missiles, the decision rule has the linear form $\dot{x}_i = k_i(x_i^* - x_i)$ where x_i^* depends on the missile duel model and the strategic objectives. $k_i > 0$ is the adjustment or *reaction coefficient*. For $x_i < x_i^*$ more weapons will be built in order to reach x_i^*; for $x_i > x_i^*$, country i will reduce its missiles (e.g., for cost-saving reasons). The equilibrium lines $x_i^* = x_i$, where the armament is constant, separate the regions of deterrence and war initiation (where one side loses second strike capability) in the (x_1, x_2)-space. The resulting dynamical equations for these decision rules are linear Richardson type equations, whose coefficients can be derived from strategic considerations. A specific type of decision rules for military expenditures, combining mutual stimulation and self-stimulation, has been used by Lambelet and Luterbacher 1979.

4 Chaos and Predictability in the Arms Race

While mathematical models such as the Richardson equations help to structure the field of conflict research and international security by identifying fundamental relationships among actors and basic system variables, they are often rather simple in dealing with the complexity of reality. Here the "fog of war" (as Clausewitz named it) comes into play, that is, the lack of knowledge and a chaotic sequence of events that occurs in real conflicts.

Because of the linearity and continuity of the Richardson equations, an analytic treatment is quite simple. During the 1980s new mathematical concepts have been developed, such as the theory of complexity, chaos and nonlinear dynamics, which may be applicable to conflicts and arms race phenomena. Contrary to the well-ordered world of Newtonian mechanics, symbolized by the predictable swinging pendulum or the regular movement of the celestial bodies, is the unpredictability of everyday-life experience. Since the 1970s the natural sciences have begun to systematically explore critical phenomena, such as self organization, discontinuous phase transitions, and catastrophe theory. Methods and results were transferred to the social sciences. Chaos became the paradigm for the turbulent transformation of the international system after 1989.

The concept of deterministic chaos was developed to study simple nonlinear phenomena with complicated dynamics, in which it is "practically impossible to predict the long-term behavior of these systems, because in practice one can only fix their initial conditions with finite accuracy, and errors increase exponentially fast."[Schuster 1988, pp. 3–4]

Many properties of chaos have been studied for the time-discrete logistic mapping $x_{t+1} = f_r(x_t) = rx_t(1 - x_t)$, where $r \in [0, 4]$ is the reaction coefficient (order parameter), whose value determines the transition from predictability to chaos. For small r the variable x_t converges against the stable *fixed point* $x^* = 0$, independent of the initial conditions. For $r > 3$ the single fixed point bifurcates into *periodic cycles.* For large times a chaotic region is reached. An *attractor* is a set of system states on which the time evolution accumulates over longer periods. For *dissipative* systems, where phase-space volumes on the attractor are separated exponentially, one speaks of *strange attractors.*

The concept of chaos as a model for arms race and war outbreak was introduced by [Saperstein 1984, 1986], to show that even simple nonlinear deterministic arms race models may lead to the breakdown of predictability, which is defined in this context as [Saperstein 1984, p. 303]

> a situation in which small perturbations of initial conditions, such as malfunctions of early-warning radar systems or irrational acts of individuals disobeying orders, lead to large unforeseen changes in the solutions to the dynamical equations of the model. There is no way then to predict the effect of the actions of any participant analyst, planner, statesman or general with any certainty.

Saperstein used a pair of nonlinear difference equations with quadratic mappings for two variables, denoting the fractions of the available resources which two countries pay annually. The problem of chaotic dynamics in arms race models was further investigated by Grossmann/Mayer-Kress 1989. The nonlinear difference equations have the following Richardson-like form but with discrete time and a damping of the fraction of expenditure x_t and y_t in year t at the upper cost limits x_m and y_m :

$$\Delta x_t = x_{t+1} - x_t = (x_m - x_t)(-k_{11}(x_t - x_s) + k_{12} \cdot y_t),$$

$$\Delta y_t = y_{t+1} - y_t = (y_m - y_t)(-k_{22}(y_t - y_s) + k_{21} \cdot x_t),$$

where $\Delta x_t, \Delta y_t$ is the change in the armament expenditure between the years t and $t + 1$, and $k_{ii}, k_{ij} > 0$ are the reaction coefficients corresponding to the defense and fatigue coefficients of Richardson. In addition, x_s, y_s represent self-establishing armament levels, comparable to Richardson's grievance terms and Intriligator's decision rules.

Factors provoking chaotic behavior are multinational interactions (more than two actors), overshooting or underestimation, hectic responses, delay in information processing or discretization. There is an important distinction between chaos and instability:

> In particular, regarding the arms race and international relations, it is wrong to identify the general onset of bounded chaos with the outbreak of a war or another global crisis. The really dangerous case is instability. By this we mean that the $x_\alpha(t)$ leave their previously bounded range rapidly and dramatically. [Grossmann/Mayer-Kress 1989, p. 702]

In the asymmetric case

$$\sigma = \frac{k_{11}k_{22}}{k_{12}k_{21}} > 1$$

is a condition for stability of a fixed point, which corresponds to Richardson's stability condition.

A nonlinear time-discrete model, using decision rules for weapons procurement, was derived in 1988 by Saperstein and Mayer-Kress to simulate the implications of ballistic missile defense systems (SDI) on the arms race between the USA and the former USSR. Two questions have been raised in this context [Saperstein 1988b, p. 41]:

- Can we reasonably predict that there will be a transition from the present offense-dominated strategic nuclear confrontation to a defensive security posture?
- Can we be reasonably assured that predictions about the transition from offense-dominated strategic security postures will actually be fulfilled? Or will the resulting world be chaotic, so sensitive to small changes (small threats) as to be 'crisis unstable'?

To answer these questions, a nonlinear model was formulated, describing the dynamical interaction between the number of intercontinental ballistic missiles (ICBMs), anti-ballistic missile (ABM) satellites, anti-ABM systems (such as anti-satellite weapons, ASAT). Each procurement step depends on the anticipated role of the weapons in a simple two-strike model of nuclear war and on the immediately preceding steps of the opponent. The "*weapons builders recursion relations*" are coupled dynamical difference equations for the deployment of the three weapon types, having the form of deterministic decision rules with nonlinear constraints due to maximum costs. The time evolution (arms race) is simulated numerically under various parameter conditions (scenarios), starting from a set of initial conditions. In most cases, the arms race remains predictable and offense-dominated. But if the production rates are increased by a factor of 10, a chaotic transition occurs [Saperstein 1988b, p. 43]:

- When the model does not turn chaotic the present offense dominance does not evolve to the desired defensive configuration.
- Conversely, when the situation evolves from offense to defense, the model turns chaotic. The model seems to indicate that if SDI evolves to the point where the defense can overwhelm the offense, the result is a crisis-unstable international system that cannot be mathematically distinguished from a system of war.

5 A Dynamic Multi-Actor Conflict Model

The Richardson and other arms race models can be embedded into a larger framework of conflict modelling based on a multi-actor dynamic game of power-value interaction, called the VCX model [Scheffran 2001]. It describes the interaction among multiple actors which influence their system environments by using power resources to pursue their objectives. For players acting upon the same system, conflicts may occur if actions and objectives are incompatible. With such a general framework it is possible to study the complexity and instability of multi-actor constellations in more detail. In particular, it is possible to express problems of war and peace, of arms race and disarmament as well as environmental conflicts and international security problems within the model framework.

5.1 The representative actor

In order to understand the interaction among multiple actors, it is important first to analyze the actions of an individual or representative actor (see Figure 3) which acts in a system environment (context or situation), represented by relevant *system variables*, defining a system state x, which in general is a vector but will be treated as a scalar in this section. The system state is observed and evaluated by the actor with a problem-specific *value function* $V(x)$, and compared to a target value V^*. The actor decides on investing its *power resources* (costs) C into a particular *action* $a(x, C)$ that changes the system state by $\Delta x = g(x, a)$ to approach the state-dependent value function $V(x)$ towards the target value V^* in repeated action cycles. The flow of power resources is allocated with a priority p to several action alternatives.

In several cases (such as buying goods on a market for a fixed price) the action a and the induced system change is not state-dependent and can thus be completely controlled by the power resources. Then by use of C an actor changes the system state $\Delta x = g(a) = C/c$ (with unit cost or price c), resulting in a value gain $V = v \cdot \Delta x = f \cdot C$ where v is the value per unit of system changes and $f = v/c$ is the efficiency of resource input with regard to the value output (also called benefit-cost ratio). If an actor aims at achieving a particular value gain $V = V^*$, this leads to the required power resources $C^* = V^*/f$.

If at a given time t the actual resource flow $C(t)$ differs from the required one $C^*(t)$, this gap can be bridged by an adaptation process among power resources according to "decision rules":

$$\Delta C(t) = C(t+1) - C(t) = \alpha(C^*(t) - C(t)) = (\alpha / f)(V^*(t) - V(t))$$

where α is the reaction strength. The underlying decision rule is to increase power resource input as long as $V < V^*$ and to decrease it if the goal is exceeded ($V > V^*$). In the satisfactory state $V = V^*$ the resource input remains constant. The dynamical system describes a single actor who pursues a goal by use of power resources.

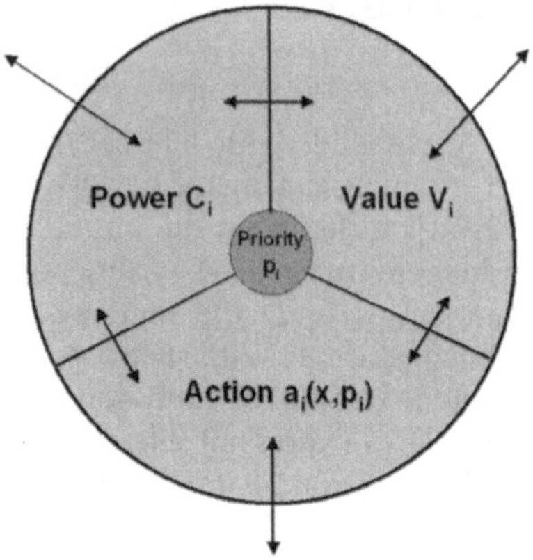

Figure 3.
The representative actor and its basic variables.

The actor has to take into consideration that its resource flow should not exceed the upper limit C^+ of available power resources, such that $0 \leq C \leq C^+$. In case of $C^+ < C^*$ the resources are insufficient to achieve the objective. In this case the actor can try to improve the action efficiency f which implies that the same objective V^* can be achieved at lower cost C^*. This may be done by switching to another action 2 with higher efficiency $f^2 > f^1$ than the first one. This implies that a higher share p of the power resources is allocated to this new option, at the cost of a lower share $1 - p$ for the old option. Then the overall efficiency of the mixed action is $f(p) = (1-p)f^1 + p \cdot f^2$.

5.2 From action to interaction

If several representative actors act upon the same system, an interaction evolves which is controlled by n actors (in the following called players) according to their strategies. During the interaction process each player P_i ($i = 1,..., n$) invests its power resources C_i into m system variables x^k ($k = 1,..., m$), and evaluates the outcome of the combined actions according to its own value criteria V_i to derive new actions in the next time step. The players can adapt the amount C_i and allocation of power resources with regard p_i^k to variables x^k, leading to an induced change $\Delta x_i^k = (p_i^k / c_i^k) C_i$, where the condition

$$\sum_{k=1}^{m} p_i^k = 1$$

is to be satisfied. Important is the learning capability of the players to adapt the resource allocation to achieve the value target V_i^*. In the linear case the value change ($i = 1,..., n$)

$$V_i = \sum_j f_{ij}(p) C_j + V_i^U$$

depends on the power resources of all players, weighted by the mutual efficiencies f_{ij} which describe the value impact of players on each other per resource unit. V_i^U includes all other value gains which can be important but are neglected in the following analysis. Now the target conditions $V_i = V_i^*$ cannot be achieved by one player alone but depends on the actions and powers of others according to (i = 1,..., n)

$$C_i^* = \frac{V_i^* - \sum_{j \neq i} f_{ij} C_j}{f_{ii}}.$$

These target conditions correspond to reaction curves (which in general are multidimensional hyperplanes) which lead to an adaptation $\Delta C_i = \alpha_i (C_i^* - C_i)$ in the space of power resources (costs) and thus a multi-player dynamical system. The intersection of the reaction curves is the cost equilibrium vector $\hat{C} = F^{-1} V^*$ (balance of power) where det F is the determinant of the interaction matrix $F = (f_{ij})_{i,j=1,...,n}$. The equilibrium coordinates can be determined by Cramer's Rule, which for two players is $\hat{C}_i = (f_{jj} V_i^* - f_{ij} V_j^*) / \det F$.

Since the interaction efficiencies $f_{ij}(p)$ between the players depend on the vector of the allocation priorities $p = (p_1,..., p_n)$ of all players, each of them can control the dynamical system by changing their own p_i (which themselves can be vectors) to achieve the respective target equilibria and stabilize or destabilize it. If the action priorities p_i are changed towards hostile relations, the equilibrium moves towards higher costs, for a cooperative relation towards lower costs (see the stylized symmetric cases in Figure 4). In asymmetric cases one player can show cooperative attitudes to another player, which in return can show hostile attitudes, and vice versa. The players can also negotiate and cooperate on the right choice of their own p_i which can be expressed as a dynamic game problem. In addition they can form coalitions by pooling some of their resources and redistribute the gains (or losses) to the individual players.

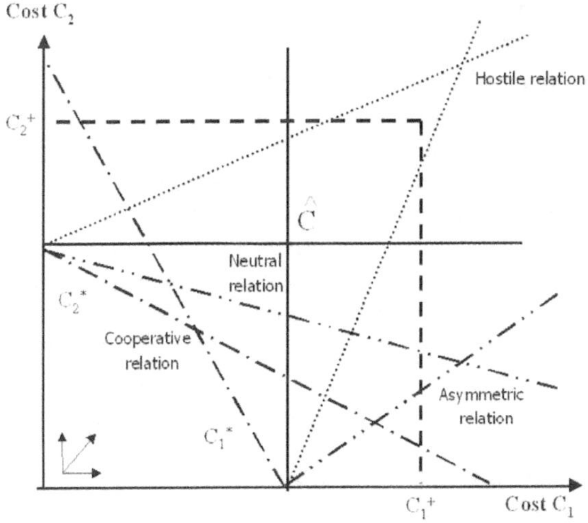

Figure 4.
Cost balance for neutral, hostile, friendly and asymmetric relations

Using the deviations from the equilibrium $y_i = C_i - \hat{C}_i$ as variables, we obtain a system of linear difference equations for the vector $y = (y_1,\dots, y_n)$

$$\Delta y(t) = y(t+1) - y(t) = -\bar{F} \cdot y(t)$$

where $\bar{F}$ is the modified interaction matrix with elements $\bar{f}_{ij} = \alpha_i f_{ij} / f_{ii}$. The dynamical system can be also expressed in the form $y(t + 1) = A y(t)$ with matrix $A = I - \bar{F}$ where I denotes the $n \times n$ unity matrix.

5.3 Stability conditions for interaction

From standard linear algebra it follows that as long as the interaction matrix F is stable, all players are able to achieve positive values, even for hostile relations with $f_{ij} < 0$. F also determines the stability of the resource dynamics $y(t+1) = (I - \bar{F})y(t)$, i.e., the conditions under which all eigenvalues are within the unit circle of the complex plane. For arbitrary f_{ij}, a set of sufficient stability conditions is [Scheffran 2001, Murata 1977]

$$\left|1 - \bar{f}_{ii}\right| + \sum_{j \neq i}^{n} \left| \bar{f}_{ij} \right| = \left|1 - \alpha_i\right| + \sum_{j \neq i}^{n} \left| \alpha_i f_{ij} / f_{ii} \right| < 1 \quad \text{for all } i = 1, 2, \dots, n.$$

For $f_{ii} > 0$ and $0 < \alpha_i \le 1$, this condition holds for

$$f_{ii} > \sum_{j \neq i} \left| f_{ij} \right|,$$

i.e., for each player positive self-efficiency should exceed the aggregate impact of cross-efficiencies by other players. For an increasing number of players, this condition is more difficult to achieve without cooperation. For $f_{ij} \le 0$ $(j = 1,\dots, n)$, we obtain the condition

$$\sum_j f_{ij} > 0 .$$

The stability of the interaction matrix determines whether the players can achieve specific value goals unilaterally or need to form coalitions and cooperate. A coalition is called stable if all members of the coalition are satisfied and have no incentive to leave the coalition. If the interaction matrix of a coalition becomes unstable, some players are not satisfied and fail to achieve their target values, i.e., the coalition breaks apart into smaller coalitions for which the interaction matrix is stable. With an increasing number of players, the number of eigenvalues of $\bar{F}$ increases and thus the chance that some are outside the unit circle and the interaction would become unstable. This relationship has been extensively discussed in ecology and population dynamics as the "complexity-stability" tradeoff. For an initial selection of arbitrary players only those players survive which are powerful enough or are part of a stable coalition.

To clarify the escalating sequence of events it is assumed that a player P_i tries to leave the power equilibrium $\hat{C}$ by changing its power use ($\delta C_i > 0$) in order to achieve a unilateral value gain $\delta V_i = f_{ii}\,\delta C_i > 0$. If its action leads to value losses $\delta V_j = f_{ji}\,\delta C_i < 0$ for other players P_j, it provokes countermeasures $\delta C_j = -f_{ji} / f_{jj}\,\delta C$ to compensate for the loss and keep V_j constant, assuming $f_{ji} < 0$. Then the outcome of P_i's action and other players' reaction after the second time step is

$$\delta V_i = f_{ii}\,\delta C_i + \sum f_{ij}\,\delta C_j = f_{ii}(1-\sigma)\,\delta C_i$$

where

$$\sigma = \sum\nolimits_j f_{ij} f_{ji} / (f_{jj}\, f_{ii}).$$

Thus, we have $\delta V_i < 0$ for $\sigma > 1$. Then, even though player P_i aims for an individual value gain, the overall social efficiency $f_{ii}^{\sigma} = f_{ii}(1-\sigma)$ due to the interaction with other players is negative and its value is worse than before. If this process continues, the interaction among the other players evolves, which may further contribute to instability (similar to an arms race), or stabilize the process by containing the first player's move. The latter can be induced externally by punishments or rewards, imposing some kind of "social regulation".

6 Striving for Security

6.1 The security-cost dynamics

In the following we apply the multi-actor framework to problems of international security. An important precondition for security is the absence of threat, i.e., the expectation that no intolerable damage is likely to occur. In mathematical terms this would imply that in a situation described by system variables $x(t)$ at time t, the expected security value $S^+(x, t) - S^-(x, t)$, combining gains and losses, is below a tolerable security threshold $S^*(x, t)$. Provided that the expected security loss S^- increases with potential damage $Y(x, t)$ and the likelihood $p(x, t)$, we obtain the net security function

$$S(x,t) = S^+(x,t) - p(x,t) \cdot Y(x,t) - S^*(x,t)$$

which is thus an indicator of relative security against threat. For $S > 0$ an actor feels relatively secure, for $S < 0$ relatively insecure. By use of power resources C (such as the defense budget) an actor can change its security state $\Delta S = f \cdot C$, where $f = s/c$ is the efficiency of the power resources with regard to the security impact. If an actor aims at achieving a particular security gain $\Delta S = V^*$, this translates into the required budget $C^* = V^* / f$. One obvious objective would be to reduce a negative security gap $S < 0$ through changes $\Delta S = -S / \tau = V^*$ where τ indicates how many time steps are required to bridge the security gap for constant ΔS. For $\tau = 1$ the gap shall be bridged in one single time step, for $\tau > 1$ the gap is expected to decay over an extended period. If at a given time t the actual budget $C(t)$ differs from the required one $C^*(t)$, the gap can be bridged by a budget adaptation process

$\Delta C_i(t) = \alpha_i(C_i^*(t) - C_i(t))$ where the reaction parameter $\alpha_i = \kappa_i C_i(C_i^+ - C_i)$ contains a logistic growth factor, reflecting the consideration that for a country with a limited budget $0 \le C_i \le C_i^+$ the dynamics is damped near the boundaries. κ_i is a constant cost reaction parameter (this model variant is called the SCX model, according to [Scheffran 1989] and [Jathe/Krabs/Scheffran 1997]).

6.2 Stability of the offense-defense competition

Throughout military history the offense-defense competition has been a major driver of the arms race, in which offenses increased the potential damage in war while defense tried to limit it. With the advent of nuclear-armed ballistic missiles any attempt to protect against this immense threat by defensive measures remained economically and technically unfeasible, depite enormous costs and efforts in programs such as the Strategic Defense Initiative (SDI) or the current Missile Defense programs of the Bush administration. One of the crucial issues is the impact of (missile) defense on strategic and international stability (for an earlier survey see [Scheffran 1989]).

The potential damage for a defender P_i grows with the offensive capability O_j of an attacker P_j, weighted with its offense efficiency o_j (representing the fraction of the existing offense units being efficiently launched) and destructive impact y_i (translating weapons into damage at the target), dimished by its own defense capability D_i, which is weighted with the defense efficiency d_i. In the simple scenario of first and second strike (which dominated Cold War thinking for a long time), each country perceives basically three options: being the first striker (index F), being the second striker (index S), no attack (index 0). The case of mutual attack (duel) in the same period is neglected here for reasons of simplicity. Assume that attacker P_i destroys some of P_j's offense capability before it can be launched in retaliatory attack. Then damage for the first and second striker is

$$Y_i^F = y_i^F[o_j(O_j - e_i(o_i^F O_i - d_j D_j)) - d_i^F D_i],$$

$$Y_i^S = y_i[o_j^F O_j - d_i D_i],$$

where o_j^F, d_j^F and y_i^F are the respective variables for the first striker while the index S for the second striker is neglected to simplify notation. e_i is the efficiency of the first striker's offense units in destroying the second striker's offense units, depending on factors such as the attacker's accuracy or the number of offense units per delivery vehicle. Now assume that each of the options is perceived with a likelihood $q_i^F, q_i = q_i^S, q_i^0$, which may also be interpreted as the weight that the decisionmakers in a country gives to the damage in a particular scenario. $q_i^k = 0$ implies that the decisionmakers completely neglect the damage in a scenario k. Then net security includes the weighted security impacts for all considered options

$$S_i = q_i^0 S_i^+ - q_i^F Y_i^F - q_i^S Y_i^S - S_i^*.$$

To norm the likelihoods one can make the additional assumption $q_i^0 = 1 - q_i^F - q_i$. To determine the impact of the military capabilities on security function S_i, we use the partial derivatives $s_{ij}^x = \partial S_i / \partial x_j$ where x_j can be either offensive or defensive capabilities of two countries P_i and P_j. For analytic treatment, we now use the special case that each country P_i ignores the option of being the first striker ($q_i^F = 0$) but feels threatened by a first strike of an opponent P_j with likelihood $q_i > 0$. Then security $S_i = (1 - q_i)S_i^+ - q_i Y_i - S_i^*$ of each player only depends on the weighted damage Y_i of being attacked and the security gains of no attack.

A country can change its security situation by applying its power resources (military budget) C_i to increase or decrease the offensive and defensive capabilities according to $\Delta O_i = (p_i^o / c_i^o)C_i, \Delta D_i = (p_i^d / c_i^d)C_i$ where $p_i^d, p_i^o = 1 - p_i^d$ are the cost shares (action priorities) for the respective military capabilities. In this case, and for constant likelihoods,[3] the overall security efficiencies are $f_{ii} = p_i^d s_{ii}^d / c_i^d = q_i y_i d_i^c p_i^d$, $f_{ij} = p_j^o s_{ij}^o / c_j^o = -q_i y_i o_j^c (1 - p_j^d$ $(i, j = 1,..., n)$ with $o_j^c = o_j / c_j^o, d_i^c = d_i / c_i^d$ and $p_j^o = 1 - p_j^d$. According to the analyis in the previous chapter, for $f_{ii} > 0$ and $f_{ij} < 0\ (j \neq i)$, sufficient conditions for stability are $f_{ii} + f_{ij} > 0$ which leads to (for $c_i^d > 0$)

$$p_i^d > \frac{o_j^c}{d_i^c}(1 - p_j^d) \equiv p_i^{d*}.$$

This defines a defense share which country P_i has to exceed to ensure stability. Below this threshold the threat due to offensive weapons is intolerable and has to be compensated by increasing defenses. The situation is unstable which implies that there are incentives for an arms race. This threshold increases with the security-cost efficiency of the offense o_j^c and decreases with the efficiency of the defense d_i^c. There is an inverse coupling between the defense shares of both countries, which implies that the threshold for one country declines the lower the offense share for the other country is. For $o_j^c = d_i^c$ the lines connect the corners (0, 1) and (1, 0).

We use a simple example with relative units. Compared to a country P_1 another country P_2 has a higher offense efficiency $o_2^c = 2o_1^c = 2$, which exceeds the defense efficiencies $d_2^c = 2d_1^c = 1$ of both sides. Then the two stability conditions are $p_1^{d*} = 4(1 - p_2^d)$ and $p_2^{d*} = 1 - p_1^d$. The results are depicted in Figure 5.

An alternative way of stabilizing the security relationship is the option of mutually disarming the offense which in the model corresponds to negative unit costs $c_i^o < 0$. In this case we have $f_{ij} > 0$ which always guarantees $f_{ii} + f_{ij} > 0$ and thus stability. The equilibrium point (balance of power) is given by

$$C_i = \frac{p_j^d \left[q_j y_j d_j^c V_i^* - q_i y_i o_j^c V_j^* \right] + q_i y_i o_j^c V_j^*}{\det F}$$

where V_i^* is the given value target for player P_i (e.g., bridging a security gap). Assume $\det F = f_{11} f_{22} - f_{12} f_{21} > 0$, and a negative bracket [...]. Then the budget of

[3] The assumption of exogenously given likehoods can be revised, making the likelihood of being attacked dependent on the force structure. See further [Scheffran 1989].

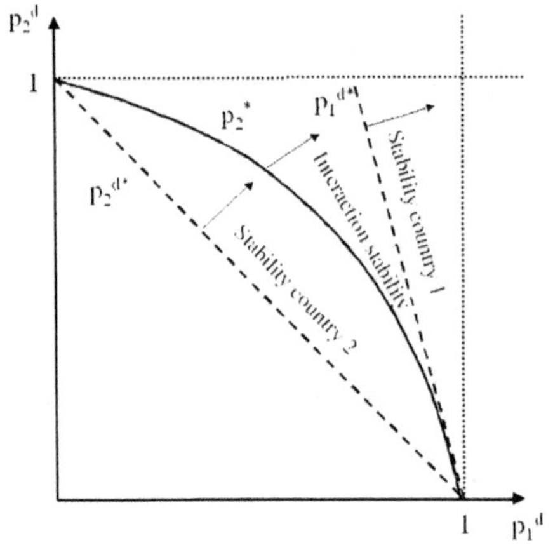

Figure 5.
Stability regions for defense shares p_1^d and p_2^d.

P_i can be reduced by increasing the defense share of $P_j(p_j^d \to 1)$; if the bracket is positive, costs can be reduced by decreasing the defense share $(p_j^d \to 0)$. The latter is the case in particular for disarming the offensive capability $(c_j^o < 0)$. The threshold between increasing and decreasing defense is given by

$$q_i = q_j \frac{y_j d_j V_i^*}{y_j o_j^c V_j^*} = q_i^*.$$

Thus for an attack likelihood $q_i < q_i^*$ there is a tendency to increase the offense share for cost reasons which depends on the perceived likelihood q_j of the opponent being attacked. The rule of thumb is that for a low perceived attack likelihood the defense share is also low. If both sides cooperate and trust each other, they don't believe in an attack by their partner and have thus less tendency to build defenses because they evaluate the other's offense as less threatening. Stability of the interaction matrix depends on the eigenvalues solving the characteristic equation:

$$\lambda_{1/2} = -(f_{11} + f_{22})/2 \pm \sqrt{(f_{11} + f_{22})^2/4 - \det F}\,.$$

Negative eigenvalues occur for $f_{11} + f_{22} > 0$ and the determinant of the interaction matrix

$$\det F = f_{11} f_{22} - f_{12} f_{21} = q_1 q_2 y_1 y_2 [d_1^c d_2^c p_1^d p_2^d - o_1^c o_2^c (1 - p_1^d)(1 - p_2^d)] > 0$$

which leads to the condition

$$p_2^d > \frac{1 - p_1^d}{1 + (do - 1) p_1^d} = p_2^*,$$

where

$$do = \frac{d_1^c d_2^c}{o_1^c o_2^c}.$$

Both the points $(p_1^d, p_2^d) = (0,1)$ and (1, 0) satisfy the equality relation $p_2^d = p_2^*$, whereas for $p_1^d = 0.5$ we achieve the threshold

$$p_2^* = \frac{1}{1+do}.$$

Our example yields $do = 1/4$ and a threshold $p_2^* = 4/5$ (see Figure 5).

The competition between offenses and defense can show a very complex and unstable dynamics which is difficult to control because of a wide range of uncertainties in the security-relevant variables [Scheffran 1989]. With an increasing number of countries whose offense-defense competition is closely linked, the potential for instability increases as the number of eigenvalues increases and thus the likelihood that the chosen parameter combination represents an unstable eigenvalue with positive real parts. With regard to the introduction of missile defense, there is the danger of a domino effect, i.e., the possibility, that Russia and China build more weapons because of the USA, that India follows China and Pakistan responds to India. Such a strong coupling can lead to an international escalation, with missile defense as a key trigger.

7 Chaos, Uncertainty, and Instability in International Security

The more variables and relations are included in a model to match the complexity of reality, the more does the combined uncertainty in model variables matter, leaving a wide range of possible outcomes. This also can play a role in the SCX conflict model outlined in the previous chapter, with the following set of equations for the variable changes of actors P_i ($i = 1,..., n$):

$$\Delta S_i = \sum_{j=1}^{n} f_{ij} C_j + U_i^S ,$$

$$\Delta C_i = -\kappa_i C_i (C_i^+ - C_i)(S_i + \tau_i \Delta S_i) ,$$

$$\Delta x_i^k = p_i^k C_i / c_i^k .$$

Here $k = 1, ..., m_i$ represents the index of system variables, κ_i is the cost reaction parameter, and U_i^S comprises all other security impacts (which are again ignored here). Each of the parameters can be time-dependent which is neglected to facilitate notation. Even in the bilateral case of two actors, chaos can emerge, as the following case shows. Here we start with anti-symmetric (antagonistic) security conditions $S_1(0) = - S_2(0) = 0.3$, which corresponds to a zero-sum situation. Both actors adapt their budgets towards achieving their satisfying security objectives $S_i = 0$, starting with an initial budget of $C_i(0) = C_i^* / 2$, which is half of the maxi-

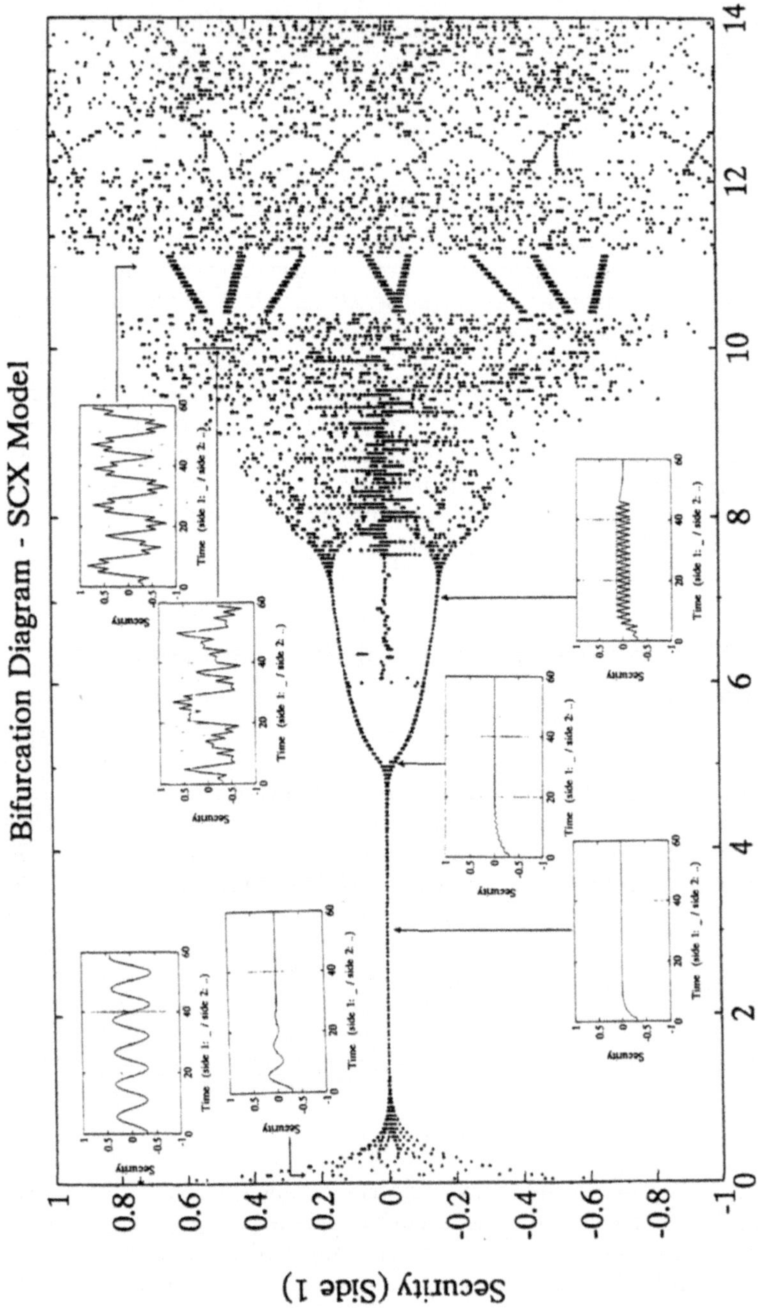

Figure 6. The security bifurcation diagram for two antagonistic actors.

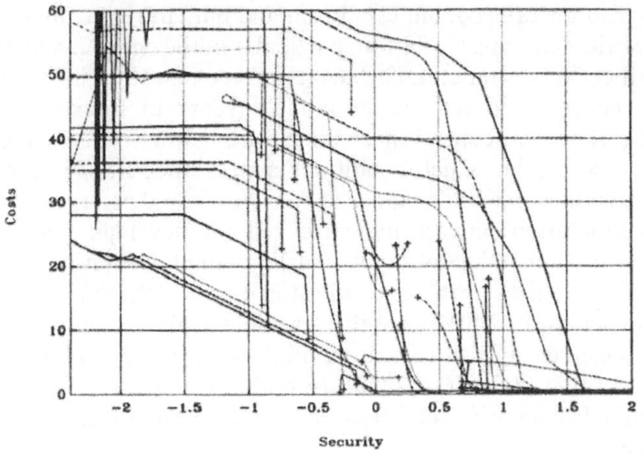

Figure 7. Dichotomy between winners and losers for a large number of actors.

mum budget $C_i^+ = 60$ cost units. Relations are hostile, represented by efficiencies $f_{ii} = -f_{ij} = 0.01$, which are chosen to be at the threshold between stability and instability. The cost reaction parameter is $\kappa_i = 0.04$. If we use the memory parameter τ_i as an order parameter and depict the security states to which the trajectory is attracted after some time, then we get a bifurcation diagram (Figure 6). For very low τ values, security oscillates around $S_1 = 0$, for increasing τ the trajectory approaches this state asymptotically (the second player shows inverse behavior). At some τ value, the first bifurcation occurs by splitting the equilibrium point into two, leading to a whole cascade. For high $\tau > 11$ the security situation is completely unpredictable because the overreaction of the actors moves the dynamics into any possible security state [Jathe/Scheffran 1995ab].

Now consider a large number of actors (in this case 30) whose interaction efficiencies f_{ij} as well as the initial conditions for security and costs (marked by +) are chosen stochastically around the previous baseline case. This implies that for some players the interactions are stable, for others unstable. Figure 7 clearly shows that a dichotomy emerges between two groups of actors. Some actors steadily win from the dynamics, forming a winning coalition, others lose by moving towards high insecurity and high costs. This polarization resembles some aspects that can be observed in real-world conflicts, such as the North-South conflict. Here also the asymmetry in maximum power resources C_i^+ plays an important role, which implies that a strong player (such as the US) can compensate unstable relations with much weaker players by use of their overwhelming force (dominance). However, this can be countered by an increased destruction efficiency f_{ij}, e.g., by a terror attack against vulnerable spots of the strong player which require only small efforts.

One way to avoid a polarization into blocks and minimize violent conflicts between them is to change the interaction efficiencies f_{ij} towards positive values, e.g., by strengthening of cooperation beyond the blocks and the formation of coalitions in order to achieve common aims. This can involve non-military security dimensions, such as environmental, economic, cultural and human rights dimensions. The model framework seems appropriate, in particular in the environmental sciences, to analyze conflicts between human behavior and ecological limits. Here systemic models (from ecology) and actor-based models (economic-social relations) are directly linked. If central control mechanisms are neither realistic nor desirable, the question arises under which conditions a spontaneous, self-organized transition to ecologically compatible modes of behavior would emerge.

One example is the cooperation between industrialized and developing countries in greenhouse gas emission reductions, e.g., via Joint Implementation and the Clean Development Mechanism of the Kyoto Protocol (see [Scheffran/Pickl 2000] and [Ipsen/Rösch/Scheffran 2001]). Another example deals with cooperative and sustainable management of fish resources, minimizing the potential for fish scarcity and severe conflict [Scheffran 2000]. In both fields the VCX model framework has been applied to identify conditions for cooperation, a task that is to be continued in the future.

References

Ashford, O. M. (ed.) (1985): *Prophet – or Professor?: The Life and Work of Lewis Fry Richardson*, Bristol 1985.

Ashford, O. M. (ed.) (1993): *Collected Papers of Lewis Fry Richardson*, Cambridge 1993.

Axelrod, R. (1984): *The Evolution of Cooperation*, Basic Publ., New York 1994.

Axelrod, R. (1997): *The Complexity of Cooperation*, Princeton University Press 1997.

Braun, M.; Coleman, C. S.; Drew, D. A. (1983): *Differential Equation Models, Modules in Applied Mathematics*, Vol. 1, New York: Springer 1983.

Carraro, C.; Filar, J. (eds.) (1995): *Control and Game-theoretic Models of the Environment*, Basel: Birkhäuser 1995.

Dockner, E.; Jorgensen, S.; Van Long, N.; Sorger, G. (eds.) (2000): *Differential games in economics and management science*, Cambridge Univ. Press 2000.

Epstein, J. M.; Axtell, R. (1997): *Growing Artificial Societies*, MIT Press, Cambridge 1997.

Gaylord, R. J.; D'Andria, L. J. (1998): *Simulating Society – A Mathematica Toolkit for Modeling Socioeconomic Behavior*, Springer/Telos, Berlin et al. 1998.

Grossmann, S.; Mayer-Kress, G. (1989): Chaos in the international arms race, *Nature* **337**, 23 February 1989, pp. 701–704.

Helbing, D. (1995): *Quantitative Sociodynamics Stochastic Methods and Models of Social Interaction Processes*, Boston/London/Dordrecht: Kluwer 1995.

Hess, G. D. (1995): An introduction to Lewis Fry Richardson and his mathematical theory of war and peace, *Conflict Management and Peace Science* **14**, no. 1, pp. 77–113.

Hofbauer, J.; Sigmund, K. (1998): *Evolutionary Games and Population Dynamics*, Cambridge University Press 1998.

Intriligator, M. D. (1975): Strategic considerations in the Richardson model of arms races, *Journal of Political Economy* **83**, no. 2, pp. 339–353.

Intriligator, M. D.; Brito, D. L. (1985): Heuristic decision rules, the dynamics of the arms race, and war initiation, in: U. Luterbacher, M. D. Ward (eds.), *Dynamic Models of International Conflict*, Boulder, Colorado, 1985, pp. 133–160.

Ipsen, D.; Rösch, R.; Scheffran, J. (2001): Cooperation in global climate policy: potentialities and limitations, *Energy Policy* **29**/4 (Jan. 2001), pp. 315–326.

Jathe, M.; Scheffran, J. (1995a): The transition to chaos in the SCX model of international security, in: G. Johannsen (ed.), *Proceedings of the IFAC Conference on Integrated Systems Engineering*, 27.-29. September 1994, IFAC Workshop Series, Oxford: Pergamon Press 1995.

Jathe, M.; Scheffran, J. (1995b): Modelling international security and stability in a complex world, in: P. Berge, R. Conte, M. Dubois, J. Thran Thanh Van (eds.), *Chaos and Complexity*, Gif sur Yvette: Editions Frontieres 1995, pp. 331–332.

Jathe, M.; Krabs, W.; Scheffran, J. (1997): Control and game theoretical treatment of a cost-security model for disarmament, *Mathematical Methods in the Applied Sciences* **20**, pp. 653–666.

Krabs, W. (1998): *Dynamische Systeme: Steuerbarkeit und chaotisches Verhalten*, Stuttgart: Teubner 1998.

Lambelet. J. C.; Luterbacher, U. (1979): Dynamics of arms races: mutual stimulation vs. self-stimulation, *Journal of Peace Science* **4**, no. 1, pp. 55–68.

Murata, Y. (1977): *Mathematics for Stability and Optimization of Economic Systems*, Academic Press, New York 1977.

von Neumann, J.; Morgenstern, O. (1944): *Theory of Games and Economic Behavior*, Princeton University Press, Princeton N.J., 1944.

Neuneck, G. (1995): *Die mathematische Modellierung von konventioneller Stabilität und Abrüstung*, Baden-Baden: Nomos 1995.

Olsder, G. J. (ed.) (1995): *New Trends in Dynamic Games and Applications*, Basel: Birkhäuser 1995.

Rapoport, A. (1960): *Fights, Games, and Debates*, Ann Arbor: The University of Michigan Press 1960.

Richardson, L. F. (1960a): *Arms and Insecurity: A Mathematical Study of the Causes and Origins of War*, Pittsburgh: Boxwood 1960.

Richardson, L. F. (1960b): *Statistics of Deadly Quarrels*. (Quincy Wright, C.C. Lienau, eds.) Pittsburgh: Boxwood Press 1960.

Saperstein, A. M. (1984): Chaos – a model for the outbreak of war, *Nature* **309**, 24 May 1984, pp. 303–305.

Saperstein, A. M. (1986): Predictability, chaos, and the transition to war, *Bulletin of Peace Proposals* **17**, no. 1, p. 879.

Saperstein, A. M.; Mayer-Kress, G. (1988a): A nonlinear dynamic model for the impact of SDI on the arms race, *J. Confl. Resol.* **32** (1988), no. 4, pp. 636–670.

Saperstein, A. M. (1988b): SDI a model for chaos, *Bulletin of the Atomic Scientists* **44**, no. 8, October 1988, pp. 40–43.

Scheffran, J. (1989): *Strategic Defense, Disarmament, and Stability – Modelling Arms Race Phenomena with Security and Costs under Political and Technical Uncertainties*, Doctoral Thesis, Marburg: IAFA No. 9, 1989.

Scheffran, J.; Pickl, S. (2000): Control and game-theoretic assessment of climate change options for joint implementation, *Annals of Operations Research* **97** (2000), pp. 203–212.

Scheffran, J. (2000): The dynamic interaction between economy and ecology cooperation, stability and sustainability for a dynamic-game model of resource conflicts, *Mathematics and Computers in Simulation* **53** (2000), pp. 371–380.

Scheffran, J. (2001): Stability and control of value-cost dynamic games, *Central European Journal of Operations Research* **9**(7) (Nov. 2001), pp. 197–225.

Schuster, H. G. (1988): *Deterministic Chaos*, 2nd edition, Weinheim: Verlag Chemie 1988.

Simaan, M.; Cruz, J. B. (1976): On the Stackelberg strategy in nonzero-sum games, in: G. Leitmann (ed.), *Multicriteria Decision Making and Differential Games*, New York: Plenum Press 1976, pp. 173–195.

Szidarovsky, F.; Li, W. (2000): A note on the stability of a Cournot-Nash equilibrium: the multiproduct case with adaptive expectations, *Journal of Mathematical Economics* **33** (2000), pp. 101–107.

Weidlich, W. (2000): *Sociodynamics – A Systematic Approach to Mathematical Modelling in the Social Sciences*, Harwood Academic Publishers 2000.

List of Contributors

Editors

Bernhelm Booß-Bavnbek is Associate Professor of Mathematics and Mathematical Modelling at the Mathematics and Physics Department of Roskilde University (Denmark). His main interest is the global analysis of partial differential equations, in particular gluing formulas for spectral invariants. He is the author of *Topology and Analysis – The Atiyah-Singer Index Formula and Gauge-Theoretical Physics* (Springer 1985, with D. Bleecker) and *Elliptic Boundary Problems for Dirac Operators* (Birkhäuser, 1993, with K. P. Wojciechowski). Email: <booss@ruc.dk>.

Jens Høyrup is Docent (~ British "Reader") at Roskilde University, Section for Philosophy and Science Studies (Denmark). His main research field is the conceptual and cultural history of pre- and early modern mathematics. He has recently published *Human Sciences: Reappraising the Humanities through History and Philosophy* (SUNY Press, 2000) and *Lengths, Widths, Surfaces: a Portrait of Old Babylonian Algebra and its Kin* (Springer, 2002). Email: <jensh@ruc.dk>.

Authors

Finn Aaserud has the equivalent of a PhD in physics from the University in Oslo and a PhD in the history of science from the Johns Hopkins University in Baltimore. From 1985 to 1989 he was Associate Historian at the Center for History of Physics of the American Institute of Physics, then in New York City, where he conducted an oral history and research project on American physicists' involvement in science policy after World War II. Since 1989 he has been director of the Niels Bohr Archive in Copenhagen. Aaserud is General Editor of the *Niels Bohr Collected Works* published by Elsevier Physics Publishing. His other publications include: *Redirecting Science: Niels Bohr, Philanthropy and the Rise of Nuclear Physics* (New York and Cambridge, 1991). Email: <aaserud@nbi.dk>.

Svend Bergstein was commissioned as an air force officer in 1971, and graduated at General Staff Course in 1980 at the Royal Danish Defence College (RDDC). Besides service in air defence units and a number of special courses, he has held the following positions: 80-82, Air Staff, Studies and Development. 82-85, RDDC, assistant faculty advisor, International Politics and Strategy. 88-92, The Defence Minister's Advisory and Analysis Group, studies in technology. 93-94, Minister for Research and Technology. 94-95, RDDC, faculty advisor, International Politics and Strategy. 95-99, defence attaché at the Danish embassies in Stockholm and Helsinki. 99-01, RDDC, head of secretariat for the reorganisation of RDDC. 2001, retired as Lieutenant Colonel. He currently works as a consultant at RDDC. Email: <svend.bergstein@mail.dk>.

Ute Bernhardt studied philosophy, computer science and psychology. She was CEO and member of the board of the German "Computer Professionals for Peace and Social Responsibility". Her works centres on privacy issues and computers and the military. She works at the Fraunhofer Gesellschaft and is a lecturer for computer science at the University of Applied Science Bonn-Rhein-Sieg. Email: <ute@kriton.bn.shuttle.de>.

Svend Clausen is head of the Operations Research and Mathematical Statistics Department, Danish Defence Research Establishment. Furthermore he is associate professor at the Informatics and Mathematical Modelling Institute, Technical University of Denmark (DTU). He is also part owner of DADCON, Danish Defence Consultants aps. His main interest is the use of mathematics for practical problem solving and he has recently proposed and introduced a new subject *Problemløsning og Modellering* at DTU. He also takes an interest in natural sciences and philosophy in general. He is author of *Kybernetik, systemer og modeller* (IMSOR, DTU 1969) and *Matematisk Modellering* (IMM, DTU 2002). Email: <sc@ddre.dk>.

Philip J. Davis, a native of Massachusetts, Harvard B.S, 1943; Ph.D. 1950, D.S.H.C., Roskilde University, 1997, has been in Applied Mathematics since 1963. He is known for his work in classical, functional, and numerical analysis. But he is also known to a larger audience as a commentator on mathematics particularly in its relation to society. Among his many nontechnical publications is the widely read and prize-winning book with Reuben Hersh, *The Mathematical Experience*. Email: <Philip_Davis@brown.edu>.

Setsuo Fukutomi was born 1919 in Sakhalin (called Karafuto when a Japanese territory). Graduated from the Imperial University of Tokyo mathematics (topological geometry). Called up for the Army in 1942 as a Soldier of the Second Class, in which respect he worked with cryptography and decoding and with finding the keys for the military cryptography of the General Staff of the Army. Demobilized with the rank of second lieutenant in 1945. Assistant at the Faculty of Sciences of the University of Tokyo. Co-editor (1947–1954) of the first edition of the *Encyclopedic Dictionary of Mathematics* (published by the Mathematical Society of Japan) together with Professor Shokiti Iyanaga. Worked to found the Mathematical Institute of the Faculty of Sciences and Human Sciences of the University of Nihon (Nihon Daigaku, Bunrigakubu). Then Professor of Mathematics at the Tokyo University for Agriculture and Technology until his retirement. Fax: +81-3-3390 3436.

Revaz Gamkrelidze is Member-Correspondent of the Russian Academy of Sciences and a Member of the Steklov Mathematical Institute, Moscow. His early interests in Mathematics were Algebraic Topology and Algebraic Geometry, where he computed the Chern classes of complex algebraic manifolds. From mid-fifties he was engaged in the Pontryagin group in developing the Optimal Control Theory. He is the co-author, together with L. S. Pontryagin, V. Boltyansky and E. Mishchenko, of the monograph *The Mathematical Theory of Optimal Processes*, Wiley, 1962, and of *Principles of Optimal Control Theory*, Plenum, 1978. Email: <gam@ipsun.ras.ru>.

Wolf Göhring studied Mathematics and Physics at the University of Saarland. There in 1960, he wrote his first computer programme. In the late 60s, early 70s he was concerned with artificial intelligence, man-machine interaction and networking. After a 15 years membership in the works council of the German National Research Institute for Computer Science (GMD) he now is studying the political economy of information and communication technology and its implication for the social development and sustainability. Email: <wolf.goehring@bi.fhg.de>.

Andrew Hodges studied mathematics at Cambridge and London universities (Ph.D. 1975). His research work in mathematical physics has been in association with Sir Roger Penrose (University of Oxford), and has mainly concerned the reformulation of quantum field theory in Penrose's twistor geometry. He is currently a Lecturer in mathematics at Wadham College, Oxford University. He is also the biographer of Alan Turing, having published the large-scale work *Alan Turing: the Enigma* in 1983, and a shorter philosophical text in 1997. Email: <andrew@synth.co.uk>.

Ib Martin Jarvad is Reader in philosophy of law at Roskilde University, Section for Philosophy and Science Studies (Denmark). He has published books and articles on constitutional law, international law, political and legal philosophy. He is now working on the history of natural law theories in law and political philosophy focussing on Hugo Grotius. Email: <jarvad@ruc.dk>.

Tinne Hoff Kjeldsen is Associate Professor of History of Mathematics at the Department of Mathematics and Physics at Roskilde University (Denmark). She studied mathematics at the University of Copenhagen (M.A. 1990) and took a Ph.D. in the history of mathematics from Roskilde University. Her research has mainly dealt with topics from the 20th century, especially within the history of game theory, mathematical programming, and the theory of systems of linear inequalities and the impact of World War II on the development of these fields. She is currently working on the historical development of the theory of convexity with focus on the significance of the different contexts in which this development took place. Email: <thk@ruc.dk>.

Helge Löfstedt is Director of Research at the Swedish Defence Research Establishment. He has conducted various analyses in the military field from a basic training in engineering, operations research and economics. His work has mainly dealt with force structuring and force planning on levels from single weapons system over composite unit structure to joint army, navy and air force structure. During that he has received experiences of military technology and weapons systems evaluation, cost-effectiveness studies and uncertainty analysis to seek balance and flexibility in solutions to meet an uncertain future. He is also experienced in using analyzing tools as manual and computer-assisted war-gaming, simulation and analytical modelling. Since 1997 he is serving as an analyst in the Swedish Ministry of Defence. Email: <helge.lofstedt@foi.se>.

Tetu Makino studied mathematical analysis at Kyoto University and Osaka City University, and took a degree of Dr. Sci. from Kyoto University in 1987 on a study of the Euler-Poisson equation which describes the evolution of gaseous stars. After teaching at Osaka Sangyo University, since 1995 he is a Professor of Mathematics at the Faculty of Engineering, Yamaguchi University. He has been studying the mathematical theory of gas dynamics, about which he gave an invited lecture at the Annual Meeting of the Mathematical Society of Japan, March 2001. His recent work is *On the Relativistic Euler Equation*, a joint work with Taiwanese mathematicians. Email: <makino@po.cc.yamaguchi-u.ac.jp>.

Elisabeth Rakus-Andersson is employed as a Docent in mathematics at the Blekinge Institute of Technology in Karlskrona – Sweden. She took her doctor's degree in applied mathematics in Poland in 1991 and the docent competence in Sweden in 2001. Besides the classical mathematical domains she researches in Fuzzy Set Theory and applies it to medicine, numerical methods and statistics, which results in many international publications and invited lectures. Her teaching duties concern the courses in mathematics for Master of Science students. She currently prepares a book about fuzzy mathematics in medicine and numerical methods due to an agreement with Springer-Verlag. Email: <era@bth.se>.

Ingo Ruhmann worked after his diploma in computer and political sciences as an scientific adviser for the German Bundestag and the European Parliament. His areas of interests are computers and the military, IT security. He was a member of the board of the German "Computer Professionals for Peace and Social Responsibility". He is a lecturer for computer science at the University of Applied Science Bonn-Rhein-Sieg. Email: <ingo.ruhmann@acm.org>.

Jesper Ryberg is Associate Professor in philosophy at Roskilde University and presently head of the Board of Studies for Philosophy and Science Studies. His main interests are moral philosophy, political philosophy and philosophy of law. He has written several works on normative ethics and applied ethics and is presently editing a forthcoming book on population ethics. Email: <ryberg@ruc.dk>.

Jürgen Scheffran received a PhD in physics and is Senior Researcher at the Potsdam Institute for Climate Impact Research (PIK) in Germany. His main fields of research are the mathematical modelling of complex systems, technology assessment, and the analysis of conflict and cooperation in energy, environment and international security. He has experience in teaching mathematical modelling and interdisciplinary studies at the universities of Marburg, Darmstadt and Potsdam. He is co-editor of two journals on peace and conflict studies, and published a number of articles and books, most recently on conflicts on biodiversity; energy conflicts; and space use and ethics. Email: <scheffran@pik-potsdam.de>.

Elmar Schmähling had served for more than 30 years in the Federal German Armed Forces before in 1990 he was forced to leave the service after heavily criticizing the nuclear doctrine of NATO security strategy and German armament policy. After a career as naval officer with positions on destroyers he had joined the central branch of the armed forces. There promoted to Rear Admiral he held the position of Chief of Military Counterintelligence Service (MAD) (1982–1983). During his last five service years he served as Chief of the Office for Studies and Exercises, a kind of think tank for the German Minister of Defence. After he had left service he joined the German and international peace movement. He has published three books on *NATO Strategy, German Military Politics*, and *The Reform of the Bundeswehr.* Email: <elmar_schmaehling@t-online.de>.

Albert N. Shiryaev, mathematician, pupil of Andrei N. Kolmogorov, Senior Researcher of the Steklov Mathematical Institute of the Russian Academy of Sciences, Corresponding Member of the Russian Academy of Sciences; Professor of the Moscow State University (since 1970) and Head of the Probability Theory Department of that university (since 1996); have been awarded the Markov Prize of the Soviet Academy of Sciences (1974), Kolmogorov Prize of the Russian Academy of Sciences (1994); President of the Bernoulli Society for Probability Theory and Mathematical Statistics (1889–1991), President of the Bachelier Financial Mathematics Society (1998–1999). D.S.H.C. of Albert Ludwigs University of Freiburg im Breisgau, Germany (2000), Professor Honoris Causa of Amsterdam University (2002). Email: <shiryaev@mi.ras.ru>.

Reinhard Siegmund-Schultze studied mathematics in Halle (Germany) and took a Ph.D. from Leipzig in the history of mathematics in 1979. Since 1978 he has been Assistant Professor at Humboldt University Berlin. After an interruption of his career for about 10 years due to the political situation in Germany he became Full Professor for history of mathematics at Agder University College in Kristiansand (Norway) in 2000. He has published articles on the history of mathematical analysis and books on mathematics in Nazi Germany, on emigration, and on the Rockefeller policies with respect to mathematics. He is currently working on applied mathematics in international contexts in the first half of the 20th century. Email: <Reinhard.Siegmund-Schultze@hia.no>.

Kathleen Broome Williams is Associate Professor of History at Bronx Community College, City University of New York, and a member of the history doctoral faculty at the City University of New York Graduate School and University Center. Her main interest is military history, particularly twentieth century naval science and technology. She is the author of Secret Weapon: High-frequency Direction Finding in the Battle of the Atlantic (Naval Institute Press, 1996) and Improbable Warriors: Women Scientists and the U.S. Navy in World War II (Naval Institute Press, 2001). Email: <kbbwilliam@aol.com>.